Steel Design

for the Civil PE and Structural SE Exams

Second Edition

Frederick S. Roland, PE, SECB, RA, CFEI, CFII

The Power to Pass®
www.ppi2pass.com

Professional Publications, Inc. • Belmont, California

Benefit by Registering This Book with PPI

- Get book updates and corrections.
- Hear the latest exam news.
- Obtain exclusive exam tips and strategies.
- Receive special discounts.

Register your book at **ppi2pass.com/register**.

Report Errors and View Corrections for This Book

PPI is grateful to every reader who notifies us of a possible error. Your feedback allows us to improve the quality and accuracy of our products. You can report errata and view corrections at **ppi2pass.com/errata**.

STEEL DESIGN FOR THE CIVIL PE AND STRUCTURAL SE EXAMS
Second Edition

Current printing of this edition: 3

Printing History

date	edition number	printing number	update
Nov 2014	2	1	New edition. Code updates. Copyright update.
May 2015	2	2	Minor cover updates.
Apr 2016	2	3	Minor corrections.

Printed in the United States of America.

PPI
1250 Fifth Avenue
Belmont, CA 94002
(650) 593-9119
ppi2pass.com

ISBN: 978-1-59126-483-5

Library of Congress Control Number: 2014953817

F E D C B A

Table of Contents

Preface and Acknowledgments

My purpose in writing this book is twofold. First, to help practicing engineers who are preparing for the civil structural Principles and Practice of Engineering (PE) exam or the structural engineering (SE) exam, both administered by the National Council of Examiners for Engineering and Surveying (NCEES). Second, to help engineering students who are learning about structural steel. This book, then, is designed to be useful as a guide for studying on your own or as a text for an introductory or intermediate class in steel design.

This second edition has been thoroughly updated to follow the fourteenth edition of AISC's *Steel Construction Manual*. The changes that AISC made to the fourteenth edition are not as far-reaching as those to the thirteenth, but they are considerable all the same, and I have revised many sections of this book to keep it up to date. At the end of the Introduction, I've summarized the most important changes to the fourteenth edition of the *Steel Construction Manual* and given pointers to where they are covered here in this book.

I want to express my thanks to C. Dale Buckner, PhD, PE, SECB, who reviewed an early draft of the first edition of this book and made many valuable suggestions for improvement. Thanks as well to the staff at PPI who worked on this new edition, including Magnolia Molcan, editorial project manager; Scott Marley, lead editor (who also typeset the book); Ralph Arcena, EIT, engineering intern; Tom Bergstrom, production associate and technical illustrator; Cathy Schrott, production services manager; and Sarah Hubbard, director of product development and implementation.

This book is dedicated to all those from whom I have learned: faculty members, supervisors, colleagues, subordinates, and my students.

Despite our best efforts, as you work through this book you may discover an error or a better way to solve a problem. I hope you will bring such discoveries to PPI's attention through their website at **ppi2pass.com/errata**. Valid corrections and improvements will be posted in the errata section of their website and incorporated into future printings of this book.

Frederick S. Roland, PE, SECB, RA, CFEI, CFII

Introduction

ABOUT THIS BOOK

The main purpose of this book, *Steel Design for the Civil PE and Structural SE Exams*, is to be a study reference for engineers and students who are preparing to take either the civil structural PE exam or the structural SE exam, both of which are given by the National Council of Examiners for Engineering and Surveying (NCEES). These exams—even the breadth section of the civil PE exam, which is more general in its scope—often contain structural questions that go beyond the basics. If you want to be prepared for all questions in steel design, this book will give you the thorough review you need.

However, anyone who wants to learn more about the most current steel design methods can benefit from this book. It can serve as a guide for those who are studying on their own or as a text in a formal course.

After a quick review of some basics in the early chapters, each chapter in turn explores in greater detail a different aspect of steel design. Among the topics covered are

- loads and load combinations
- analysis methods
- design of beams, columns, and plate girders
- design of members under combined stresses
- design of composite members
- bolted and welded connections

Many examples are included with detailed, step-by-step solutions to show you how to attack various kinds of problems and apply the relevant AISC criteria. The principles, equations, and information that you'll learn in this book are those you will need to solve the kinds of problems in structural steel that you're most likely to encounter on the civil structural PE and structural SE exams.

WHAT YOU'LL NEED AS YOU STUDY

Steel Design for the Civil PE and Structural SE Exams is designed to complement and be used with PPI's *Civil Engineering Reference Manual* (CERM) or *Structural Engineering Reference Manual* (STRM). If you're studying for the civil PE exam or the structural SE exam, then your basic text should be CERM or STRM, respectively.

As you study this book, you will also need to have by your side the *Steel Construction Manual*, fourteenth edition, published by the American Institute of Steel Construction (hereafter referred to as the *AISC Manual*). An earlier edition will not suffice, as the changes introduced in the fourteenth edition are considerable. (The most important differences between the thirteenth and fourteenth editions are summarized the end of this Introduction.) This book explains and clarifies those aspects of the *AISC Manual* that are most likely to come up during the civil PE and structural SE exams. But it isn't a substitute, and the text will frequently assume that you can refer to the *AISC Manual* when needed.

In this book, references to chapters, sections, figures, tables, and equations in the *AISC Manual* are so labeled, such as "*AISC Manual* Table 4-13" or "AISC Eq. J10-4." References that don't specify a source refer to this book; for example, "Figure 6.2" will be found in Chap. 6 of this book.

HOW TO USE THIS BOOK

Each chapter in this book treats a different topic. If you only want to brush up on a few specific subjects, you may want to study only those particular chapters. However, later chapters frequently build on concepts and information that have been set out in earlier chapters, and the book is most easily studied by reading the chapters in order.

The civil PE and structural SE exams are open book, so as you study it is a good idea to mark pages in both the *AISC Manual* and this book that contain important information, such as tables, graphs, and commonly used equations, for quick reference during the exam. (Some states don't allow removable tabs in books brought into the exam. Check with your state board, or use permanent tabs.) Become as familiar as possible with this book and with the *AISC Manual*. Remember that preparation and organization are just as important to passing the PE and SE exams as knowledge is.

Throughout the book, example problems illustrate how to use the standard design principles, methods, and formulas to tackle common situations you may encounter on your exam. Take your time with these, and make sure you understand each example before moving ahead. Keep in mind, though, that in actual design situations there are often several correct solutions to the same problem.

In the last chapter, you'll find 37 practice problems. These are multiple-choice problems similar in scope, subject matter, and difficulty to problems you'll encounter on the breadth and depth sections of the civil PE exam or the breadth sections of the structural SE exam. These problems cover the full range of steel design topics and show the variety of approaches needed to solve them. The topics covered by the problems are listed in Table I.1 at the end of this introduction.

When you feel comfortable with the principles and methods taught by the example problems, work these practice problems under exam conditions. Try to solve them without referring to the solutions, and limit yourself to the tools and references you'll have with you during the actual exam—an NCEES-approved calculator, pencil and scratch paper, and the references you plan to bring.

After studying this book, you should be able to solve most common problems in structural steel, both on the exams and in real design applications.

TWO DESIGN METHODS: LRFD AND ASD

Steel design problems on the PE and SE exams may be solved using either the load and resistance factor design (LRFD) method or the allowable strength design (ASD) method. You should plan to use whichever method is most familiar to you. If your classes in school emphasized one method, or if you routinely use one method at your job, then you should use that method on the exam.

This book covers both methods. The principles that underlie the two methods are explained and compared in Chap. 3. Throughout the book, wherever the LRFD and ASD methods use different equations for a calculation, they are both given and explained.

A particularly useful feature of this book is that example problems and practice problems are not given separate LRFD and ASD solutions. Instead, a single solution is presented for each problem, and when a step or a calculation is different in the two methods, the two versions are displayed side by side. This makes it very easy to compare the LRFD and ASD methods and see where they are similar and where they differ. In some solutions, you'll find that the LRFD and ASD methods are substantially the same, differing in only one or two calculations along the way. In others, you'll find that every calculation from beginning to end is different. Whichever method you are studying, your understanding of both methods and how they are related will increase as you use this book.

ABOUT THE EXAMS

The NCEES PE exam in civil engineering is an eight-hour exam consisting of two four-hour sections, which are separated by a one-hour lunch period. Each section contains 40 multiple-choice problems, and you must answer all problems in a section to receive full credit. There are no optional questions. The breadth section is taken in the morning by all examinees, and may include general steel problems. In the afternoon, you must select one of five depth sections: water resources and environmental, geotechnical, transportation, construction, or structural. The structural depth section covers a range of structural engineering topics including loads, analysis, mechanics of materials, materials, member design, design criteria, and other topics.

The structural engineering (SE) exam is a 16-hour exam given in two parts, each part consisting of two four-hour sections separated by a one-hour lunch period. The first part—vertical forces (gravity/other) and incidental lateral—is given on a Friday. The second part—lateral forces (wind/earthquake)—takes place on a Saturday.

Each part contains a breadth section, which is given in the morning, and a depth section, given in the afternoon. Each breadth section contains 40 multiple-choice problems that cover a range of structural engineering topics specific to vertical or lateral forces. Each depth section contains either three or four essay (design) problems. For each of the depth sections, you have a choice between two subject areas, bridges and buildings, but you must choose the same area for both depth sections. That is, if you choose to take the buildings depth section during the first part, you must also take the buildings depth section during the second part.

According to NCEES, the vertical forces (gravity/other) and incidental lateral depth section in buildings covers loads, lateral earth pressures, analysis methods, general structural considerations (e.g., element design), structural systems integration (e.g., connections), and foundations and retaining structures. The depth section in bridges covers gravity loads, superstructures, substructures, and lateral loads other than wind and seismic. It may also require pedestrian bridge and/or vehicular bridge knowledge.

The lateral forces (wind/earthquake) depth section in buildings covers lateral forces, lateral force distribution, analysis methods, general structural considerations (e.g., element design), structural systems integration (e.g., connections), and foundations and retaining structures. The depth section in bridges covers gravity loads, superstructures, substructures, and lateral forces. It may also require pedestrian bridge and/or vehicular bridge knowledge.

ABOUT THE STEEL CONSTRUCTION MANUAL, FOURTEENTH EDITION

AISC published the fourteenth edition of its *Steel Construction Manual* (usually shortened to *AISC Manual* in this book) in 2011. Although the format of the fourteenth edition is similar to that of the thirteenth (2005), there are many differences, ranging from minor to significant, between the two editions. Many of these changes are contained in Part 16, *Specification for Structural Steel Buildings* (or *AISC Specification* for short). Using different editions of the *AISC Manual* interchangeably can lead to undesirable results, including over- or underdesigned structural elements.

The 2012 edition of the *International Building Code* incorporates the provisions set forth in the fourteenth edition of the *AISC Manual*. Where authorities having jurisdiction have adopted the 2012 IBC, the use of the fourteenth edition of the *AISC Manual* is mandated.

Perhaps the most significant change in the fourteenth edition is that Chap. C of the *AISC Specification*, Design for Stability, now specifies that required strengths (ASD or LRFD) are to be determined by the direct analysis method instead of by the effective length method. Discussion of the effective length method has been moved to App. 7 of the *AISC Specification* as an alternative that may be used if all other requirements of Chap. C are met.

In the direct analysis method, notional lateral loads are applied to the frame to account for the initial imperfection of the members and out-of-plumbness of the structure. In this method, the effective length of each member is equal to the member's true length, so that the effective length factor, K, is always equal to 1.0. (The direct analysis method is discussed in this book's Chap. 7, Steel Column Design. Chapter 13 includes practice problems for both methods. For problems on the direct analysis method, use the actual length of the member; for problems on the effective length method, multiply the member length by the appropriate effective length factor, K.)

Other significant changes to the fourteenth edition include the following.[1]

- In Chap. B, width-to-thickness ratio requirements have been reorganized into two tables. Table B4.1a is for members subject to axial compression, and Table B4.1b is for members subject to flexure. (Width-to-thickness requirements are discussed in this book's Chap. 5, Steel Beam Design.)
- In Chap. D, some modifications have been made to the table of shear lag factors, Table D3.1. Requirements for gross and net section areas have been moved from Chap. D to Chap. B. (Shear lag factors and gross and net section areas are discussed in Chap. 4, Tension Member Design.)
- In Chap. F, the equation for the lateral-torsional buckling modification factor, C_b (Eq. 5.3 in this book), has been modified. (This and related subjects are discussed in Chap. 5, Steel Beam Design.)
- Chapter G establishes a minimum required moment of inertia (Eq. 11.33 in this book) for stiffeners with tension-field action. (Stiffeners are discussed in Chapter 11, Plate Girders, and the moment of inertia requirement is illustrated in Ex. 11.1.)
- Chapter H adds a new section discussing the rupture of flanges with holes subject to tension under combined axial force and flexure. (Combined forces are discussed in Chap. 8, Combined Stress Members.)
- Chapter I has been extensively reorganized, and new provisions for the design of composite members have been added. (These are discussed in Chap. 12, Composite Steel Members.)
- In Chap. J, high-strength bolts are now classified as Class A or Class B rather than by ASTM specification number. The shear strength capacity of high-strength bolts has been increased by approximately 12.5%. (High-strength bolts and the new classifications are discussed in Chap. 9, Bolted Connections.)
- In Chap. K, new provisions for the design of hollow structural section (HSS) and box member connections have been added, and the complicated requirements have been reorganized into a series of tables for easier reference. (HSS and box member connections are discussed in Chap. 10, Welded Connections, and the solution to Ex. 10.4 illustrates how to use the tables in Chap. K.)

Minor changes made in the fourteenth edition include the following.

- The W36×800 shape has been deleted.
- Four HP18 sections have been added.
- Six HP16 sections have been added.
- For about one-third of the W shapes and about two-thirds of the angles, small changes have been made in cross-sectional area.

---

[1]For a more detailed description, see Eric J. Bolin, Thomas J. Dehlin, and Louis F. Geschwindner, "A Comparison between the 2005 and 2010 AISC Specification," *Engineering Journal,* First Quarter 2013.

Table I.1 *List of Practice Problems in Chap. 13*

List of Figures

List of Tables

Codes Used to Prepare This Book

ACI 318: *Building Code Requirements for Structural Concrete*, 2011. American Concrete Institute, Farmington Hills, MI.

AISC 325: *Steel Construction Manual*, 14th ed., 2011. American Institute of Steel Construction, Chicago, IL. (Includes ANSI/AISC 360: *Specification for Structural Steel Buildings*, 2010, and *Specification for Structural Joints Using High-Strength Bolts*, 2009. American Institute of Steel Construction, Chicago, IL.)

ASCE 7: *Minimum Design Loads for Buildings and Other Structures*, 2010. American Society of Civil Engineers, Reston, VA.

IBC: *International Building Code*, 2012. International Code Council, Washington, DC.

TMS 402/ACI 530/ASCE 5 and TMS 602/ACI 530.1/ASCE 6: *Building Code Requirements and Specifications for Masonry Structures*, 2011. American Concrete Institute, Farmington Hills, MI.

1 Structural Steel

1. THE DEVELOPMENT OF STRUCTURAL METALS

Since the mid-1890s, structural steel has been the principal metal used in the construction of bridges and buildings. Before this time, however, other metals such as cast iron and wrought iron were favored.

Cast iron was developed in China in the sixth century B.C. It was introduced to western Europe in the 15th century, where it was used mostly for weaponry, including cannons and shot. In the 18th century, new manufacturing techniques made cast iron cheap enough and available in large enough quantities to become a practical building material, and in the late 1770s, the first cast-iron bridges were built in England. Because cast iron is strong in compression but relatively brittle, these bridges were usually arch-shaped to minimize tensile stresses.

By the 1820s, mills had begun rolling rails for railroads. At first, most of these rails were made from wrought iron. Wrought iron had been manufactured in western Europe since the Middle Ages, but in the 1820s it was not yet widely used in building. As processes improved, however, wrought iron became more plentiful and of better quality, and around 1840, wrought iron began to replace cast iron in building. Cast iron was effectively abandoned as a structural material by the end of the century, due in part to the catastrophic collapses of a number of cast-iron railway bridges between the 1840s and 1890s.

In the 1850s, improvements in the manufacturing process made steel production faster and cheaper. Steel, which had previously been expensive and thus used mainly for small items such as knives, became practical for use as a building material.

The rolling of wrought-iron rails evolved into the rolling of I-shaped beams by the 1870s. At first, these beams were manufactured in both wrought iron and steel, but steel could be produced with less effort and in greater quantity. Shapes rolled in steel gradually replaced the wrought-iron shapes, and steel almost completely dominated construction by 1900.

2. THE STANDARDIZATION OF STEEL

In 1896, the Association of American Steel Manufacturers began standardizing the rolling of beams and establishing their regular depths and weights. These beams were called American standard beams or *I-beams,* and eventually became known as S-beams (for *standard*). The inside surfaces of the flanges of S-beams have a slope of approximately 16.7%. Sizes for S-beams are given in true depths rather than nominal depths. S-beams range in depth from 3 in to 24 in, and in weight from 5.7 lbf to 121 lbf per linear foot.

In 1900, the National Steel Fabricators Association, which later became the American Institute of Steel Construction (AISC), in conjunction with the American Society for Testing and Materials (ASTM), began standardizing the configuration of shapes, weights, tensile strengths, and yield strengths for various structural steel products. The AISC published *Steel Construction* in 1923, which established a basic allowable working stress of 18,000 psi (pounds per square inch) for rolled steel. This value remained in effect through several revisions of the code until 1936, when it was increased to 20,000 psi.

With the development of new, stronger steels, higher allowable working stresses became warranted. With the sixth edition of the *Manual of Steel Construction*, published by AISC in 1963, the basic allowable working stress was increased to 24,000 psi for steels with a yield strength of 36,000 psi. By changing the chemical composition of the steel, factories could produce a variety of types of steel. Yield points ranged from 33,000 psi to 50,000 psi, depending on the steel's chemical composition and thickness. Later, through further development of the manufacturing process, steels were produced with yield points up to 100,000 psi depending on alloy composition, thickness of material, and heat treatment.

Structural steels are commonly referred to by the designations given by the ASTM, which are based on a steel's characteristics and chemical composition. Generally, structural steels are divided into three groups: carbon steels, high-strength low-alloy (HSLA) steels, and quenched and tempered alloy steels. *Carbon steels* are usually divided into four categories according to the percentage of carbon they contain.

- Low carbon steel contains less than 0.15% carbon.
- Mild carbon steel contains between 0.15% and 0.29% carbon.
- Medium carbon steel contains between 0.30% and 0.59% carbon.
- High carbon steel contains between 0.60% and 1.70% carbon.

One of the most common structural steels, ASTM A36, has a yield point of 36 ksi (kips per square inch) and belongs to the mild carbon category. ASTM A36 steel was used extensively in rolled shapes from 1963 until 2000, when ASTM specification A992, for steel with a yield point of 50 ksi (an HSLA steel), supplanted A36 steel for wide-flange (W shape) rolled beams.

Regardless of type, there are three important characteristics that remain constant for all steels. They are

- modulus of elasticity: $E = 29{,}000$ ksi
- shear modulus of elasticity: $G = 11{,}200$ ksi
- coefficient of linear expansion: $\varepsilon = 0.00065$ per 100°F change in temperature.[1]

[1]This is the value as given in the AISC *Steel Construction Manual.* The coefficient of linear expansion is also sometimes expressed as 6.5×10^{-6} per °F change in temperature.

It is important when using these constants to make sure the units (pounds per inch or kips per inch) are consistent within the requirements of the formulas. This is particularly true when taking a root of a constant or raising a constant to a power.

There are two common methods of designing steel structures. The method of design specified in the first eight editions of the AISC *Manual of Steel Construction* was *allowable stress design*, an older form of what is now known as *allowable strength design* (ASD). In 1986, AISC gave support to a newer, alternative method of design when it published the first edition of the *Manual of Steel Construction: Load and Resistance Factor Design* (LRFD). From 1986, both methods of design and analysis were widely used and permitted for use by building codes, but until 2005, they continued to be covered in separate manuals. In 1989, a ninth edition of the earlier book was published under the title *Manual of Steel Construction: Allowable Stress Design*. Second and third editions of the LRFD Manual were published in 1998 and 2001, respectively.

Supplement No. 1 to the *Specification for Structural Steel Buildings* was approved on December 17, 2001, and altered some provisions in the *Manual of Steel Construction: Allowable Stress Design*. AISC published *Errata List, September 4, 2001* for editorial corrections made to the *Load and Resistant Factor Design Specifications for Structural Steel Buildings*, which was dated December 27, 1999, and contained in the third edition of the LRFD Manual.

In 2005, the AISC published the *Steel Construction Manual*, which for the first time incorporated both load and resistance factor design and allowable strength design, the latter being a modified form of allowable stress design. All versions of the AISC *Manual of Steel Construction*, whether for ASD or LRFD, are treated as earlier editions of the *Steel Construction Manual*.

3. STRUCTURAL SHAPES

Structural steel comes in a variety of shapes. Most shapes are designated by a letter that indicates the shape series, followed by the nominal depth of the member, and the unit weight per linear foot.

There are four series of shapes that are collectively referred to as I- or H-beams because their cross-sectional shape resembles those uppercase letters. These are the W, M, S, and HP series, with the W series being the most commonly used.

WT, MT, and ST sections are T-shaped members produced by cutting W, M, and S members longitudinally down the center of the web to make two T-shaped members of equal size.

Table 1.1 gives information on some commonly used I- and T-beams. Table 1.2 gives information on some channels, angles, and hollow structural shapes (HSS) and pipes.

***Table 1.1** W, M, S, HP, WT, MT, and ST Series Shapes*

series	example	depth range (in)	weight range (lbf/ft)	description
W	W18 × 50	4–44	8.5–798	wide-flange sections • inside and outside faces of flanges are parallel
M	M8 × 6.5	4–12	3.7–11.8	miscellaneous beams • section proportions do not conform to requirements of W, S, or HP sections • inside and outside faces of flanges are parallel
S	S12 × 35	3–24	5.7–121	American standard beams • inside face of flange slopes 16.66%
HP	HP10 × 57	8–14	36–117	H-piles, or bearing piles • inside and outside faces of flanges are parallel • web and flange thickness are nominally equal, as are beam depth and flange width
WT MT ST	WT15 × 74 MT7 × 9 ST4 × 11.5	2–22 2–6 1.5–12	6.5–296.5 3.0–5.9 2.85–60.5	structural tees • fabricated by cutting W, M, and S sections longitudinally along the web center

Table 1.2 *Channels, Angles, and Hollow Structural Shapes and Pipes*

series	example	depth range (in)	weight range (lbf/ft)	description
C	C10 × 20	3–15	3.5–50	American standard channels • inside face of flange slopes 16.66%
MC	MC8 × 20	6–18	12–58	miscellaneous channels • section proportions do not conform to requirements of American standard channels • inside face of flange slopes 16.66%
L	L4 × 3 × $^1/_4$	2 × 2 × $^1/_8$ to 8 × 8 × $1^1/_8$	1.67–57.2	angles (L shapes) • equal and unequal legs • long leg is always listed first
HSS	HSS8 × 4 × $^1/_4$	1.25–20	1.77–127	rectangular and square hollow structural sections • designated by long face × short face × wall thickness
HSS round	HSS4 × 0.125	4–20	5.18–104	round hollow structural sections • designated by diameter × wall thickness
pipe	3 in std. pipe 3 in X strong pipe 3 in XX strong pipe	0.5–12	0.582–72.5	standard steel pipe (std.) extra strong pipe (X strong) double extra strong pipe (XX strong)

2 Loads and Load Combinations

Nomenclature

D	dead load	lbf
E	earthquake (seismic) load	lbf
L	live load	lbf
L_r	roof live load	lbf
R	rain load	lbf
R	strength	lbf
S	snow load	lbf
W	wind load	lbf

Subscripts

a	required (ASD)
u	required (LRFD)

1. GENERAL

When designing a structure, the types and magnitudes of loads that will be imparted to that element or structure must be considered. The loads may act individually or in a variety of combinations. Therefore, it is important to determine the individual load or the combination of loads that will produce the maximum load on the element being designed.

Most codes incorporate the types, magnitudes, and combinations of loads specified in ASCE Standard 7, *Minimum Design Loads for Buildings and Other Structures* (ASCE 7), published by the American Society of Civil Engineers (ASCE). For example, the AISC *Steel Construction Manual,* published by the American Institute of Steel Construction (AISC), no longer gives information about loads and load combinations, but has incorporated, by reference, the loads and load combinations specified in ASCE 7.

2. LOAD TYPES

ASCE 7 Table 4.1 gives the minimum live loads, both uniformly distributed and concentrated, that are to be used in the design of buildings and other structures. ASCE 7 Table C3-1 gives a list of uniform dead loads for common building materials.

Table 1607.1 of the *International Building Code* (IBC) also lists the minimum uniformly distributed live loads and the minimum concentrated live loads that various structures must be designed for. Though the IBC is based on ASCE 7, there are some differences between the two; therefore, when calculating loads it is important to use only the code that is specified.

3. LOAD COMBINATIONS

Unless otherwise specified by a local code with jurisdiction over a project, the load combinations given in the AISC *Steel Construction Manual* (*AISC Manual*) should be used in designing steel structures. The AISC basic load combinations are derived from ASCE 7, and include combinations for both allowable stress design (ASD) and load and resistance factor design (LRFD). The number of combinations can be extensive, taking into consideration wind direction, unbalanced snow loads, or any number of other variables. It is not unusual for computer printouts with 20 to 30 or even more load combinations to be generated.

Fortunately, not all structural members will be subjected to every type of load. Therefore, a number of terms may drop out of the load combination formulas.

Load Combinations for Allowable Strength Design

The following are the basic load combinations used with ASD, as given in ASCE 7 Sec. 2.4.1.

$$D \quad (2.1)$$

$$D+L \quad (2.2)$$

$$D+(L_r \text{ or } S \text{ or } R) \quad (2.3)$$

$$D+0.75L+0.75(L_r \text{ or } S \text{ or } R) \quad (2.4)$$

$$D+(0.6W \text{ or } 0.7E) \quad (2.5)$$

$$D+0.75L+0.75(0.6W)+0.75(L_r \text{ or } S \text{ or } R) \quad (2.6)$$

$$D+0.75L+0.75(0.7E)+0.75S \quad (2.7)$$

$$0.6D+0.6W \quad (2.8)$$

$$0.6D+0.7E \quad (2.9)$$

In 2001, *Supplement No. 1 to the Specifications for Structural Steel Buildings* eliminated a provision from the ninth edition of the *Manual of Steel Construction: Allowance Stress Design* that permitted an increase of 33% in allowable stress for any load or load combination incorporating a wind load, W, or earthquake load, E. (This provision had never applied to the LRFD method.)

Example 2.1
Calculating Load Using ASD

The loads on a steel beam consist of 15 kips due to dead load and 41 kips due to live load. No other loads need to be considered. Using ASD, calculate the required strength of the beam.

Solution

The required strength is the greatest value among the combinations in Eq. 2.1 through Eq. 2.9. Equation 2.2 gives

$$\begin{aligned} D+L &= 15 \text{ kips} + 41 \text{ kips} \\ &= 56 \text{ kips} \end{aligned}$$

Because all other variables are zero, no other combination has a sum greater than $D + L$. The required strength is therefore given by the combination $R_a = D + L =$ 56 kips.

Load Combinations for Load and Resistance Factor Design

The following are the basic load combinations used with load and resistance factor design (LRFD), as given in ASCE 7 Sec. 2.3.2.

$$1.4D \qquad 2.10$$

$$1.2D+1.6L+0.5(L_r \text{ or } S \text{ or } R) \qquad 2.11$$

$$1.2D+1.6(L_r \text{ or } S \text{ or } R)+(L \text{ or } 0.5W) \qquad 2.12$$

$$1.2D+1.0W+L+0.5(L_r \text{ or } S \text{ or } R) \qquad 2.13$$

$$1.2D+1.0E+L+0.2S \qquad 2.14$$

$$0.9D+1.0W \qquad 2.15$$

$$0.9D+1.0E \qquad 2.16$$

Example 2.2
Calculating Load Using LRFD

The loads on a steel beam consist of 15 kips due to dead load and 41 kips due to live load. No other loads need to be considered. Using LRFD, calculate the required strength of the beam.

Solution

The required strength is the greatest value among the combinations in Eq. 2.10 through Eq. 2.16. Equation 2.10 gives

$$\begin{aligned}1.4D &= (1.4)(15 \text{ kips}) \\ &= 21 \text{ kips}\end{aligned}$$

As all variables but D and L are zero, Eq. 2.11 can be reduced to

$$\begin{aligned}1.2D+1.6L &= (1.2)(15 \text{ kips})+(1.6)(41 \text{ kips}) \\ &= 83.6 \text{ kips}\end{aligned}$$

Because all other variables are zero, none of the other combinations can have a sum greater than $1.2D + 1.6L$. The required strength is therefore given by the combination $R_u = 1.2D + 1.6L = 83.6$ kips.

4. MOVING LOADS

The loads listed in the previous section are generally considered to be static loads or applied as static loads. In addition to these, *moving loads*, such as vehicles on bridges and traveling cranes on or in buildings, may also need to be determined. For example, Sec. 4.9 of ASCE 7 specifies that, to allow for induced vertical impact or vibration force, the maximum wheel loads of a powered crane shall be increased by the percentages given in Table 2.1.

Table 2.1 *Increase for Vertical Impact Force from Crane Load*

crane type	percentage of increase
powered monorail	25%
powered bridge, cab operated	25%
powered bridge, remotely operated	25%
powered bridge, pendant operated	10%
hand-geared bridge, trolley, and hoist	no increase

Source: ASCE 7 Sec. 4.9.3

Powered cranes are also considered to create lateral and longitudinal forces on their runway beams (ASCE 7 Sec. 4.9). The lateral force on a runway beam is taken as 20% of the sum of the crane's rated capacity and the weight of the hoist and trolley. The longitudinal force on a runway beam is taken as 10% of the crane's maximum wheel loads. Both forces are assumed to act horizontally at the beam surface, the lateral force acting in either direction perpendicular to the beam and the longitudinal force acting in either direction parallel to the beam.

5. IMPACT LOADS

When a live load will impart a greater-than-ordinary impact load to a structure, the increased load is usually taken into consideration by increasing the weight of the equipment by a certain amount. Table 2.2 gives the percentages of increase that ASCE 7 Sec. 4.6 specifies should be applied to the static load to account for induced vertical impact or vibration force from various causes.

Table 2.2 *Increase for Impact Load*

load	percentage of increase
elevators and elevator machinery	100%
reciprocating and power-driven machinery	50%
hangers supporting floors or machinery	33%
light machinery, driven by a shaft or motor	20%

Source: ASCE 7 Sec. 4.6

3 Design and Analysis Methods for Structural Steel

Nomenclature

D	dead load	lbf
F_u	tensile strength	lbf/in^2
F_y	minimum yield stress	lbf/in^2
FS	factor of safety	–
L	live load	lbf
Q_i	nominal effect of load of type i	lbf
R_a	required strength (ASD)	lbf
R_n	nominal strength	lbf
R_u	required strength (LRFD)	lbf

Symbols

γ	effective load factor	–
γ_i	load factor for load of type i	–
Γ	section property	in^2, in^3, in^4
ϕ	resistance factor (LRFD)	–
Ω	safety factor (ASD)	–

1. ALLOWABLE STRENGTH DESIGN (ASD)

Before steel design was formalized, a variety of methods were used to design steel structures. Some were not much more sophisticated than trial and error, but others involved running tests in which steel members were loaded until they failed, and then using the results to determine a maximum allowable safe load for each size and type of member. This approach was the forerunner to *allowable stress design*, which later became *allowable strength design*.

In 1923, the American Institute of Steel Construction (AISC) formalized the procedures for designing structural steel members. By then, enough testing had been performed that results were consistently predictable. A steel member can fail in a number of different ways, such as by buckling or by rupturing. For each such *failure mode*, the amount of stress that would cause a member to fail in that way could be determined. A safety factor was assigned to each failure mode based on its effects on a structure and its occupants.

Allowable stress design was stated in terms of keeping induced stresses less than allowable stresses, each allowable stress being equivalent to either the yield stress or tensile stress of the steel member, depending on the failure mode being considered, divided by the appropriate safety factor for that mode.

- calculated stress ≤ allowable stress
- allowable stress = (yield stress, F_y, or tensile stress, F_u) ÷ appropriate safety factor, FS

These relationships are combined and described by the following equation.

$$\frac{\sum Q_i}{\Gamma} \leq \frac{F_y \text{ or } F_u}{\text{FS}} \qquad 3.1$$

In Eq. 3.1, Q_i is the nominal effect of a load of type i and Γ is the appropriate section property (such as gross area, net area, effective area, and so on).

Since the introduction of other design methods based on ultimate strength (such as load and resistance factor design—see Sec. 2), AISC has moved to base this method on strength as well. With the thirteenth edition of the *AISC Manual*, allowable stress design has been replaced with *allowable strength design* (ASD). This is very similar to the older method and uses the same load combinations, but the provisions are expressed in terms of forces and moments rather than stresses. In the older method, calculated design stresses cannot exceed the specified allowable design stress; in allowable strength design, calculated design loads cannot exceed the calculated strength capacity. The term "strength" describes the load capacities now listed in the *AISC Manual*'s tables more accurately than "stress" would.

In this book, the abbreviation ASD always stands for *allowable strength design*. However, the term *allowable stress design* is still used by other authorities, including ASCE 7.

2. LOAD AND RESISTANCE FACTOR DESIGN (LRFD)

By the time AISC published the ninth edition of the *Manual of Steel Construction: Allowable Stress Design*, it had already published the first edition of the *Manual of Steel Construction: Load and Resistance Factor Design* with the intention that the LRFD method would eventually replace the allowable stress design method.

The purpose of developing the LRFD method was to establish a theoretically more consistent and accurate safety factor, based both on variations in load (that is, the left side of Eq. 3.1) and on variations in load capacity (the right side of Eq. 3.1). On the left side of the equation, the size of each load factor varies with the type of load and how predictable it is; for example, live loads are more difficult to predict accurately than dead loads, so the load factor is larger for live loads than for dead loads. On the right side of the equation, the size of the resistance factor is a function of the limit states for the various modes of failure and the normal variances in steel manufacture.

The result is a more efficient use of steel. Using LRFD instead of ASD can often reduce the weight of needed structural steel members by 5% to 15%. Whether such a reduction can in fact be made depends on the live-to-dead load ratio and other design criteria such as serviceability. For example, the need to limit beam deflection may demand a heavier or deeper beam than strength requirements alone would call for.

The general requirement of LRFD is that the required strength is less than or equal to the design strength. This can be stated as follows.

$$\sum \gamma_i Q_i \leq \phi R_n \qquad 3.2$$

The left side of Eq. 3.2 is the sum of the applied load types—each load type, Q_i, multiplied by its applicable load factor, γ_i. The right side of Eq. 3.2 represents the nominal load capacity, R_n, multiplied by the applicable resistance factor, ϕ.

3. DESIGN BASIS

The term *design basis* is used to designate the method (LRFD or ASD) used in the design or analysis of the structure.

For LRFD, the *required strength* is determined by combining factored loads (nominal load × respective load factor) in the combinations given in ASCE 7 Sec. 2.3 (Eq. 2.9 through Eq. 2.15 in this book). The critical (that is, controlling or governing) load combination is the one that gives the greatest total load; this load is equivalent to the required strength.

The required strength, R_u, must be less than or equal to the *design strength*, ϕR_n. This is expressed by Eq. 3.3.

$$R_u \leq \phi R_n \quad \text{[AISC Eq. B3-1]} \qquad 3.3$$

For ASD, the *required strength* is determined by combining nominal loads in the combinations given in ASCE 7 Sec. 2.4 (Eq. 2.1 through Eq. 2.8 in this book). The critical (that is, controlling or governing) load combination is the one that gives the greatest total load; this load is equivalent to the required strength.

The required strength, R_a, must be less than or equal to the *allowable strength*, R_n/Ω. This is expressed by Eq. 3.4.

$$R_a = \frac{R_n}{\Omega} \quad \text{[AISC Eq. B3-2]} \qquad 3.4$$

In both ASD and LRFD methods, the term *limit state* refers to the design limit for a failure mode that could occur, based both on the member's properties and the load conditions.

4. DEFLECTION AND ELONGATION CALCULATIONS

When calculating beam deflections or elongation of tension or compression members, the *service load* (unfactored load) should be used. Using factored loads will result in a value that is too large.[1]

5. EFFECTIVE LOAD FACTOR

In 1986, AISC calibrated the LRFD with the allowable stress design method at $L/D = 3.0$. The effective load factor, γ, is found by setting the LRFD load combination equal to the equivalent allowable stress design load combination.

$$1.2D + 1.6L = \gamma(L + D) \qquad 3.5$$

Dividing both sides by D and replacing L/D with 3.0, Eq. 3.5 becomes

$$1.2 + 1.6\left(\frac{L}{D}\right) = \gamma\left(\frac{L}{D} + \frac{D}{D}\right)$$

$$1.2 + (1.6)(3.0) = \gamma(3.0 + 1.0)$$

$$\gamma = 1.5 \qquad 3.6$$

Therefore, when the live load is 3.0 times the dead load, the effective load factor will be 1.5, and LRFD and ASD will result in identical answers.

[1]This is important to remember when using the LRFD method, as using factored loads is an easy mistake to make.

4 Tension Member Design

Nomenclature

a	shortest distance from edge of pin hole to edge of member, measured parallel to direction of force	in
A	area	in^2
b	width	in
c	shortest distance from edge of pin hole to cut edge of corner cropped at 45°, measured perpendicular to cut	in
d	depth or diameter	in
D	dead load	lbf
F	strength or stress	lbf/in^2
g	transverse spacing (gage) between centers of fastener gage lines	in
L	length	in
L	live load	lbf
L/r	slenderness ratio for tension members	–
n	number of items	–
P	force or load	lbf
r	radius of gyration	in
R	required strength (ASD)	lbf
s	longitudinal spacing (pitch) between centers of consecutive holes	in
t	thickness	in
U	reduction coefficient	–
U	shear lag factor	–
w	width	in
$\overline{x}$	connection eccentricity (see *AISC Manual* Table D3.1)	in
$\overline{x}$, $\overline{y}$	horizontal or vertical distance from outer edge of leg or flange to centroid (see *AISC Manual* Table 1-7 and Table 1-8)	in

Symbols

ϕ	resistance factor (LRFD)	–
Ω	safety factor (ASD)	–

Subscripts

a	required (ASD)
e	effective
f	flange
g	gross
h	holes
n	net or nominal
pb	projected bearing
sf	shear on failure path
t	tensile or tension
u	required (LRFD) or ultimate tensile
y	y-axis or yield

1. INTRODUCTION

The design of members for tension is covered in Chap. D of the *AISC Specification.*[1] Chapter D is divided into the following sections.

D1	Slenderness Limitations
D2	Tensile Strength
D3	Effective Net Area
D4	Built-Up Members
D5	Pin-Connected Members
D6	Eyebars

Among structural steel members, the member in pure tension is probably the easiest to design and analyze. Part 5 of the *AISC Manual* contains many tables to assist in the design and analysis of tension members. These tables list the tensile yield strength for the member's gross area, A_g, and the tensile rupture strength for an effective net area, A_e, equal to $0.75A_g$. The table values are conservative as long as the actual effective net area is at least 75% of the gross area. If the effective net area is less than 75% of the gross area, the tensile rupture strength of the member will have to be calculated.

A *pure tension member* is a member that is subjected to axial forces that create uniform tensile stresses across the member's entire cross section. Figure 4.1 shows a member in pure tension. A tension member can consist of a single element (a rod, bar, plate, angle, or W, M, S, or C shape) or a built-up section. Pin-connected members and eyebars are also used for tension members. Tension members can be found in trusses (as chord and web members), in hangers used to support machinery, in lateral-load bracing, and in platforms, stairs, and mezzanines.

---

[1]The AISC *Steel Construction Manual* (*AISC Manual*) is divided into Part 1 through Part 16. The AISC *Specification for Structural Steel Buildings* (referred to in this book as the *AISC Specification*) is Part 16 of the *AISC Manual,* and it is further divided into 14 chapters designated Chap. A through Chap. N.

Figure 4.1 *Member in Pure Tension*

The slenderness ratio for tension members, L/r, has no maximum limit. For tension members other than rods and hangers in tension, however, it is preferable that L/r be no greater than 300.

$$\frac{L}{r} \le 300 \quad \text{[AISC Sec. D1]} \tag{4.1}$$

For single-angle members, the least radius of gyration, r, may be about the z-axis rather than the x- or y-axis.

2. TENSION MEMBER LIMIT STATES

In designing a tension member, there are generally two strength limit states to consider. The first limit state is for yielding on the gross section of the member. The second limit state is for rupture across the member's effective net section; this applies when there are holes in the member or when there is a change in the member's cross-sectional area. (A third limit state is possible, involving a serviceability issue such as excessive elongation for the particular installation.)

When calculating a member's design tensile strength, $\phi_t P_n$ (in LRFD), or a member's allowable tensile strength, P_n/Ω_t (in ASD), both limit states—tensile yielding and tensile rupture—must be considered. In either case, the quantity shall be the lower of the values obtained for the two states.

Tensile yielding on the gross section is calculated with Eq. 4.2.

$$P_n = F_y A_g \quad \text{[AISC Eq. D2-1]} \tag{4.2}$$

A_g is the member's gross cross-sectional area and F_y is the steel's minimum yield stress. P_n is modified by a factor of $\phi_t = 0.90$ (for LRFD) or $\Omega_t = 1.67$ (for ASD).

Tensile rupture on the net section is calculated with Eq. 4.3.

$$P_n = F_u A_e \quad \text{[AISC Eq. D2-2]} \tag{4.3}$$

A_e is the member's effective net area, and F_u is the steel's minimum tensile stress. P_n is modified by a factor of $\phi_t = 0.75$ (for LRFD) or $\Omega_t = 2.00$ (for ASD).

For safety, a tension member should fail by tensile yielding before it fails by tensile rupture. In simpler terms, the member should stretch before it breaks. To ensure that the member's failure is ductile rather than brittle, a tension member should be designed so that its limit state for yielding will be reached before its limit state for rupture. The following formulas ensure ductile failure.

$$0.9F_yA_g \le 0.75F_uA_e \quad \text{[LRFD, AISC Part 5]} \qquad 4.4$$

$$0.6F_yA_g \le 0.5F_uA_e \quad \text{[ASD, AISC Part 5]} \qquad 4.5$$

Both expressions reduce to

$$\frac{A_e}{A_g} \le \frac{0.90F_y}{0.75F_u} = 1.2\left(\frac{F_y}{F_u}\right)$$

In Table 5-1 through Table 5-8 in the *AISC Manual*, where $A_e = 0.75A_g$, tensile rupture rather than tensile yielding may be the governing design value.

3. NET AREA

In designing a tension member, both the member's net cross-sectional area, A_n, and its effective net cross-sectional area, A_e, must be calculated. Where the tension load is transmitted directly to each of the cross-sectional elements by fasteners or welds, the net area is identical to the effective net area. When this is not the case, the net area must be reduced by the applicable reduction coefficient, U, as described later in this chapter. The net area is equal to the member's gross area less the area of the hole or holes in a line that is perpendicular to the axis of the member, A_h, (and, therefore, that is also perpendicular to the force being applied to the member).

$$A_n = A_g - A_h \quad \text{[AISC Part B3.13]} \qquad 4.6$$

To facilitate insertion, the nominal diameter of a hole for a standard bolt or rivet is $^1/_{16}$ in larger than the diameter of the bolt or rivet itself. At the same time, the net width of a hole for a bolt or rivet must be taken as $^1/_{16}$ in greater than the hole's nominal diameter to allow for possible peripheral edge damage caused when punching the hole. It follows from these two requirements that the effective width of a hole will be $^1/_8$ in larger than the diameter of the bolt or rivet.

$$\begin{aligned} A_h &= n_{\text{holes}} t d_{\text{hole}} \\ &= n_{\text{holes}} t \left(d_{\text{bolt}} + 0.125 \text{ in}\right) \end{aligned} \qquad 4.7$$

Combining Eq. 4.6 and Eq. 4.7, the net area can be calculated as

$$A_n = A_g - n_{\text{holes}}\left(d_{\text{bolt}} + 0.125 \text{ in}\right)t \qquad 4.8$$

Example 4.1

Net Area of a Bar or Plate

The steel bars shown are subject to a tensile load, P.

Section properties	Material properties	Bolt properties
$w = 6$ in	ASTM A36 bars	$d_{\text{bolt}} = {}^3/_4$ in
$t = {}^1/_2$ in	$F_y = 36$ ksi	
	$F_u = 58$ ksi	

Determine the gross area and net area of each section and the design tensile strength (LRFD) and allowable tensile strength (ASD) of the member with two holes in it.

Solution

From Eq. 4.6,

$$A_n = A_g - A_h$$

For the bar with one hole,

$$A_g = tw = (0.5 \text{ in})(6 \text{ in}) = 3 \text{ in}^2$$

The effective width of the hole is ${}^1/_8$ in larger than the diameter of the bolt, so, using Eq. 4.7,

$$\begin{aligned} A_h &= n_{\text{holes}} t (d_{\text{bolt}} + 0.125 \text{ in}) \\ &= (1)(0.5 \text{ in})(0.75 \text{ in} + 0.125 \text{ in}) \\ &= 0.44 \text{ in}^2 \end{aligned}$$

Therefore,

$$A_n = A_g - A_h = 3 \text{ in}^2 - 0.44 \text{ in}^2 = 2.56 \text{ in}^2$$

For the bar with two holes,

$$\begin{aligned} A_g &= tw = (0.5\text{ in})(6\text{ in}) \\ &= 3\text{ in}^2 \end{aligned}$$

The effective width of the hole, d_{hole}, is $^1/_8$ in larger than the diameter of the bolt, so using Eq. 4.7,

$$\begin{aligned} A_h &= n_{\text{holes}} t(d_{\text{bolt}} + 0.125\text{ in}) \\ &= (2)(0.5\text{ in})(0.75\text{ in} + 0.125\text{ in}) \\ &= 0.88\text{ in}^2 \end{aligned}$$

Therefore,

$$A_n = A_g - A_h = 3\text{ in}^2 - 0.88\text{ in}^2 = 2.12\text{ in}^2$$

For each limit state, determine the design tensile strength, $\phi_t P_n$ (LRFD), and the allowable tensile strength, P_n/Ω_t (ASD), of the member with two holes. For the yield limit state,

$$\begin{aligned} P_n &= F_y A_g = \left(36\ \frac{\text{kips}}{\text{in}^2}\right)(3\text{ in}^2) \\ &= 108\text{ kips} \end{aligned}$$

LRFD	ASD
$\phi_t P_n = (0.90)(108\text{ kips})$ $= 97.2\text{ kips}$	$\dfrac{P_n}{\Omega_t} = \dfrac{108\text{ kips}}{1.67} = 64.67\text{ kips}$

For the rupture limit state,

$$\begin{aligned} P_n &= F_u A_n = \left(58\ \frac{\text{kips}}{\text{in}^2}\right)(2.12\text{ in}^2) \\ &= 122.96\text{ kips} \end{aligned}$$

LRFD	ASD
$\phi_t P_n = (0.75)(122.96\text{ kips})$ $= 92.22\text{ kips}$	$\dfrac{P_n}{\Omega_t} = \dfrac{122.96\text{ kips}}{2.00}$ $= 61.48\text{ kips}$

In both LRFD and ASD, the rupture limit is lower than the yield limit. Therefore, the governing limit state is rupture on the net area of the member, and this is not a ductile failure. The design tensile strength (LRFD) is 92.22 kips, and the allowable tensile strength (ASD) is 61.48 kips.

Example 4.2

Net Area of an Angle

The steel angle shown is fabricated from A36 stock and is subject to a tensile load, P.

For angles, the gage for holes in opposite adjacent legs is the sum of the gages from the back of the angles, less the thickness of the angle. When the load is transmitted directly to each cross-sectional element by connectors, the effective net area is equal to the net area.

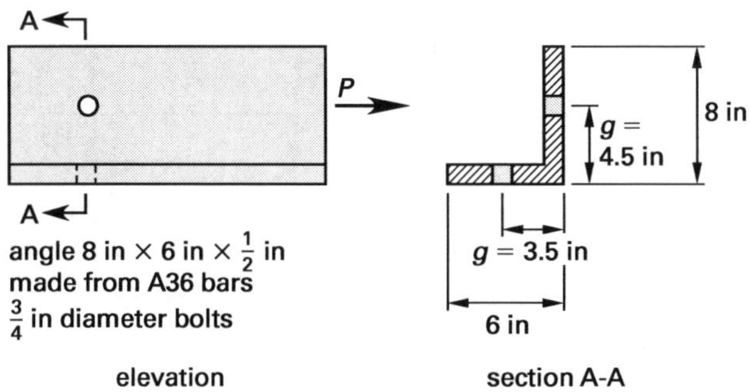

Section properties

$t = {}^1/_2$ in

$d_{\text{bolt}} = {}^3/_4$ in

Material properties

$F_y = 36$ ksi

$F_u = 58$ ksi

Determine the angle's gross area, net area, design tensile strength (LRFD), and allowable tensile strength (ASD).

Solution

The effective gross width of the angle is

$$\begin{aligned} w_e &= w_1 + w_2 - t \\ &= 8 \text{ in} + 6 \text{ in} - 0.5 \text{ in} \\ &= 13.5 \text{ in} \end{aligned}$$

The gross area of the angle is

$$\begin{aligned} A_g &= tw_e = (0.5 \text{ in})(13.5 \text{ in}) \\ &= 6.75 \text{ in}^2 \end{aligned}$$

(If this were a rolled angle rather than a built-up one, the tabulated gross area from the *AISC Manual* would be used.) The effective width of the hole, d_{hole}, is ${}^1/_8$ in larger than the diameter of the bolt. Using Eq. 4.7,

$$\begin{aligned} A_h &= n_{\text{holes}} t (d_{\text{bolt}} + 0.125 \text{ in}) \\ &= (2)(0.5 \text{ in})(0.75 \text{ in} + 0.125 \text{ in}) \\ &= 0.88 \text{ in}^2 \end{aligned}$$

Therefore,

$$A_n = A_g - A_h = 6.75 \text{ in}^2 - 0.88 \text{ in}^2 = 5.87 \text{ in}^2$$

For each limit state, determine the design tensile strength, $\phi_t P_n$ (LRFD), and the allowable tensile strength, P_n/Ω_t (ASD), of the steel angle. For the yield limit state,

$$\begin{aligned} P_n &= F_y A_g = \left(36 \ \frac{\text{kips}}{\text{in}^2}\right)\left(6.75 \text{ in}^2\right) \\ &= 243 \text{ kips} \end{aligned}$$

LRFD	ASD
$\phi_t P_n = (0.90)(243 \text{ kips}) = 218.7 \text{ kips}$	$\dfrac{P_n}{\Omega_t} = \dfrac{243 \text{ kips}}{1.67} = 145.5 \text{ kips}$

For the rupture limit state,

$$\begin{aligned} P_n &= F_u A_n = \left(58 \ \frac{\text{kips}}{\text{in}^2}\right)\left(5.87 \text{ in}^2\right) \\ &= 340.46 \text{ kips} \end{aligned}$$

LRFD	ASD
$\phi_t P_n = (0.75)(340.46 \text{ kips}) = 255.35 \text{ kips}$	$\dfrac{P_n}{\Omega_t} = \dfrac{340.46 \text{ kips}}{2.00} = 170.23 \text{ kips}$

In both LRFD and ASD, the yield limit is lower than the rupture limit, so the governing limit state is yielding on the gross area of the member. The design tensile strength (LRFD) is 218.7 kips, and the allowable tensile strength (ASD) is 145.5 kips.

Calculating the effective width of a channel is similar to the method used to calculate the angle. To obtain the effective width of a channel, add the width of the two flanges to the depth of the web and subtract twice the average thickness of the flanges.

4. NET AREA FOR A CHAIN OF HOLES

If a chain of holes runs in a diagonal or zigzag line across a member, the net area of the member is

$$A_n = A_g - \sum d_{\text{hole}} t + \sum \left(\frac{s^2}{4g}\right)_{ij} t \quad [\text{AISC Sec. B4.3b}] \qquad 4.9$$

In other words, the net area is equal to the gross area minus the area of the holes, plus the quantity $s^2/4g$ for each gage line. The line between two consecutive holes in the chain is a *gage line* if it is diagonal (neither parallel nor perpendicular) to the direction of the load. In the quantity $s^2/4g$, s is the longitudinal spacing (or *pitch*) between the centers of the two holes, and g is the transverse spacing (or *gage*) between the centers of the two holes. (See Fig. 4.2.)

Figure 4.2 *Net Area for a Diagonal or Zigzag Chain of Holes*

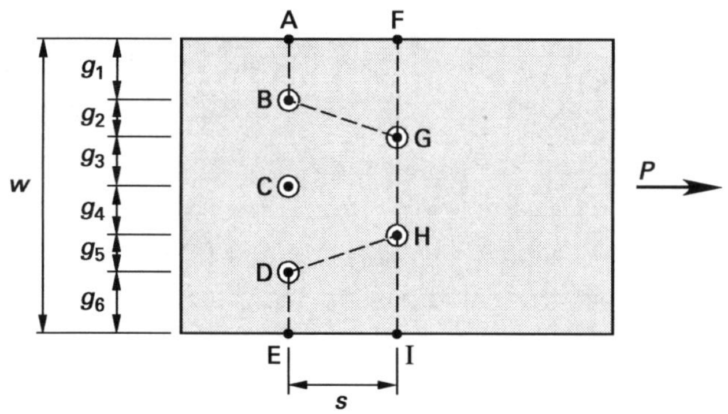

Example 4.3

Net Area for a Chain of Holes

The steel plate shown in Fig. 4.2 is fabricated from A36 stock and is subjected to a tensile load, P.

Section properties	Material properties	Bolt properties
$w = 10$ in	$F_y = 36$ ksi	$d_{\text{bolt}} = {}^3/_4$ in
$t = {}^1/_2$ in	$F_u = 58$ ksi	$s = 2$ in
		$g_1 = g_6 = 2$ in
		$g_2 = g_3 = g_4 = g_5 = 1.5$ in

Determine the plate's gross area and critical net area. Also, determine the design tensile strength (LRFD) and allowable tensile strength (ASD).

Solution

In Fig. 4.2, rupture will occur through the net area of a chain of holes, either F–G–H–I or A–B–G–H–D–E. (Failure could not occur first through chain A–B–C–D–E because the stress in the material is dissipated as the load is transferred from where it is applied to the far end of the connection.) Use Eq. 4.9 to determine which chain will fail first. The area having the least value is the critical net area and will govern the design.

$$A_g = wt = (10\text{ in})(0.5\text{ in}) = 5\text{ in}^2$$

For chain F–G–H–I, from Eq. 4.8,

$$\begin{aligned} A_n &= A_g - n_{\text{holes}}(d_{\text{bolt}} + 0.125\text{ in})t \\ &= 5.0\text{ in}^2 - (2)(0.75\text{ in} + 0.125\text{ in})(0.5\text{ in}) \\ &= 4.125\text{ in}^2 \end{aligned}$$

Chain A–B–G–H–D–E is staggered, so Eq. 4.9 is needed.

$$\begin{aligned}A_n &= A_g - \sum d_{\text{hole}}t + \sum\left(\frac{s^2}{4g}\right)_{ij} t \\ &= 5.0\ \text{in}^2 - (4)(0.875\ \text{in})(0.5\ \text{in}) + (2)\left(\frac{(2\ \text{in})^2}{(4)(1.5\ \text{in})}\right)(0.5\ \text{in}) \\ &= 3.92\ \text{in}^2\end{aligned}$$

Therefore, A–B–G–H–D–E, with a smaller net area of 3.92 in^2, governs for the limit state of tensile rupture on the net area. As indicated in the following table, the governing limit state is yielding on the gross area of the member.

For the yield limit state,

$$\begin{aligned}P_n &= F_yA_g = \left(36\ \frac{\text{kips}}{\text{in}^2}\right)(5\ \text{in}^2) \\ &= 180\ \text{kips}\end{aligned}$$

LRFD	ASD
$\phi_tP_n = (0.90)(180\ \text{kips}) = 162\ \text{kips}$	$\frac{P_n}{\Omega_t} = \frac{180\ \text{kips}}{1.67} = 107.78\ \text{kips}$

For the rupture limit state,

$$\begin{aligned}P_n &= F_uA_n = \left(58\ \frac{\text{kips}}{\text{in}^2}\right)(3.92\ \text{in}^2) \\ &= 227.36\ \text{kips}\end{aligned}$$

LRFD	ASD
$\phi_tP_n = (0.75)(227.36\ \text{kips}) = 170.52\ \text{kips}$	$\frac{P_n}{\Omega_t} = \frac{227.36\ \text{kips}}{2.00} = 113.68\ \text{kips}$

In both LRFD and ASD, the yield limit is lower than the rupture limit. Therefore, the governing limit state is yielding on the gross area of the member. The design tensile strength (LRFD) is 162 kips, and the allowable tensile strength (ASD) is 107.78 kips.

5. REDUCTION COEFFICIENTS FOR EFFECTIVE NET AREA

When a tension member is connected to a supporting member in such a way that stress is not uniformly distributed, some of the tension member's load-carrying capacity is lost. This phenomenon is called *shear lag*. A common cause of shear lag is a tensile load transmitted by bolts, rivets, or welds through some but not all of the cross-sectional elements of the member.

For example, Fig. 4.3 shows an angle connected to its support by only one of its legs. As shown, part of the member is not contributing fully to the angle's load-carrying capacity.

Figure 4.3 *Shear Lag Effect Shown on Angle*

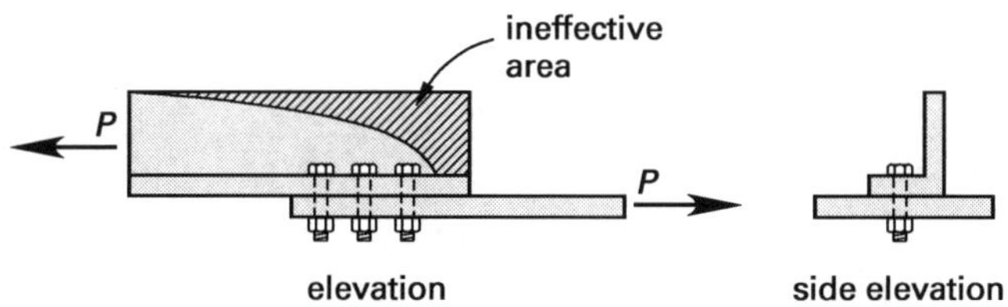

When this is the case, calculations of load-carrying capacity are based not on the member's net cross-sectional area but on a smaller value, the *effective net area*, A_e, which is obtained by multiplying the net area by a *shear lag factor*, U. For bolted sections, the shear lag factor is applied to the net section, A_n.

$$A_e = A_n U \quad \text{[AISC Eq. D3-1]} \tag{4.10}$$

For welded connections, the factor is applied to the gross section, A_g.

$$A_e = A_g U \tag{4.11}$$

As the length of the connection is increased, the shear lag effects diminish.

AISC Specification Table D3.1 describes eight ways of joining members in tension and gives a corresponding shear lag factor for each. Six of these are shown in Table 4.1. (Cases 5 and 6 are omitted here because they involve connections with hollow structural sections in tension.)

***Table 4.1** Shear Lag Factors for Connections to Tension Members*

description	shear lag factor
case 1: For all tension members. Tension load is transmitted directly to each cross-sectional element by fasteners or welds (except where case 3 or 4 applies).	$U = 1.0$
case 2: For all tension members except plates and HSS. Tension load is transmitted to some but not all cross-sectional elements by fasteners or longitudinal welds. (L is the length of the connection and $\overline{x}$ is the connection eccentricity. Case 7 may be used as an alternative for W, M, S, and HP sections.)	$U = 1 - \frac{\overline{x}}{L}$
case 3: For all tension members. Tension load is transmitted by transverse welds to some but not all cross-sectional elements.	$U = 1.0$ A = area of directly connected elements
case 4: For plates only. Tension load is transmitted by longitudinal welds only. (L is the length of the weld and w is the width of the welded member.)	$U = 1.0 \quad [L \ge 2w]$ $U = 0.87 \quad [2w > L \ge 1.5w]$ $U = 0.75 \quad [1.5w > L \ge w]$
case 7: For W, S, M, and HP shapes, and for tees cut from these shapes. (b_f is the flange width and d is the depth of the member. If U can also be calculated as in case 2, the larger value may be used.)	
• flange is connected with at least three fasteners per line in direction of load	$U = 0.9 \quad [b_f \ge \frac{2}{3}d]$ $U = 0.85 \quad [b_f < \frac{2}{3}d]$
• web is connected with at least four fasteners per line in direction of load	$U = 0.70$
case 8: For single angles. (If U can also be calculated as in case 2, the larger value may be used.)	
• at least four fasteners per line in direction of load	$U = 0.80$
• two or three fasteners per line in direction of load	$U = 0.60$

Example 4.4
Effective Net Area for W Shape Tension Member

A steel I-shaped member, a W8 × 21, is subject to a tensile load, *P*, as shown.

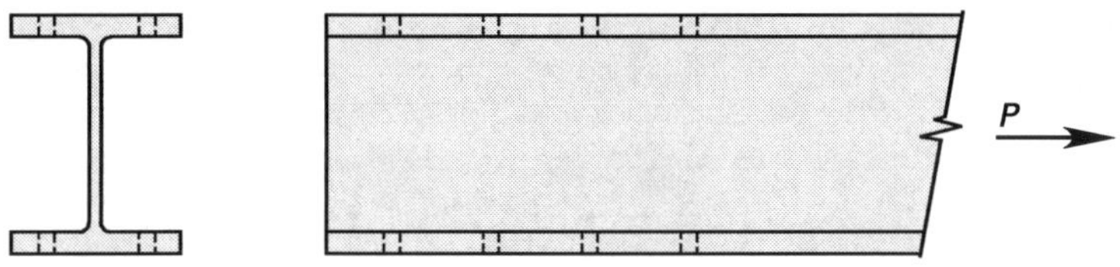

side elevation　　elevation

Section properties	Material properties	Bolt properties
$A_g = 6.16\ \text{in}^2$	ASTM A992 steel	$d_{bolt} = {}^3/_4$ in
$b_f = 5.27$ in	$F_y = 50$ ksi	end distance = 1.25 in
$t_f = 0.40$ in	$F_u = 65$ ksi	spacing = 3.0 in
$d = 8.28$ in		
$r_x = 1.26$ in		
$\bar{y} = 0.831$ in [for WT4 × 10.5, AISC Table 1-8]		

Determine the effective net area. Also, determine the design tensile strength (LRFD) and the allowable tensile strength (ASD).

Solution

Calculate the shear lag factor, *U*. Cases 2 and 7 in Table 4.1 both apply; therefore, it is permissible to take the larger value. Check case 2, considering the member as two WT shapes. The length of the connection is (3)(3.0 in) = 9.0 in.

$$U = 1 - \frac{\bar{x}}{L} = 1 - \frac{0.831\ \text{in}}{9.0\ \text{in}} = 0.91$$

Check case 7 with the flange containing three or more fasteners per line in the direction of loading.

$$\tfrac{2}{3}d = \left(\frac{2}{3}\right)(8.28\ \text{in}) = 5.52\ \text{in} \quad \left[> b_f = 5.27\ \text{in, so } U = 0.85\right]$$

For case 2, $U = 0.91$; for case 7, $U = 0.85$. Use the larger value of $U = 0.91$. From Eq. 4.8, for the net area,

$$\begin{aligned} A_n &= A_g - n_{\text{holes}}\left(d_{\text{bolt}} + 0.125\ \text{in}\right)t \\ &= 6.16\ \text{in}^2 - (4)(0.75\ \text{in} + 0.125\ \text{in})(0.40\ \text{in}) \\ &= 4.760\ \text{in}^2 \end{aligned}$$

From Eq. 4.10, the effective net area is

$$A_e = UA_n = (0.91)(4.760 \text{ in}^2) = 4.33 \text{ in}^2$$

Calculate the design and allowable tensile strengths of the member. For the yield limit state,

$$P_n = F_y A_g = \left(50 \ \frac{\text{kips}}{\text{in}^2}\right)(6.16 \text{ in}^2) = 308 \text{ kips}$$

LRFD	ASD
$\phi_t P_n = (0.90)(308 \text{ kips})$ $= 277.20 \text{ kips}$	$\frac{P_n}{\Omega_t} = \frac{308 \text{ kips}}{1.67}$ $= 184.43 \text{ kips}$

For the rupture limit state,

$$P_n = F_u A_e = \left(65 \ \frac{\text{kips}}{\text{in}^2}\right)(4.33 \text{ in}^2) = 281.45 \text{ kips}$$

LRFD	ASD
$\phi_t P_n = (0.75)(281.45 \text{ kips})$ $= 211.09 \text{ kips}$	$\frac{P_n}{\Omega_t} = \frac{281.45 \text{ kips}}{2.00}$ $= 140.73 \text{ kips}$

In both LRFD and ASD, the rupture limit is lower than the yield limit. Therefore, the governing limit state is rupture on the net area of the member. The design tensile strength (LRFD) is 211.09 kips, and the allowable tensile strength (ASD) is 140.73 kips.

6. LOAD AND RESISTANCE FACTOR DESIGN

The basic design requirement for load and resistance factor design (LRFD) is that each structural component's design strength, ϕR_n, must meet or exceed its required strength, R_u. The required strength is equal to the critical (i.e., greatest) combination of applicable service loads, with each load multiplied by its appropriate load factor.

When designing tension members to resist yielding on the gross area, Eq. 4.12 is used. In this case, the resistance factor for tension, ϕ_t, is 0.90.

$$R_u \leq \phi_t R_n \quad \text{[AISC Eq. B3-1]} \qquad 4.12$$

The nominal strength is

$$R_n = F_y A_g \quad \text{[AISC Eq. D2-1]} \qquad 4.13$$

Combining Eq. 4.12 and Eq. 4.13,

$$R_u \le \phi_t F_y A_g \quad 4.14$$

The minimum gross area required is

$$A_g \ge \frac{R_u}{\phi_t F_y} \quad 4.15$$

When designing tension members to resist rupture on the effective net area, the resistance factor is $\phi_t = 0.75$. The nominal strength is

$$R_n = F_u A_e \quad \text{[AISC Eq. D2-2]} \quad 4.16$$

Combining Eq. 4.12 and Eq. 4.16,

$$R_u \le \phi_t F_u A_e \quad 4.17$$

The minimum effective net area required is

$$A_e \ge \frac{R_u}{\phi_t F_u} \quad 4.18$$

7. ALLOWABLE STRENGTH DESIGN

The basic design requirement for allowable strength design (ASD) is that each structural component's design strength, R_n/Ω, meets or exceeds its required strength, R_a. The required strength is equal to the critical (i.e., greatest) combination of applicable service loads.

When designing tension members to resist yielding on the gross area, Eq. 4.19 is used. The safety factor for tension, Ω_t, is in this case equal to 1.67.

$$R_a \le \frac{R_n}{\Omega_t} \quad \text{[AISC Eq. B3-2]} \quad 4.19$$

Substituting Eq. 4.13 into Eq. 4.19,

$$R_a \le \frac{F_y A_g}{\Omega_t} \quad 4.20$$

The minimum gross area required is therefore

$$A_g \ge \frac{\Omega_t R_a}{F_y} \quad 4.21$$

When designing tension members to resist rupture on the effective net area, the basic design requirement is

$$R_a \leq \frac{R_n}{\Omega_t} \quad \text{[AISC Eq. B3-2]} \tag{4.22}$$

The safety factor for tension, Ω_t, is 2.0 here. Combining this with Eq. 4.16,

$$R_a \leq \frac{F_u A_e}{\Omega_t} \tag{4.23}$$

The minimum net effective area required is therefore

$$A_e \geq \frac{\Omega_t R_a}{F_u} \tag{4.24}$$

Example 4.5

Tension Member Design to Resist Yielding and Rupture

The steel I-shaped member shown is made from ASTM A992 steel (F_y = 50 ksi, F_u = 65 ksi) and is subject to the following tensile loads: D = 71 kips and L = 213 kips. Four $^3/_4$ in diameter bolts are to be placed in a row in the flanges perpendicular to the axis of the load; bolt spacing is 3 in.

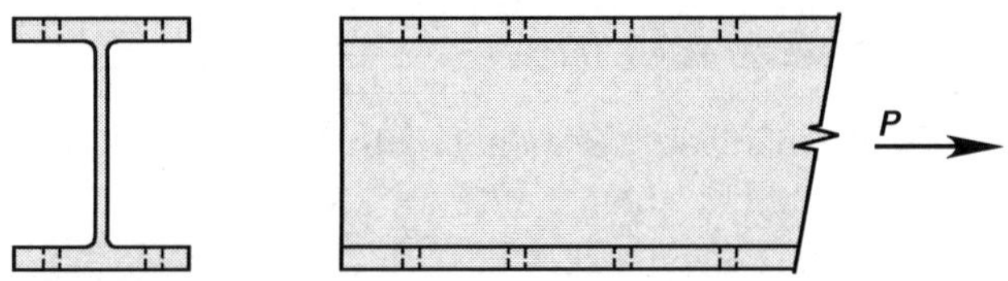

Determine the minimum gross area, A_g, the minimum effective net area, A_e, the design tensile strength, $\phi_t P_n$, and the allowable tensile strength, P_n/Ω_t. Select the lightest W8 section that will support the design loads.

Solution

Calculate the required tensile strength.

LRFD	ASD
$R_u = 1.2D + 1.6L$ $= (1.2)(71 \text{ kips}) + (1.6)(213 \text{ kips})$ $= 426 \text{ kips}$	$R_a = D + L$ $= 71 \text{ kips} + 213 \text{ kips}$ $= 284 \text{ kips}$

Calculate the gross area, A_g, required to resist yielding.

LRFD	ASD
$A_g = \frac{R_u}{\phi_t F_y} = \frac{426 \text{ kips}}{(0.90)\left(50 \frac{\text{kips}}{\text{in}^2}\right)}$ $= 9.47 \text{ in}^2$	$A_g = \frac{\Omega_t R_a}{F_y} = \frac{(1.67)(284 \text{ kips})}{50 \frac{\text{kips}}{\text{in}^2}}$ $= 9.49 \text{ in}^2$

Calculate the net effective area, A_e, required to resist rupture.

LRFD	ASD
$A_e = \frac{R_u}{\phi_t F_u} = \frac{426 \text{ kips}}{(0.75)\left(65 \frac{\text{kips}}{\text{in}^2}\right)}$ $= 8.74 \text{ in}^2$	$A_e = \frac{\Omega_t R_a}{F_u} = \frac{(2.0)(284 \text{ kips})}{65 \frac{\text{kips}}{\text{in}^2}}$ $= 8.74 \text{ in}^2$

The following W8 sections meet the requirement for $A_g \geq 9.49 \text{ in}^2$: W8 × 35, W8 × 40, and W8 × 48. Find the lightest of these sections that meets the required effective net area. For the W8 × 35,

$$A_g = 10.3 \text{ in}^2, d = 8.12 \text{ in}, b_f = 8.02 \text{ in}, t_f = 0.495 \text{ in}$$

Calculate the areas of the holes in the flanges.

$$A_h = 4d_{\text{hole}}t_f = (4)(0.875 \text{ in})(0.495 \text{ in}) = 1.73 \text{ in}^2$$

Calculate the net area.

$$A_n = A_g - A_h = 10.3 \text{ in}^2 - 1.73 \text{ in}^2 = 8.57 \text{ in}^2$$

The net area of the W8 × 35 is less than the required effective net area, $A_e = 8.74 \text{ in}^2$, so this is not OK. Try the next lightest member. For the W8 × 40,

$$A_g = 11.7 \text{ in}^2, d = 8.25 \text{ in}, b_f = 8.07 \text{ in}, t_f = 0.560 \text{ in}$$

Calculate the areas of the holes in the flanges.

$$A_h = 4d_h t_f = (4)(0.875 \text{ in})(0.560 \text{ in}) = 1.96 \text{ in}^2$$

Calculate the net area.

$$A_n = A_g - A_h = 11.7 \text{ in}^2 - 1.96 \text{ in}^2 = 9.74 \text{ in}^2$$

The net area of the W8 × 40 is greater than the required effective net area, $A_e = 8.74 \text{ in}^2$; therefore, calculate the shear lag factor, U, based on *AISC Specification* Table D3.1, case 2. The dimension $\overline{x}$ for a W member is obtained from *AISC Manual*

Table 1-8 for a WT with half the depth and weight of the W member. For a WT4 × 20, $\overline{x} = 0.735$ in.

$$U = 1 - \frac{\overline{x}}{L} = 1 - \frac{0.735 \text{ in}}{9.0 \text{ in}} = 0.92$$

Calculate the shear lag factor again, based on *AISC Specification* Table D3.1, case 7.

$$\tfrac{2}{3}d = \left(\tfrac{2}{3}\right)(8.25 \text{ in}) = 5.50 \text{ in}$$
$$b_f = 8.07 \text{ in} \quad \left[> \tfrac{2}{3}d, \text{ so } U = 0.90\right]$$

Case 2 gives the greater value for U and therefore governs. Use Eq. 4.10 to calculate the net effective area, A_e, for a W8 × 40.

$$A_e = A_n U = \left(9.74 \text{ in}^2\right)(0.92)$$
$$= 8.96 \text{ in}^2 \quad \left[> 8.74 \text{ in}^2, \text{ so OK}\right]$$

The net effective area of the W8 × 40 section is greater than required. Therefore, the W8 × 40 section is the lightest W8 section meeting the requirements.

Example 4.6

Tension Member Analysis to Resist Yielding and Rupture

A steel angle is bolted to a gusset plate with three $^3/_4$ in diameter bolts as shown.

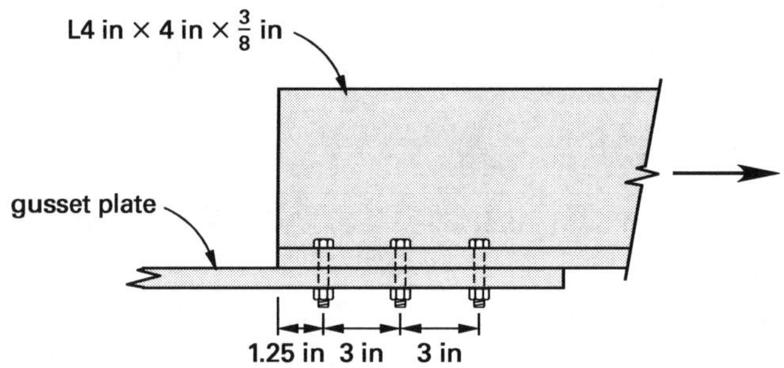

The bolt spacing is 3 in and the end spacing is 1.25 in. The plate and angle are of ASTM A36 steel, with a specified minimum yield stress of 36 ksi and a specified minimum tensile stress of 58 ksi. The angle is 4 in × 4 in × $^3/_8$ in. The gross area is 2.86 in^2. The dimensions $\overline{x}$ and $\overline{y}$ are both equal to 1.13 in. Determine the load capacity of the angle.

Solution

Use Eq. 4.8 to calculate the net area.

$$A_n = A_g - \left(d_{\text{bolt}} + 0.125 \text{ in}\right)t$$
$$= 2.86 \text{ in}^2 - (0.75 \text{ in} + 0.125 \text{ in})(0.375 \text{ in})$$
$$= 2.53 \text{ in}^2$$

Calculate the shear lag factor, U, as the larger of the values permitted from *AISC Specification* Table D3.1. Based on case 8, $U = 0.60$. Based on case 2,

$$U = 1 - \frac{\bar{x}}{L} = 1 - \frac{1.13 \text{ in}}{6 \text{ in}} = 0.81$$

The larger value of U is 0.81. From Eq. 4.10, the net effective area is

$$A_e = A_n U = (2.53 \text{ in}^2)(0.81) = 2.05 \text{ in}^2$$

For the yield limit state,

$$\begin{aligned} P_n &= F_y A_g = \left(36 \ \frac{\text{kips}}{\text{in}^2}\right)(2.86 \text{ in}^2) \\ &= 102.96 \text{ kips} \end{aligned}$$

LRFD	ASD
$\phi_t P_n = (0.90)(102.96 \text{ kips})$ $= 92.66 \text{ kips}$	$\frac{P_n}{\Omega_t} = \frac{102.96 \text{ kips}}{1.67} = 61.65 \text{ kips}$

For the rupture limit state,

$$\begin{aligned} P_n &= F_u A_e = \left(58 \ \frac{\text{kips}}{\text{in}^2}\right)(2.05 \text{ in}^2) \\ &= 118.90 \text{ kips} \end{aligned}$$

LRFD	ASD
$\phi_t P_n = (0.75)(118.96 \text{ kips})$ $= 89.22 \text{ kips}$	$\frac{P_n}{\Omega_t} = \frac{118.96 \text{ kips}}{2.00} = 59.48 \text{ kips}$

In both LRFD and ASD, the rupture limit is lower than the yield limit. Therefore, the governing limit state is rupture on the net area of the angle. The load capacity for LRFD is 89.22 kips and for ASD is 59.48 kips.

AISC Manual Table 5-2 provides the following data for the angle.

$$A_e = 0.75 A_g = (0.75)(2.86 \text{ in}^2) = 2.15 \text{ in}^2$$

$$\begin{aligned} \phi_t P_n &= 92.7 \text{ kips} \quad [\text{yielding}] \\ &= 93.5 \text{ kips} \quad [\text{rupture}] \end{aligned}$$

$$\begin{aligned} \frac{P_n}{\Omega_t} &= 61.7 \text{ kips} \quad [\text{yielding}] \\ &= 62.4 \text{ kips} \quad [\text{rupture}] \end{aligned}$$

The calculated rupture values are less than the tabulated values because the actual effective net area is less than the $0.75A_g$ assumed in the table.

8. PIN-CONNECTED MEMBERS

Pin-connected members are occasionally used for tension members with very large dead loads. It is recommended that they not be used where there is sufficient variation in the live load to cause wearing of the pin holes.

The design of pin-connected members is governed by the geometry and physical dimensions of the member as well as limit states for strength. The governing strength design limit state is the lowest value of the following: tensile rupture, shear rupture, bearing on the member, and yielding.

Tensile rupture on the net effective area is calculated with Eq. 4.25.

$$P_n = 2tb_eF_u \quad \text{[AISC Eq. D5-1]} \tag{4.25}$$

The effective width is

$$b_e = 2t + 0.63 \text{ in} \leq b \tag{4.26}$$

The effective width may not be more than the actual distance from the edge of the hole to the edge of the part, measured in the direction normal to the applied force (b in Fig. 4.4). For tensile rupture, P_n is modified by a factor of $\phi_t = 0.75$ (for LRFD) or $\Omega_t = 2.0$ (for ASD).

Shear rupture on the effective area is calculated with Eq. 4.27.

$$P_n = 0.6F_uA_{\text{sf}} \quad \text{[AISC Eq. D5-2]} \tag{4.27}$$

The shear area on the failure path is

$$A_{\text{sf}} = 2t\left(a + \frac{d}{2}\right) \tag{4.28}$$

In Eq. 4.28, a is the shortest distance from the edge of the pin hole to the edge of the member, measured parallel to the direction of the force. For shear rupture, P_n is modified by a factor of $\phi_t = 0.75$ (for LRFD) or $\Omega_t = 2.0$ (for ASD).

Bearing strength on the thickness of the pin-plate member is calculated with Eq. 4.29.

$$R_n = 1.8F_yA_{\text{pb}} \quad \text{[AISC Eq. J7-1]} \tag{4.29}$$

A_{pb} is the projected bearing area. P_n is modified by a factor of $\phi_{\text{sf}} = 0.75$ (for LRFD) or $\Omega_{\text{sf}} = 2.0$ (for ASD).

Yielding on the gross area of the pin-plate member is calculated with Eq. 4.2. P_n is modified by a factor of $\phi_t = 0.90$ (for LRFD) or $\Omega_t = 1.67$ (for ASD).

The geometry and nomenclature for a pin-connected plate are shown in Fig. 4.4. *AISC Specification* Sec. D5.2 includes the following dimensional requirements.

$$a \geq 1.33 b_e \quad (4.30)$$

$$w \geq 2b_e + d_{\text{pin}} \quad (4.31)$$

$$c \geq a \quad (4.32)$$

$$d_{\text{pin hole}} \leq d_{\text{pin}} + \tfrac{1}{32} \text{ in} \quad (4.33)$$

Figure 4.4 *Nomenclature for Pin Connection*

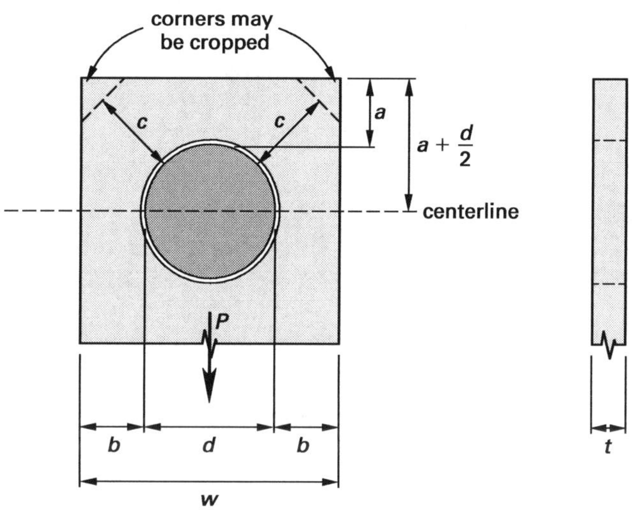

Example 4.7

Pin-Connected Tension Member

A pin-connected steel tension member supports loads of D = 18 kips and L = 6 kips. The pin diameter is 1.25 in, and the diameter of the hole for the pin is $^1/_{32}$ in larger than the pin diameter. The member is made of ASTM A36 steel, with a specified minimum yield stress of 36 ksi and a specified minimum tensile stress of 58 ksi. The member's width, w, is 5.25 in, and its thickness, t, is 0.625 in. The dimensions labeled a and c in Fig. 4.4 are 2.5 in and 3.0 in, respectively. The pin can be assumed to be satisfactory for supporting loads.

Determine whether the geometry of the pin-connected member meets the requirements of the code and whether the member will support the imposed loads.

Solution

From Eq. 4.26, the effective width is

$$\begin{aligned} b_e &= 2t + 0.63 \\ &= (2)(0.625\text{ in}) + 0.63\text{ in} \\ &= 1.88\text{ in} \end{aligned}$$

Check dimensional properties for conformance to *AISC Specification* requirements using Eq. 4.30 through Eq. 4.33.

$$\begin{aligned} a &\geq 1.33b_e \\ 2.5\text{ in} &\geq (1.33)(1.88) \\ &\geq 2.50\text{ in} \quad [\text{OK}] \\ w &\geq 2b_e + d \\ 5.25\text{ in} &\geq (2)(1.88\text{ in}) + 1.25\text{ in} \\ &\geq 5.01\text{ in} \quad [\text{OK}] \\ c &\geq a \\ 3.0\text{ in} &\geq 2.5\text{ in} \quad [\text{OK}] \end{aligned}$$

AISC Specification requirements are met. Calculate the required tensile strength.

LRFD	ASD
$P_u = 1.2D + 1.6L$ $= (1.2)(18\text{ kips}) + (1.6)(6\text{ kips})$ $= 31.20\text{ kips}$	$P_a = D + L$ $= 18\text{ kips} + 6\text{ kips}$ $= 24\text{ kips}$

Calculate the available tensile rupture strength on the effective net area using Eq. 4.25.

$$\begin{aligned} P_n &= 2tb_eF_u \\ &= (2)(0.625\text{ in})(1.88\text{ in})\left(58\ \frac{\text{kips}}{\text{in}^2}\right) \\ &= 136.30\text{ kips} \end{aligned}$$

LRFD	ASD
$\phi_t P_n = (0.75)(136.30\text{ kips})$ $= 102.23\text{ kips}$	$\dfrac{P_n}{\Omega_t} = \dfrac{136.30\text{ kips}}{2.00} = 68.15\text{ kips}$

Calculate the available shear rupture strength. From Eq. 4.28, the shear area on the failure path is

$$A_{sf} = 2t\left(a+\frac{d}{2}\right)$$
$$= (2)(0.625 \text{ in})\left(2.50 \text{ in}+\frac{1.25 \text{ in}}{2}\right)$$
$$= 3.91 \text{ in}^2$$

From Eq. 4.27, the available shear rupture strength is

$$P_n = 0.6F_uA_{sf}$$
$$= (0.6)\left(58 \frac{\text{kips}}{\text{in}^2}\right)(3.91 \text{ in}^2)$$
$$= 136.07 \text{ kips}$$

The required strength is

LRFD	ASD
$\phi_t P_n = (0.75)(136.07 \text{ kips})$ $= 102.05 \text{ kips}$	$\frac{P_n}{\Omega_t} = \frac{136.07 \text{ kips}}{2.00} = 68.04 \text{ kips}$

Calculate the available bearing strength. The projected bearing area is

$$A_{pb} = td = (0.625 \text{ in})(1.25 \text{ in}) = 0.78 \text{ in}^2$$

From Eq. 4.29, the nominal bearing strength is

$$R_n = 1.8F_yA_{pb}$$
$$= (1.8)\left(36 \frac{\text{kips}}{\text{in}^2}\right)(0.78 \text{ in}^2)$$
$$= 50.54 \text{ kips}$$

The available bearing strength is

LRFD	ASD
$\phi_t P_n = (0.75)(50.54 \text{ kips})$ $= 37.91 \text{ kips}$	$\frac{P_n}{\Omega_t} = \frac{50.54 \text{ kips}}{2.00} = 25.27 \text{ kips}$

The gross area is

$$A_g = bt = (5.25 \text{ in})(0.625 \text{ in}) = 3.28 \text{ in}^2$$

Use Eq. 4.2 to calculate the available tensile yielding strength.

$$P_n = F_y A_g = \left(36\ \frac{\text{kips}}{\text{in}^2}\right)(3.28\ \text{in}^2) = 118.08\ \text{kips}$$

LRFD	ASD
$\phi_t P_n = (0.90)(118.08\ \text{kips})$ $= 106.27\ \text{kips}$	$\dfrac{P_n}{\Omega_t} = \dfrac{118.08\ \text{kips}}{2.00} = 59.04\ \text{kips}$

The available tensile strength is governed by the bearing strength limit state.

LRFD	ASD
$(\phi_t P_n)_{\text{bearing}} = 37.91\ \text{kips}$ $[> P_u = 31.20\ \text{kips, so OK}]$	$\left(\dfrac{P_n}{\Omega_t}\right)_{\text{bearing}} = 25.27\ \text{kips}$ $[> P_a = 24\ \text{kips, so OK}]$

Code requirements are met, and the design strength and the allowable strength exceed the required strength. The pin-connected member is acceptable.

5 Steel Beam Design

Nomenclature

a	clear distance between transverse stiffeners	in
a'	distance from end of cover plate	in
A	area	in^2
b	one-half the full flange width, b_f, of an I-shaped member or tee	in
b_f	flange width	in
B	factor for lateral-torsional buckling	–
BF	bending factor	lbf
c	height factor	–
C_b	lateral-torsional buckling modification factor (beam buckling coefficient)	–
C_v	web shear coefficient	–
C_w	warping constant	in^6
d	depth	in
D	dead load	lbf
D	outside diameter	in
E	modulus of elasticity	lbf/in^2
F	strength or stress	lbf/in^2
G	shear modulus of elasticity	lbf/in^2
h	distance between flanges	in
h	effective height	in
h_o	distance between flange centroids	in
I	moment of inertia	in^4
J	torsional constant	in^4
k_v	web plate buckling coefficient	–
L	length	in
L	live load	lbf
L_b	length between braces or braced points	in
L_p	limiting unbraced length for full plastic moment	in
L_r	limiting unbraced length for inelastic lateral-torsional buckling	in
L_v	distance from maximum to zero shear force	in

M	flexural strength, moment, or moment strength	in-lbf
M_A	absolute value of moment at quarter point of unbraced section	in-lbf
M_B	absolute value of moment at centerline of unbraced section	in-lbf
M_C	absolute value of moment at three-quarter point of unbraced section	in-lbf
M_{max}	absolute value of maximum moment in unbraced section	in-lbf
M_r	available moment strength	in-lbf
r	radius of gyration	in
r_{ts}	effective radius of gyration of the compression flange	in
S	elastic section modulus	in^3
S	snow load	lbf
t	thickness	in
V	shear force or shear strength	lbf
w	load per unit length	lbf/in
w	width	in
W	wind load	lbf
$\bar{y}$	vertical distance from edge of member to centroid	in
Y_t	hole reduction coefficient	–
Z	plastic section modulus	in^3

Symbols

Δ	deflection	in
λ	width-to-thickness ratio	–
λ_p	limiting width-to-thickness ratio for compactness	–
λ_r	limiting width-to-thickness ratio for noncompactness	–
ϕ	resistance factor	–
Ω	safety factor	–

Subscripts

a	required (ASD)
b	flexural (bending)
c	compression flange
cr	critical
D	dead load

e	effective
f	flange
g	gross
L	live load
n	net or nominal
p	plastic bending
req	required
T	total
u	required (LRFD) or ultimate tensile
v	shear
w	web
x	about x-axis
y	about y-axis or yield

1. INTRODUCTION

Beams, the most prevalent members in a structure, are designed for flexure and shear. Chapter F of the *AISC Specification* provides the requirements for flexural design, and Chap. G covers shear design. These two chapters include the flexure and shear requirements for plate girders and other built-up flexural members. Chapter F is divided into the following sections.

F1	General Provisions
F2	Doubly Symmetric Compact I-Shaped Members and Channels Bent About Their Major Axis
F3	Doubly Symmetric I-Shaped Members with Compact Webs and Noncompact or Slender Flanges Bent About Their Major Axis
F4	Other I-Shaped Members with Compact or Noncompact Webs Bent About Their Major Axis
F5	Doubly Symmetric and Singly Symmetric I-Shaped Members with Slender Webs Bent About Their Major Axis
F6	I-Shaped Members and Channels Bent About Their Minor Axis
F7	Square and Rectangular HSS Box-Shaped Members
F8	Round HSS
F9	Tees and Double Angles Loaded in the Plane of Symmetry
F10	Single Angles
F11	Rectangular Bars and Rounds
F12	Unsymmetrical Shapes
F13	Proportions of Beams and Girders

The specific requirements of *AISC Specification* Sec. F2 through Sec. F13 are based on the way steel members with various shapes react to forces applied to them. Some equations may apply to more than one section, whereas others are specific to one type of member or loading.

A *beam* primarily supports transverse loads (loads that are applied at right angles to the longitudinal axis of the member). Beams, then, are primarily subjected to flexure, or bending. It's easy to design a beam that is loaded about a single strong or weak axis and through the member's shear center, eliminating torsion. When the loading conditions vary from these, however, beam design becomes more complex.

Such members as girders, rafters, purlins, and girts are sometimes referred to as beams because they, too, are primarily flexural members. *Rafters* and *purlins* are members used to support a roof; rafters run parallel to the slope of the roof, while purlins run perpendicular. *Girts* run horizontally between columns and support the building envelope siding but not a floor or roof. Figure 5.1 shows how these members are used.

Figure 5.1 *Building Frame Showing Rafters, Purlins, and Girts*

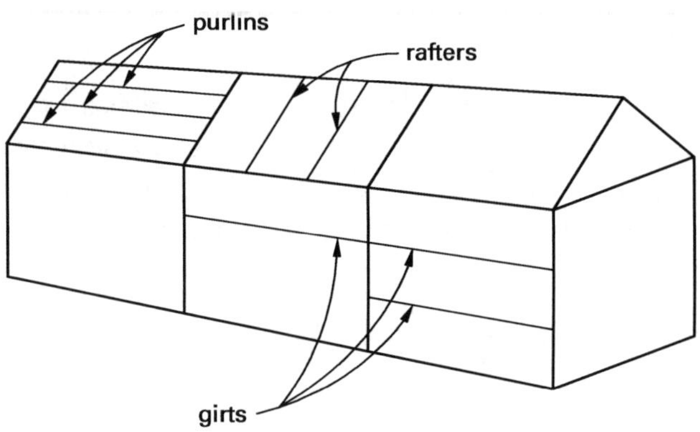

The term *girder* is frequently used to mean a beam that supports other beams. However, the *AISC Manual* uses *girder* to designate an I-shaped member fabricated from plate steel. This book follows AISC usage.

Beams subjected to flexural loads about their major and minor axes simultaneously, and beams subjected to axial compressive or tensile loads in addition to flexural loads, are covered in Chap. H of the *AISC Specification* and in Chap. 8 of this book. The design of plate girders is covered in Chap. 11 of this book.

2. LIMIT STATES

The following are the limit states that should be checked when designing beams.

- yielding
- lateral-torsional buckling
- flange local buckling
- web local buckling

- tension flange yielding
- local leg buckling
- local buckling
- shear

Local buckling can be prevented by using established limits on slenderness ratios for various elements, such as the flanges and webs of the members. Depending on its slenderness ratio, each element is classified (from the lowest ratio to the highest) as compact (C), noncompact (NC), or slender (S). If the flanges and webs are compact, the limit state for the entire member will be reached before local buckling occurs.

3. REQUIREMENTS FOR COMPACT SECTION

Chapter F of the *AISC Specification* classifies flexural members on the basis of

- member type (W, S, M, HP, C, L, or HSS)
- axis of bending (major or minor)
- flange and web slenderness ratios (compact, noncompact, or slender)

There are 11 classes in all, and they are discussed in Sec. F2 through Sec. F12. These cases are described in Table 5.1, along with the applicable section of the *AISC Specification* and the limit states that should be checked for each case. (*AISC Specification* Sec. F13 covers some additional considerations regarding the proportions of beams and girders.)

Whether a member is classified as a compact or noncompact section is a function of the width-to-thickness ratio of its projecting flanges and the height-to-thickness ratio of its web. These ratios were established to ensure that the member would fail in overall yielding before it would fail in local flange or web buckling. Table B4.1b in the *AISC Specification* provides the following limiting width-thickness ratios for compression elements.

For flanges of I-shaped rolled beams and channels in flexure, the member is compact if

$$\lambda_p = \frac{b}{t_f} \leq 0.38\sqrt{\frac{E}{F_y}} \quad [\text{case 10}] \qquad 5.1$$

For webs in flexural compression, the member is compact if

$$\lambda_p = \frac{h}{t_w} \leq 3.76\sqrt{\frac{E}{F_y}} \quad [\text{case 15}] \qquad 5.2$$

Most current ASTM A6 W, S, M, C, and MC shapes have compact flanges for F_y less than or equal to 50 ksi. The exceptions are W21 × 48, W14 × 99, W14 × 90, W12 × 65, W10 × 12, W8 × 31, W8 × 10, W6 × 15, W6 × 9, W6 × 8.5, and M4 × 6. All current ASTM A6 W, S, M, C, and MC shapes have compact webs for F_y less than or equal to 65 ksi.

Table 5.1 *Selection Table for the Application of AISC Chap. F Sections*

AISC section	cross section	flange slenderness	web slenderness	limit states*
F2	doubly symmetrical compact I-shaped members and channels bent about their major axis	C	C	Y, LTB
F3	doubly symmetrical I-shaped members with compact webs and noncompact or slender flanges bent about their major axis	NC, S	C	LTB, FLB
F4	other I-shaped members with compact or noncompact webs bent about their major axis	C, NC, S	C, NC	Y, LTB, FLB, TFY
F5	doubly symmetrical and singly symmetrical I-shaped members with slender webs bent about their major axis	C, NC, S	S	Y, LTB, FLB, TFY
F6	I-shaped members and channels bent about their minor axis	C, NC, S	–	Y, FLB
F7	square and rectangular hollow structural sections (HSS)	C, NC, S	C, NC	Y, FLB, WLB
F8	round hollow structural sections (HSS)	–	–	Y, LB
F9	tees and double angles loaded in plane of symmetry	C, NC, S	–	Y, LTB, FLB
F10	single angles	–	–	Y, LTB, LLB
F11	rectangular bars and rounds	–	–	Y, LTB
F12	unsymmetrical shapes	–	–	all limit states

*Y, yielding; LTB, lateral-torsional buckling; FLB, flange local buckling; WLB, web local buckling; TFY, tension flange yielding; LLB, leg local buckling; LB, local buckling; C, compact; NC, noncompact; S, slender

4. SERVICEABILITY CRITERIA

The criterion that governs the design of a beam often turns out to be the need to keep the beam from deflecting so far that it interferes with the purpose and usefulness of the building. According to the *AISC Manual*, the serviceability of a structure must not be impaired by any deflections caused by appropriate combinations of service loads. The *AISC Manual* no longer gives specific limits on deflections, leaving those decisions up to the engineer, the end user of the structure, and the applicable building codes. In most of the United States, the limits given by the *International Building Code* (IBC) will apply.

A beam deflection criterion is usually expressed as limiting deflection to the length of the span, L, divided by a specified constant; for example, $L/360$ or $L/600$. ACI 530, *Building Code Requirements for Masonry Structures*, limits the deflection of a beam that supports masonry to a maximum of the lesser of $L/600$ and 0.3 in. For some overhead traveling cranes, the deflection is limited to $L/1000$. Table 1604.3 in the IBC provides the specific deflection limitations given in Table 5.2.

Table 5.2 *Deflection Limitations in the International Building Code*

construction	live load, L	snow load, S, or wind load, W	dead load plus live load, $D+L$
roof members			
supporting plaster ceiling	$L/360$	$L/360$	$L/240$
supporting nonplaster ceiling	$L/240$	$L/240$	$L/180$
not supporting ceiling	$L/180$	$L/180$	$L/120$
floor members	$L/360$	–	$L/240$
exterior walls and interior partitions			
with brittle finishes	–	$L/240$	–
with nonbrittle finishes	–	$L/120$	–
farm buildings	–	–	$L/180$
greenhouses	–	–	$L/120$

Other serviceability criteria that must be considered include building drift, vibration, wind-induced motion, expansion and contraction, and connection slip. Chapter L of the *AISC Specification* does not give specific guidance for these criteria, stating only that "under appropriate service load combinations, serviceability issues shall not impair the serviceability of the structure."

The IBC, however, does set seismic design requirements for allowable story drift, incorporating by reference the requirements of ASCE 7, *Minimum Design Loads for Buildings and Other Structures*. These requirements, found in ASCE 7 Table 12.12-1, must be met wherever the IBC applies.

5. LATERAL-TORSIONAL BUCKLING

The moment gradient can be considered when the unbraced length of a beam exceeds the limiting length for full plastic moment, L_p. Under these conditions, the available strength of the beam can be adjusted using the *lateral-torsional buckling modification factor*, C_b (also called the *beam bending coefficient*). Whether the LRFD or ASD method of design is being used, C_b is calculated as in Eq. 5.3.

$$C_b = \frac{12.5M_{\text{max}}}{2.5M_{\text{max}} + 3M_A + 4M_B + 3M_C} \quad \text{[AISC Eq. F1-1]} \qquad 5.3$$

Table 5.3 and Table 5.4 give values of C_b for some common conditions. In all cases, C_b may be taken conservatively as 1.0, though under certain loading conditions this may produce ultraconservative designs. C_b must be taken as 1.0 for cantilevers and overhangs where the free end is unbraced, in accordance with *AISC Specification* Sec. F1. C_b must also be taken as 1.0 for tees with the stems in compression, in accordance with *AISC Commentary* Sec. F9.

Table 5.3 *Values for Lateral-Torsional Buckling Modification Factors for Simply Supported Beams with Concentrated Loads*

loading	lateral bracing along span	lateral-torsional buckling modification factors, C_b
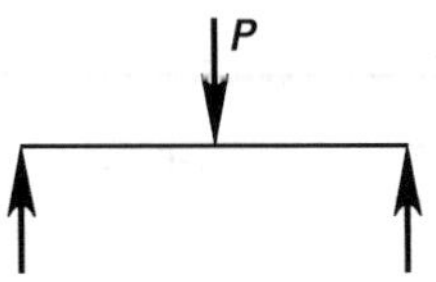	no bracing, load at midpoint	1.32
	bracing at load point	1.67 1.67
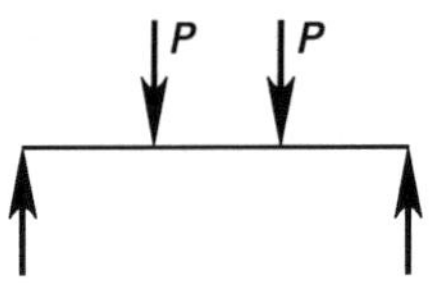	no bracing, loads at third points	1.14
	bracing at load points, loads symmetrically placed	1.67 1.0 1.67
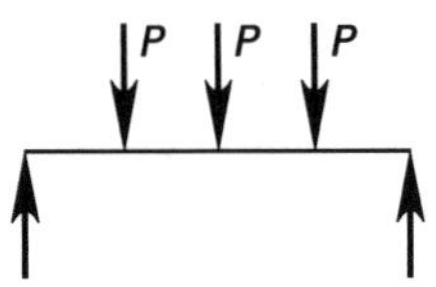	no bracing, loads at quarter points	1.14
	bracing at load points, loads at quarter points	1.67 1.11 1.11 1.67

Lateral bracing must be provided at points of support per *AISC Specification* Chap. F.

Table 5.4 *Values for Lateral-Torsional Buckling Modification Factors for Simply Supported Beams with Uniform Loads*

loading	lateral bracing along span	lateral-torsional buckling modification factors, C_b
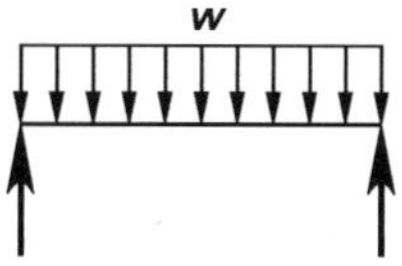 	none	1.14
	at midpoint	1.30 1.30
	at third points	1.45 1.01 1.45
	at quarter points	1.52 1.06 1.06 1.52
	at fifth points	1.56 1.12 1.00 1.12 1.56

M_{max} is the absolute value of the maximum moment in the unbraced section, and M_A, M_B, and M_C are the absolute values of the moments at the quarter point, centerline, and three-quarter point of the unbraced section, respectively.

The nominal flexural strength values in Table 3-2 through Table 3-10 in the *AISC Manual* are based on a value of 1.0 for C_b. As a result, these tables give conservative values for the nominal flexural strength. The value of the adjusted available flexural strength, C_bM_n, can vary from 1.0 to 3.0. However, regardless of the value of C_b, the moment capacity may never be increased to more than the full plastic moment, M_p.

6. FLEXURAL REQUIREMENTS

Flexural stresses are frequently the controlling design criteria for beams. The flexural resistance capacity of a beam is a function of all the following: steel yield point, axis of bending, compact or noncompact section, and the distance between lateral bracing points of the compression flange of the beam, L_b.

The type and extent of the bracing of a beam's compression flange determines its potential failure modes—yielding, inelastic torsional buckling, or elastic torsional buckling—and consequently the appropriate formulas to use for designing or analyzing beams. These three modes are commonly referred to as flexural zones 1, 2, and 3. These zones are illustrated in Fig. 5.2. (See also Sec. 7, Sec. 8, and Sec. 9.)

Figure 5.2 *Moment Capacity Based on Unbraced Length*

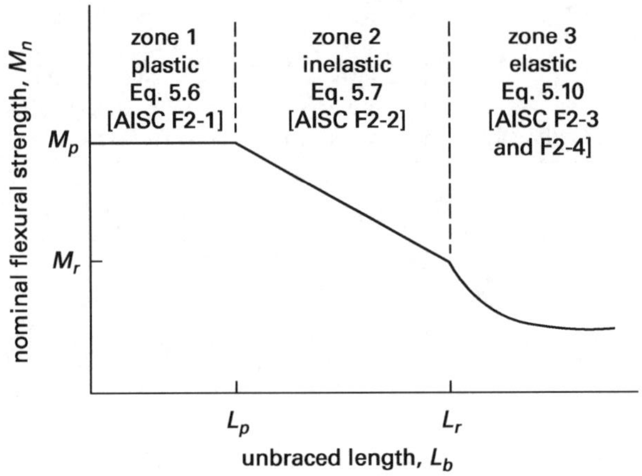

In flexural zone 1 (i.e., yielding), the compression flange of the beam is braced laterally at distances less than or equal to L_p, the limiting length for plastic bending. The lateral bracing prevents the compression flange from buckling, thereby enabling the maximum stress to reach the yield stress and develop the full plastic moment of the section.

In flexural zone 2 (i.e., inelastic torsional buckling), the compression flange is braced laterally at distances greater than L_p but less than or equal to L_r, the limiting length for inelastic torsional buckling. With bracing spaced at this distance, inelastic torsional buckling occurs before the yield stress is reached. M_r is the moment strength available when L_b equals L_r for service loads where the extreme fiber reaches the yield stress, F_y, including the residual stress.

In flexural zone 3 (i.e., elastic torsional buckling), the compression flange is braced at distances greater than L_r. This results in elastic torsional buckling.

Regardless of the flexural zone, the basic design formulas for designing flexural members for strength are based on design requirements given in *AISC Specification* Chap. B, and shown in Eq. 5.4 and Eq. 5.5.

For LRFD, the nominal strength, M_n, when multiplied by a resistance factor, ϕ_b (given in *AISC Specification* Chap. F), must be greater than or equal to the required strength, M_u. The quantity $\phi_b M_n$ is also known as the *design strength*. Equation 5.4 is derived from *AISC Specification* Eq. B3-1.

$$M_u \le \phi_b M_n \quad [\text{LRFD}] \tag{5.4}$$

For ASD, the nominal strength, M_n, when divided by a safety factor, Ω_b (given in *AISC Specification* Chap. F), must be greater than or equal to the required strength, M_a. The quantity M_n/Ω_b is also known as the *allowable strength.* Equation 5.5 is derived from *AISC Specification* Eq. B3-2.

$$M_a \leq \frac{M_n}{\Omega_b} \quad [\text{ASD}] \tag{5.5}$$

The nominal flexural strength, M_n, is taken as the lower of the values obtained according to the limit states of yielding (plastic moment) and lateral-torsional buckling in accordance with *AISC Specification* Chap. F.

All W, S, M, C, and MC shapes, with the exceptions listed in Sec. 3 of this chapter, qualify as compact sections. For this reason, the most important factor in determining a beam's capacity to resist an imparted load is the distance between lateral supports of the compression flange. Increasing the unbraced distance, L_b, between lateral supports decreases the load-carrying capacity of a beam.

7. ZONE 1, PLASTIC BENDING: $L_b \leq L_p$

Beams that are laterally supported at distances less than or equal to L_p and bent about the strong axis are referred to as zone 1 bending (plastic bending zone). These are the easiest to design because in zone 1 the flexural member can obtain the full plastic moment and will not be subjected to lateral-torsional buckling. Therefore, only the limit state of yielding applies, and Eq. 5.6 is used.

$$M_n = M_p = F_y Z_x \quad [\text{AISC Eq. F2-1}] \tag{5.6}$$

Because lateral-torsional buckling will not occur under these conditions, a beam selection can be determined directly by determining the required plastic section modulus, Z_x.

For LRFD, when $M_u \leq 0.90F_yZ_x$,

$$Z_{x,\text{req}} = \frac{M_u}{0.90F_y} \tag{5.7}$$

For ASD, when $M_a \leq F_yZ_x/1.67$,

$$Z_{x,\text{req}} = \frac{1.67M_a}{F_y} \tag{5.8}$$

Example 5.1

Zone 1 Bending

The 30 ft beam shown is laterally supported for its entire length. The beam supports a uniform dead load including the beam weight of 0.30 kip/ft and a uniform live load of 0.70 kip/ft.

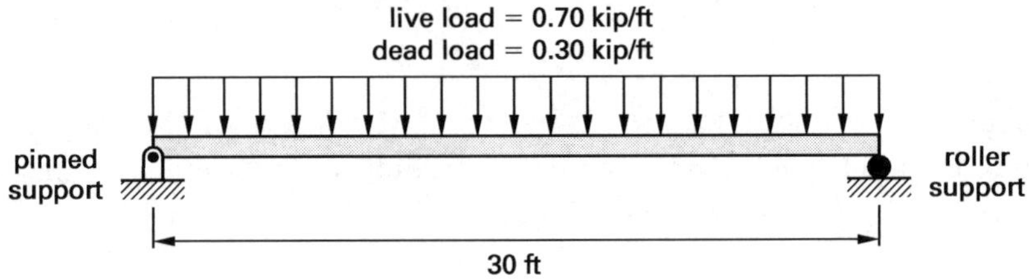

Material properties

ASTM A992 steel

$F_y = 50$ ksi

$F_u = 65$ ksi

Select the most economical beam that complies with the deflection criteria for floor beams in the IBC.

Solution

Calculate the required flexural strength. The loading is

LRFD	ASD
$w_u = 1.2w_D + 1.6w_L$ $= (1.2)\left(0.30\ \frac{\text{kip}}{\text{ft}}\right) + (1.6)\left(0.70\ \frac{\text{kip}}{\text{ft}}\right)$ $= 1.48\ \text{kips/ft}$	$w_a = w_D + w_L$ $= 0.30\ \frac{\text{kip}}{\text{ft}} + 0.70\ \frac{\text{kip}}{\text{ft}}$ $= 1.00\ \text{kip/ft}$

The moment is

LRFD	ASD
$M_u = \frac{w_u L^2}{8}$ $= \frac{\left(1.48\ \frac{\text{kips}}{\text{ft}}\right)(30\ \text{ft})^2}{8}$ $= 166.50\ \text{ft-kips}$	$M_a = \frac{w_a L^2}{8}$ $= \frac{\left(1.00\ \frac{\text{kip}}{\text{ft}}\right)(30\ \text{ft})^2}{8}$ $= 112.50\ \text{ft-kips}$

Calculate the moment of inertia required to comply with the IBC, remembering that deflections are calculated on service loads and not factored loads. The allowable deflection and required moment of inertia for the live load are

$$\begin{aligned}\Delta_L &= \frac{L}{360} \\ &= \frac{(30\text{ ft})\left(12\ \frac{\text{in}}{\text{ft}}\right)}{360} \\ &= 1.00\text{ in}\end{aligned}$$

$$\begin{aligned}I_{\text{req}} &= \frac{5wL^4}{384E\Delta_L} \\ &= \frac{(5)\left(0.70\ \frac{\text{kip}}{\text{ft}}\right)(30\text{ ft})^4\left(12\ \frac{\text{in}}{\text{ft}}\right)^3}{(384)\left(29{,}000\ \frac{\text{kips}}{\text{in}^2}\right)(1.00\text{ in})} \\ &= 439.91\text{ in}^4 \quad \left(440\text{ in}^4\right)\end{aligned}$$

For the total load,

$$\begin{aligned}\Delta_T &= \frac{L}{240} \\ &= \frac{(30\text{ ft})\left(12\ \frac{\text{in}}{\text{ft}}\right)}{240} \\ &= 1.50\text{ in}\end{aligned}$$

$$\begin{aligned}I_{\text{req}} &= \frac{5wL^4}{384E\Delta_T} \\ &= \frac{(5)\left(1.00\ \frac{\text{kip}}{\text{ft}}\right)(30\text{ ft})^4\left(12\ \frac{\text{in}}{\text{ft}}\right)^3}{(384)\left(29{,}000\ \frac{\text{kips}}{\text{in}^2}\right)(1.50\text{ in})} \\ &= 418.97\text{ in}^4\end{aligned}$$

Therefore, I_{req} is governed by the live load deflection, and the minimum moment of inertia is 440 in^4.

Because the beam is laterally supported for its full length, the unbraced length of the beam is 0 ft and the full plastic moment, M_p, will be obtained. Therefore, C_b is 1.0, and the beam is in the plastic range, zone 1 bending.

Calculate the required plastic section modulus.

LRFD	ASD
$M_u \le \phi_b M_n$ $\le \phi_b F_y Z_x$ $Z_{x,\text{req}} = \dfrac{M_u}{\phi_b F_y}$ $= \dfrac{(166.50 \text{ ft-kips})\left(12 \dfrac{\text{in}}{\text{ft}}\right)}{(0.90)\left(50 \dfrac{\text{kips}}{\text{in}^2}\right)}$ $= 44.40 \text{ in}^3$	$M_a \le \dfrac{M_n}{\Omega_b}$ $\le \dfrac{F_y Z_x}{\Omega_b}$ $Z_{x,\text{req}} = \dfrac{M_a \Omega_b}{F_y}$ $= \dfrac{(112.50 \text{ ft-kips})\left(12 \dfrac{\text{in}}{\text{ft}}\right)(1.67)}{50 \dfrac{\text{kips}}{\text{in}^2}}$ $= 45.09 \text{ in}^3$

Basing the selection on strength, for a W14 × 30 (see *AISC Manual* Table 3-2),

LRFD	ASD
$Z_x = 47.3 \text{ in}^3 > 44.40 \text{ in}^3$ $\phi_b M_{px} = 177 \text{ ft-kips} > 166.50 \text{ ft-kips}$ $I_x = 291 \text{ in}^4 < 440 \text{ in}^4$ [no good]	$Z_x = 47.3 \text{ in}^3 > 45.09 \text{ in}^3$ $\dfrac{M_{px}}{\Omega_b} = 118 \text{ ft-kips} > 112.50 \text{ ft-kips}$ $I_x = 291 \text{ in}^4 < 440 \text{ in}^4$ [no good]

Basing the selection on I_{req} for deflection, for a W18 × 35 (see *AISC Manual* Table 3-2),

LRFD	ASD
$Z_x = 66.5 \text{ in}^3 > 44.40 \text{ in}^3$ $\phi_b M_{px} = 249 \text{ ft-kips} > 166.50 \text{ ft-kips}$ $I_x = 510 \text{ in}^4 > 440 \text{ in}^4$ [OK]	$Z_x = 66.5 \text{ in}^3 > 45.09 \text{ in}^3$ $\dfrac{M_{px}}{\Omega_b} = 166 \text{ ft-kips} > 112.50 \text{ ft-kips}$ $I_x = 510 \text{ in}^4 > 440 \text{ in}^4$ [OK]

The design of the member is controlled by deflection rather than by yielding.

8. ZONE 2, INELASTIC BENDING: $L_p < L_b \le L_r$

In zone 2, the flexural member is subjected to inelastic lateral-torsional buckling, and this limit state is applicable in the design or analysis of the member. Therefore, the governing nominal flexural strength, as calculated with Eq. 5.9, must be less than or equal to the full plastic moment.

$$M_n = C_b\left(M_p - \left(M_p - 0.7F_yS_x\right)\left(\frac{L_b - L_p}{L_r - L_p}\right)\right) \le M_p \quad \text{[AISC Eq. F2-2]} \qquad 5.9$$

Calculating the nominal flexural strength using Eq. 5.9 can be simplified with Eq. 5.10 and Eq. 5.11.

$$\phi_b M_n = C_b\left(\phi_b M_{px} - \text{BF}\left(L_b - L_p\right)\right) \le \phi_b M_{px} \quad \text{[LRFD]} \qquad 5.10$$

$$\frac{M_n}{\Omega_b} = C_b\left(\frac{M_{px}}{\Omega_b} - \text{BF}\left(L_b - L_p\right)\right) \le \frac{M_{px}}{\Omega_b} \quad \text{[ASD]} \qquad 5.11$$

The bending factor (BF) for a specific beam depends on the beam's properties and on whether the LRFD or ASD method is being used. Values for bending factors are given in the following *AISC Manual* tables.

- AISC Table 3-2 and Table 3-6: wide-flange (W) shapes
- AISC Table 3-7: I-shaped (S) shapes
- AISC Table 3-8: channel (C) shapes
- AISC Table 3-9: channel (MC) shapes

When the unbraced length of a beam exceeds the limiting length for plastic bending, L_p, but is no greater than the limiting length for inelastic bending, L_r, the beam is subject to inelastic lateral-torsional buckling. It is not possible to select such a member by calculating a required plastic section modulus. The easiest way to select such a member is to calculate the required flexural strength (M_u for LRFD or M_a for ASD) and use *AISC Manual* Table 3-10 with the required flexural strength and the unbraced length.

Example 5.2

Zone 2 Bending

A W21 × 50[c] beam is 40 ft in length, and is laterally supported at its ends and quarter points.[1] The beam supports a uniform dead load including the beam weight of 0.30 kip/ft and a uniform live load of 0.70 kip/ft.

[1]The superscript c on the beam designation in the *AISC Manual* indicates that the member is slender for compression with $F_y = 50$ ksi. Either the flange width-to-thickness or web height-to-thickness ratio exceeds the upper limit for a noncompact element, λ_r, for uniform compression as specified in *AISC Specification* Table B4.1b.

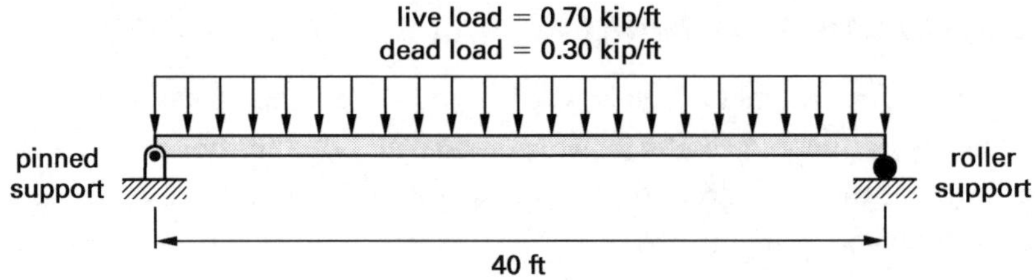

Section properties		Material properties
$A = 14.7\ \text{in}^2$	$I_y = 24.9\ \text{in}^4$	ASTM A992 steel
$d = 20.8\ \text{in}$	$S_y = 7.64\ \text{in}^3$	$F_y = 50\ \text{ksi}$
$t_w = 0.380\ \text{in}$	$Z_y = 12.2\ \text{in}^3$	$F_u = 65\ \text{ksi}$
$b_f = 6.53\ \text{in}$	$r_{ts} = 1.64\ \text{in}$	
$t_f = 0.535\ \text{in}$	$h_o = 20.3\ \text{in}$	
$I_x = 984\ \text{in}^4$	$J = 1.14\ \text{in}^4$	
$S_x = 94.5\ \text{in}^3$	$C_w = 2570\ \text{in}^6$	
$Z_x = 110\ \text{in}^3$		

Determine whether the W21 × 50 steel beam is satisfactory and whether it meets the deflection criteria for floor beams set forth in the IBC.

Solution

Calculate the design loads.

LRFD	ASD
$w_u = 1.2w_D + 1.6w_L$ $= (1.2)\left(0.30\ \frac{\text{kip}}{\text{ft}}\right) + (1.6)\left(0.70\ \frac{\text{kip}}{\text{ft}}\right)$ $= 1.48\ \text{kips/ft}$	$w_a = w_D + w_L$ $= 0.30\ \frac{\text{kip}}{\text{ft}} + 0.70\ \frac{\text{kip}}{\text{ft}}$ $= 1.00\ \text{kip/ft}$

Calculate the required strengths.

LRFD	ASD
$M_u = \frac{w_u L^2}{8} = \frac{\left(1.48\ \frac{\text{kips}}{\text{ft}}\right)(40\ \text{ft})^2}{8}$ $= 296\ \text{ft-kips}$	$M_a = \frac{w_a L^2}{8} = \frac{\left(1.00\ \frac{\text{kip}}{\text{ft}}\right)(40\ \text{ft})^2}{8}$ $= 200\ \text{ft-kips}$

From *AISC Manual* Table 3-6, $L_p = 4.59$ ft, $L_r = 13.6$ ft, BF = 12.1 kips (ASD) and 18.3 kips (LRFD), $M_p/\Omega_b = 274$ ft-kips, and $\phi_b M_p = 413$ ft-kips. The actual unbraced length, L_b, is 10 ft. Therefore, $L_p < L_b \le L_r$ is true, and the beam's failure mode is zone 2. This means that it is subject to inelastic bending and lateral-torsional buckling.

Calculate the available flexural strength. For LRFD, from Eq. 5.10,

$$\begin{aligned}
\phi_b M_n &= C_b\left(\phi_b M_{px} - \text{BF}\left(L_b - L_p\right)\right) \le \phi_b M_{px} \\
&= (1.0)\left(413 \text{ ft-kips} - (18.3 \text{ kips})(10 \text{ ft} - 4.59 \text{ ft})\right) \\
&= 314 \text{ ft-kips} \quad \begin{bmatrix} \le \phi_b M_{px} = 413 \text{ ft-kips} \\ > M_u = 296 \text{ ft-kips} \end{bmatrix}
\end{aligned}$$

For ASD, from Eq. 5.11,

$$\begin{aligned}
\frac{M_n}{\Omega_b} &= C_b\left(\frac{M_{px}}{\Omega_b} - \text{BF}\left(L_b - L_p\right)\right) \le \frac{M_{px}}{\Omega_b} \\
&= (1.0)\left(274 \text{ ft-kips} - (12.1 \text{ kips})(10 \text{ ft} - 4.59 \text{ ft})\right) \\
&= 209 \text{ ft-kips} \quad \begin{bmatrix} \le M_{px}/\Omega_b = 274 \text{ ft-kips} \\ > M_a = 200 \text{ ft-kips} \end{bmatrix}
\end{aligned}$$

The preceding calculations are based on a beam bending coefficient, C_b, of 1.0. The beam is satisfactory. The beam bending coefficient for a uniformly loaded beam braced at the quarter points is 1.06 for the two quarter lengths adjacent to the midspan of the beam. (See Table 5.2.)

For LRFD, the beam is capable of supporting a maximum factored bending moment of

$$C_b\left(\phi_b M_n\right) = (1.06)(314 \text{ ft-kips}) = 332.84 \text{ ft-kips} \quad \left[< \phi_b M_{px}\right]$$

For ASD, the beam is capable of supporting a maximum allowable bending moment due to service loads of

$$C_b\left(\frac{M_n}{\Omega_b}\right) = (1.06)(209 \text{ ft-kips}) = 221.05 \text{ ft-kips} \quad \left[< M_{px}/\Omega_b\right]$$

Calculate the moment of inertia required to comply with the IBC. Deflections are calculated on service loads and not factored loads. The allowable deflection for the live load is

$$\Delta_L = \frac{L}{360} = \frac{(40 \text{ ft})\left(12 \frac{\text{in}}{\text{ft}}\right)}{360} = 1.33 \text{ in}$$

The required moment of inertia for the live load is

$$I_{\text{req}} = \frac{5w_L L^4}{384E\Delta_L}$$

$$= \frac{(5)\left(0.70\ \frac{\text{kip}}{\text{ft}}\right)(40\ \text{ft})^4\left(12\ \frac{\text{in}}{\text{ft}}\right)^3}{(384)\left(29{,}000\ \frac{\text{kips}}{\text{in}^2}\right)(1.33\ \text{in})}$$

$$= 1045.37\ \text{in}^4$$

For the total load,

$$\Delta_T = \frac{L}{240} = \frac{(40\ \text{ft})\left(12\ \frac{\text{in}}{\text{ft}}\right)}{240} = 2.00\ \text{in}$$

$$I_{\text{req}} = \frac{5w_T L^4}{384E\Delta_T}$$

$$= \frac{(5)\left(1.00\ \frac{\text{kip}}{\text{ft}}\right)(40\ \text{ft})^4\left(12\ \frac{\text{in}}{\text{ft}}\right)^3}{(384)\left(29{,}000\ \frac{\text{kips}}{\text{in}^2}\right)(2.00\ \text{in})}$$

$$= 993.10\ \text{in}^4$$

The required moment of inertia for the specified serviceability criteria is controlled by the larger of the required values, 1045.37 in^4. The moment of inertia of a W21 × 50 is 984 in^4. This is not enough to meet the serviceability criteria. There are three possible resolutions to this problem.

- Select a section with a larger moment of inertia.
- Specify that the beam be fabricated with a camber (the deflection should be calculated to determine the amount of camber required).
- Accept the beam based on engineering judgment.

Calculate the deflections for a W21 × 50. For the live load,

$$\Delta_L = \frac{5w_L L^4}{384EI}$$

$$= \frac{(5)\left(0.70\ \frac{\text{kip}}{\text{ft}}\right)(40\ \text{ft})^4\left(12\ \frac{\text{in}}{\text{ft}}\right)^3}{(384)\left(29{,}000\ \frac{\text{kips}}{\text{in}^2}\right)(984\ \text{in}^4)}$$

$$= 1.41\ \text{in}$$

For the total load,

$$\begin{aligned}\Delta_T &= \frac{5w_TL^4}{384EI} \\ &= \frac{(5)\left(1.00\ \frac{\text{kip}}{\text{ft}}\right)(40\ \text{ft})^4\left(12\ \frac{\text{in}}{\text{ft}}\right)^3}{(384)\left(29{,}000\ \frac{\text{kips}}{\text{in}^2}\right)(984\ \text{in}^4)} \\ &= 2.02\ \text{in}\end{aligned}$$

Specify a one inch camber, thereby reducing the live load deflection to 0.41 in and the total load deflection to 1.02 in.

9. ZONE 3, ELASTIC BENDING: $L_b > L_r$

When the unbraced length of a beam exceeds the limiting length for inelastic bending, L_r, the beam is subject to elastic lateral-torsional buckling, and this is the applicable limit state in the design and analysis of the beam.

It is not possible to select a beam for which $L_b > L_r$ by calculating a required plastic section modulus. The easiest way to select such a beam is by calculating the required flexural strength (M_u for LRFD, M_a for ASD) and using *AISC Manual* Table 3-10 with the required flexural strength and unbraced length.

The governing nominal flexural strength will be less than or equal to the full plastic moment.

$$M_n = F_{\text{cr}}S_x \le M_p \quad [\text{AISC Eq. F2-3}] \qquad 5.12$$

The critical stress, F_{cr}, in Eq. 5.12 is

$$F_{\text{cr}} = \frac{C_b\pi^2E}{\left(\frac{L_b}{r_{ts}}\right)^2}\sqrt{1+0.078\left(\frac{Jc}{S_xh_o}\right)\left(\frac{L_b}{r_{ts}}\right)^2} \quad [\text{AISC Eq. F2-4}] \qquad 5.13$$

In Eq. 5.13, the square root factor can be conservatively taken as equal to 1.0. r_{ts} is the effective radius of gyration and is

$$r_{ts}^2 = \frac{\sqrt{I_yC_w}}{S_x} \quad [\text{AISC Eq. F2-7}] \qquad 5.14$$

However, r_{ts} can be approximated accurately and conservatively as the radius of gyration of the compression flange plus $^1/_6$ of the web.

$$r_{ts} = \frac{b_f}{\sqrt{12\left(1+\dfrac{ht_w}{6b_ft_f}\right)}} \tag{5.15}$$

The limiting lengths L_p and L_r are

$$L_p = 1.76r_y\sqrt{\frac{E}{F_y}} \quad \text{[AISC Eq. F2-5]} \tag{5.16}$$

$$L_r = 1.95r_{ts}\left(\frac{E}{0.7F_y}\right) \times\sqrt{\frac{Jc}{S_xh_o}+\sqrt{\left(\frac{Jc}{S_xh_o}\right)^2+6.76\left(\frac{0.7F_y}{E}\right)^2}} \quad \text{[AISC Eq. F2-6]} \tag{5.17}$$

For a doubly symmetrical I-shaped member, c = 1.0. For a channel,

$$c = \frac{h_o}{2}\sqrt{\frac{I_y}{C_w}} \quad \text{[AISC Eq. F2-8b]} \tag{5.18}$$

Example 5.3

Zone 3 Bending

The W8 × 18 steel beam shown is 14.5 ft long and is laterally supported only at the beam ends.

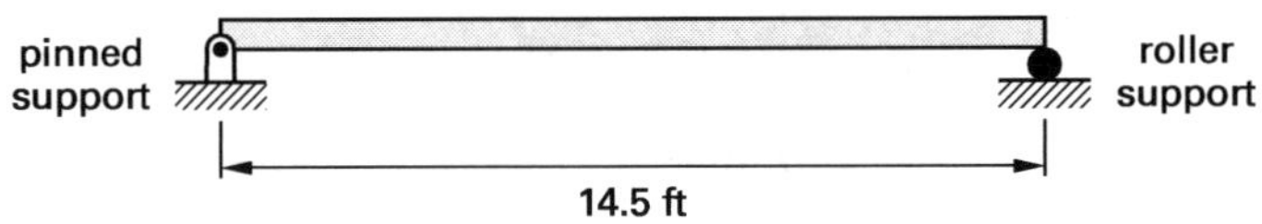

Section properties

A = 5.26 in^2	S_x = 15.2 in^3	h_o = 7.81 in
d = 8.14	Z_x = 17.0 in^3	J = 0.172 in^4
t_w = 0.230 in	I_y = 7.97 in^4	C_w = 122 in^6
b_f = 5.25 in	S_y = 3.04 in^3	L_p = 4.34 ft
t_f = 0.330 in	Z_y = 4.66 in^3	L_r = 13.5 ft
I_x = 61.9 in^4	r_{ts} = 1.43 in	

Material properties

ASTM A992 steel

$F_y = 50$ ksi

$F_u = 65$ ksi

Determine the available moment capacity of the beam.

Solution

The unbraced length, L_b, is 14.5 ft, which exceeds L_r. The beam is thus in zone 3 bending and is subject to elastic lateral-torsional buckling. Calculate the nominal moment capacity. From Eq. 5.18,

$$c = \frac{h_o}{2}\sqrt{\frac{I_y}{C_w}} = \frac{7.81 \text{ in}}{2}\sqrt{\frac{7.97 \text{ in}^4}{122 \text{ in}^6}} = 1.00$$

From Eq. 5.13, the critical stress is

$$\begin{aligned} F_{\text{cr}} &= \frac{C_b \pi^2 E}{\left(\frac{L_b}{r_{ts}}\right)^2}\sqrt{1 + 0.078\left(\frac{Jc}{S_x h_o}\right)\left(\frac{L_b}{r_{ts}}\right)^2} \\ &= \frac{(1)\pi^2\left(29{,}000 \ \frac{\text{kips}}{\text{in}^2}\right)}{\left(\frac{(14.5 \text{ ft})\left(12 \ \frac{\text{in}}{\text{ft}}\right)}{1.43 \text{ in}}\right)^2}\sqrt{\begin{array}{l} 1 + (0.078)\left(\frac{(0.172 \text{ in}^4)(1.00)}{(15.2 \text{ in}^3)(7.81 \text{ in})}\right) \\ \quad \times \left(\frac{(14.5 \text{ ft})\left(12 \ \frac{\text{in}}{\text{ft}}\right)}{1.43 \text{ in}}\right)^2 \end{array}} \\ &= 31.61 \text{ ksi} \end{aligned}$$

From Eq. 5.12, the nominal moment capacity is the lesser of

$$M_n \le \begin{cases} F_{\text{cr}} S_x = \dfrac{\left(31.61 \ \frac{\text{kips}}{\text{in}^2}\right)(15.2 \text{ in}^3)}{12 \ \frac{\text{in}}{\text{ft}}} = 40.04 \text{ ft-kips} \quad [\text{controls}] \\ M_p = F_y Z_x = \dfrac{\left(50 \ \frac{\text{kips}}{\text{in}^2}\right)(17.0 \text{ in}^3)}{12 \ \frac{\text{in}}{\text{ft}}} = 70.83 \text{ ft-kips} \end{cases}$$

Calculate the available flexural strength using LRFD and ASD.

LRFD	ASD
$\phi_b M_n = (0.90)(40.04 \text{ ft-kips}) = 36.04 \text{ ft-kips}$	$\frac{M_n}{\Omega_b} = \frac{40.04 \text{ ft-kips}}{1.67} = 23.98 \text{ ft-kips}$

These calculated moment capacities are comparable to the 36.0 ft-kips and 24.0 ft-kips obtained from *AISC Manual* Table 3-11. The lateral-torsional buckling modification factor, C_b, for a uniformly loaded beam braced at the ends only is 1.14. Taking the modification factor into consideration, the calculated available strength is as follows.

LRFD	ASD
$C_b(\phi_b M_n) = (1.14)(36.04 \text{ ft-kips}) = 41.09 \text{ ft-kips}$	$C_b\left(\frac{M_n}{\Omega_b}\right) = (1.14)(23.98 \text{ ft-kips}) = 27.34 \text{ ft-kips}$

The calculated available strengths are less than $\phi_b M_{px}$ and M_{px}/Ω_b, respectively.

10. WEAK AXIS BENDING: I- AND C-SHAPED MEMBERS

When a beam is bent about its weak axis, lateral-torsional buckling will not occur. Therefore, the beam will fail in yielding or flange local buckling. The nominal flexural strength, M_n, is the lower value obtained according to the limit states of yielding and flange local buckling. The yielding limit state will govern the design, provided the flanges are compact.

All current ASTM A6 W, S, M, C, and MC shapes except the following have compact flanges at $F_y \leq 50$ ksi: W21 × 48, W14 × 99, W14 × 90, W12 × 65, W10 × 12, W8 × 31, W8 × 10, W6 × 15, W6 × 9, W6 × 8.5, and M4 × 6.

The nominal flexural strength for yielding is

$$M_n = M_p = F_y Z_y \leq 1.6 F_y S_y \quad \text{[AISC Eq. F6-1]} \tag{5.19}$$

For sections with noncompact flanges, the nominal flexural strength is

$$M_n = M_p - (M_p - 0.7 F_y S_y)\left(\frac{\lambda - \lambda_{pf}}{\lambda_{rf} - \lambda_{pf}}\right) \quad \text{[AISC Eq. F6-2]} \tag{5.20}$$

In Eq. 5.20, for I-shaped members, λ is the ratio $b/t_f = b_f/2t_f$; for C-shaped members, λ is the ratio b/t_f where b is the full nominal dimension of the flange.

For sections with slender flanges, the nominal flexural strength is

$$M_n = F_{cr} S_y \quad \text{[AISC Eq. F6-3]} \tag{5.21}$$

For Eq. 5.21, the critical flexural stress is

$$F_{cr} = \frac{0.69E}{\left(\frac{b_f}{2t_f}\right)^2} \quad \text{[AISC Eq. F6-4]} \qquad 5.22$$

Example 5.4

Compact W Shape, Weak Axis Bending

A W10 × 30 steel beam is 20 ft long and is subjected to bending about its weak axis only.

Section properties		Material properties
$A = 8.84\ \text{in}^2$	$Z_x = 36.6\ \text{in}^3$	ASTM A992 steel
$b_f = 5.81\ \text{in}$	$I_y = 16.7\ \text{in}^4$	$F_y = 50\ \text{ksi}$
$t_f = 0.510\ \text{in}$	$S_y = 5.75\ \text{in}^3$	$F_u = 65\ \text{ksi}$
$I_x = 170\ \text{in}^4$	$Z_y = 8.84\ \text{in}^3$	
$S_x = 32.4\ \text{in}^3$		

Calculate the nominal flexural strength for yielding about the weak axis.

Solution

The W10 × 30 has compact flanges, or else it would be noted as an exception in *AISC Manual* Table 1-1. This could also be proved by determining the width-to-thickness ratio, $\lambda = b/t$, of the flanges.

Use Eq. 5.19 to find the nominal flexural strength for yielding.

$$M_n \le \begin{cases} F_yZ_y = \dfrac{\left(50\ \dfrac{\text{kips}}{\text{in}^2}\right)\left(8.84\ \text{in}^3\right)}{12\ \dfrac{\text{in}}{\text{ft}}} = 36.83\ \text{ft-kips} \quad \text{[controls]} \\ 1.6F_yS_y = \dfrac{(1.6)\left(50\ \dfrac{\text{kips}}{\text{in}^2}\right)\left(5.75\ \text{in}^3\right)}{12\ \dfrac{\text{in}}{\text{ft}}} = 38.33\ \text{ft-kips} \end{cases}$$

Calculate the available flexural strength.

LRFD	ASD
$\phi_bM_n = (0.90)(36.83\ \text{ft-kips}) = 33.15\ \text{ft-kips}$	$\dfrac{M_n}{\Omega_b} = \dfrac{36.83\ \text{ft-kips}}{1.67} = 22.05\ \text{ft-kips}$

AISC Manual Table 3-4 lists the following values: $\phi_b M_{py}$ = 33.2 ft-kips and M_{py}/Ω_b = 22.1 ft-kips.

Example 5.5

Noncompact W Shape, Weak Axis Bending

A steel W12 × 65[f] beam is 30 ft long and is subjected to bending about its weak axis only.[2]

Section properties		Material properties
$A = 19.1 \text{ in}^2$	$Z_x = 96.8 \text{ in}^3$	ASTM A992 steel
$b_f = 12.0$ in	$I_y = 174 \text{ in}^4$	$F_y = 50$ ksi
$t_f = 0.605$ in	$S_y = 29.1 \text{ in}^3$	$F_u = 65$ ksi
$I_x = 533 \text{ in}^4$	$Z_y = 44.1 \text{ in}^3$	
$S_x = 87.9 \text{ in}^3$		

Calculate the nominal flexural strength for yielding about the weak axis.

Solution

As indicated by the superscript in *AISC Manual* Table 1.1, The W12 × 65 has noncompact flanges at F_y = 50 ksi. This could also be proved by determining the slenderness ratio, $\lambda = b/t$, of the flanges.

Check whether the flanges are compact, noncompact, or slender, using *AISC Specification* Table B4.1b, case 10.

$$\lambda_{pf} = 0.38\sqrt{\frac{E}{F_y}} = 0.38\sqrt{\frac{29{,}000 \ \frac{\text{kips}}{\text{in}^2}}{50 \ \frac{\text{kips}}{\text{in}^2}}}$$
$$= 9.15$$

$$\lambda_{rf} = 1.0\sqrt{\frac{E}{F_y}} = 1.0\sqrt{\frac{29{,}000 \ \frac{\text{kips}}{\text{in}^2}}{50 \ \frac{\text{kips}}{\text{in}^2}}}$$
$$= 24.08$$

$$\lambda = \frac{b_f}{2t_f} = \frac{12.0 \text{ in}}{0.605 \text{ in}}$$
$$= 9.92 \quad [\lambda_p < \lambda < \lambda_r, \text{ so noncompact}]$$

[2]The superscript f on the beam designation means the shape exceeds the compact limit for flexure for F_y = 50 ksi, as noted in *AISC Manual* Table 1-1.

The flanges are noncompact but not slender, and the nominal moment capacity is calculated with Eq. 5.20. First, use Eq. 5.19 to calculate M_p.

$$M_p \le \begin{cases} F_y Z_y = \dfrac{\left(50\ \dfrac{\text{kips}}{\text{in}^2}\right)\left(44.1\ \text{in}^3\right)}{12\ \dfrac{\text{in}}{\text{ft}}} = 183.75\ \text{ft-kips} \quad [\text{controls}] \\ 1.6F_y S_y = \dfrac{(1.6)\left(50\ \dfrac{\text{kips}}{\text{in}^2}\right)\left(29.1\ \text{in}^3\right)}{12\ \dfrac{\text{in}}{\text{ft}}} = 194.00\ \text{ft-kips} \end{cases}$$

From Eq. 5.20,

$$\begin{aligned} M_n &= M_p - \left(M_p - 0.7F_y S_y\right)\left(\frac{\lambda - \lambda_{pf}}{\lambda_{rf} - \lambda_{pf}}\right) \\ &= 183.75\ \text{ft-kips} - \left(183.75\ \text{ft-kips} - \frac{(0.7)\left(50\ \dfrac{\text{kips}}{\text{in}^2}\right)\left(29.1\ \text{in}^3\right)}{12\ \dfrac{\text{in}}{\text{ft}}}\right) \\ &\quad \times\left(\frac{9.92-9.15}{24.08-9.15}\right) \\ &= 178.65\ \text{ft-kips} \end{aligned}$$

Calculate the available flexural strength using LRFD and ASD.

LRFD	ASD
$\phi_b M_n = (0.90)(178.65\ \text{ft-kips})$ $= 160.79\ \text{ft-kips}$	$\dfrac{M_n}{\Omega_b} = \dfrac{178.65\ \text{ft-kips}}{1.67}$ $= 106.98\ \text{ft-kips}$

AISC Manual Table 3-4 lists $\phi_b M_{py}$ as 161 ft-kips and M_{py}/Ω_b as 107 ft-kips.

11. SQUARE AND RECTANGULAR HSS AND BOX MEMBERS

Square and rectangular hollow structural sections (HSS) and doubly symmetrical box-shaped members bent about either axis may have compact or noncompact webs, and compact, noncompact, or slender flanges, as defined in *AISC Specification* Table B4.1b. (See Table 5.5.)

Table 5.5 *Compactness Criteria for Square and Rectangular HSS (F_y = 46 ksi)*

	compression	flexure		shear
nominal wall thickness	nonslender up to flange width of (in)	compact up to flange width of (in)	compact up to web width of (in)	C_v = 1.0 up to web depth of (in)
$^5/_8$ in	20	18	20	20
$^1/_2$ in	16	14	20	20
$^3/_8$ in	12	10	20	20
$^5/_{16}$ in	10	9	18	18
$^1/_4$ in	8	7	14	14
$^3/_{16}$ in	6	5	10	10
$^1/_8$ in	4	3.5	7	7

(Multiply in by 25.4 to obtain mm.)

Source: *AISC Specification* Table B4.1b

Square and rectangular HSS bent about the minor axis are not subject to lateral-torsional buckling. These closed cross sections have a high resistance to torsion, and therefore the unbraced lengths L_p and L_r are large in comparison to I-shaped members. The length of L_r is so large that deflection will almost always govern the design before an unbraced length of L_r is reached.

The nominal flexural strength, M_n, is governed by the lowest of the values given by three limit states: yielding (plastic moment), flange local buckling, and web local buckling under pure flexure. The values of M_n for these limit states are obtained from the following equations.

For *yielding*,

$$M_n = M_p = F_y Z \quad \text{[AISC Eq. F7-1]} \tag{5.23}$$

Flange local buckling does not apply to compact sections. For sections with noncompact flanges,

$$M_n = M_p - (M_p - F_y S) \times \left(3.57\left(\frac{b}{t}\right)\sqrt{\frac{F_y}{E}} - 4.0\right) \le M_p \quad \text{[AISC Eq. F7-2]} \tag{5.24}$$

For sections with slender flanges,

$$M_n = F_y S_e \quad \text{[AISC Eq. F7-3]} \tag{5.25}$$

The effective section modulus, S_e, is calculated taking the effective width of the compression flange as

$$b_e = 1.92 t_f \sqrt{\frac{E}{F_y}} \left(1 - \frac{0.38}{\frac{b}{t_f}} \sqrt{\frac{E}{F_y}} \right) \le b \quad \text{[AISC Eq. F7-4]} \qquad 5.26$$

Web local buckling does not apply to compact sections. For sections with noncompact webs,

$$M_n = M_p - (M_p - F_y S) \times \left(0.305 \left(\frac{h}{t_w} \right) \sqrt{\frac{F_y}{E}} - 0.738 \right) \le M_p \quad \text{[AISC Eq. F7-5]} \qquad 5.27$$

Example 5.6

Compact Rectangular HSS Member

Determine the design flexural strength and allowable flexural strength of an HSS6 × 4 × 1/4 member with the following properties.

Section properties

$A = 4.30 \text{ in}^2$

$t = 0.233 \text{ in}$

$I_x = 20.9 \text{ in}^4$

$S_x = 6.96 \text{ in}^3$

$Z_x = 8.53 \text{ in}^3$

$I_y = 11.1 \text{ in}^4$

$S_y = 5.56 \text{ in}^3$

$Z_y = 6.45 \text{ in}^3$

$b/t = 14.2$

$h/t = 22.8$

Material properties

ASTM A500, grade B steel

$F_y = 46 \text{ ksi}$

$F_u = 58 \text{ ksi}$

Solution

Determine whether the member is compact for flexure. For the flanges (using *AISC Specification* Table B4.1b, case 17),

$$\lambda_{pf} = 1.12 \sqrt{\frac{E}{F_y}} = 1.12 \sqrt{\frac{29{,}000 \ \frac{\text{kips}}{\text{in}^2}}{46 \ \frac{\text{kips}}{\text{in}^2}}} = 28.12 \quad [> b/t, \text{ so compact}]$$

The flanges are compact. For the webs (using *AISC Specification* Table B4.1b, case 19),

$$\lambda_{pw} = 2.42 \sqrt{\frac{E}{F_y}} = 2.42 \sqrt{\frac{29{,}000 \ \frac{\text{kips}}{\text{in}^2}}{46 \ \frac{\text{kips}}{\text{in}^2}}} = 60.76 \quad [> h/t, \text{ so compact}]$$

The webs are compact. (Whether flanges and webs are compact could also be found from Table 5.5. For a wall thickness of $^1/_4$ in, Table 5.5 shows that the flanges would be compact up to a width of 7 in and the webs would be compact up to a width of 14 in.)

Use Eq. 5.23 to calculate the nominal moment capacity based on the limit state for yielding.

$$M_n = M_p = F_y Z_x = \left(46\ \frac{\text{kips}}{\text{in}^2}\right)\left(\frac{8.53\ \text{in}^3}{12\ \frac{\text{in}}{\text{ft}}}\right)$$

$$= 32.70\ \text{ft-kips}$$

Calculate the available flexural strength.

LRFD	ASD
$\phi_b M_n = (0.90)(32.70\ \text{ft-kips})$ $= 29.43\ \text{ft-kips}$	$\frac{M_n}{\Omega_b} = \frac{32.70\ \text{ft-kips}}{1.67}$ $= 19.58\ \text{ft-kips}$

AISC Manual Table 3-12 lists $\phi_b M_{nx}$ as 29.4 ft-kips for LRFD and M_{nx}/Ω_b as 19.6 ft-kips for ASD.

Example 5.7

Noncompact Rectangular HSS Member

Determine the design flexural strength and allowable flexural strength of an HSS16 × 8 × $^1/_4$ member with the following properties.

Section properties		Material properties
$A = 10.8\ \text{in}^2$	$I_y = 127\ \text{in}^4$	ASTM A500, grade B steel
$t = 0.233\ \text{in}$	$S_y = 31.7\ \text{in}^3$	$F_y = 46\ \text{ksi}$
$I_x = 368\ \text{in}^4$	$Z_y = 35.0\ \text{in}^3$	$F_u = 58\ \text{ksi}$
$S_x = 46.1\ \text{in}^3$	$b/t = 31.3$	
$Z_x = 56.4\ \text{in}^3$	$h/t = 65.7$	

Solution

Determine whether the member is compact. For the flanges, using *AISC Specification* Table B4.1b, case 17,

$$\lambda_{pf} = 1.12\sqrt{\frac{E}{F_y}} = 1.12\sqrt{\frac{29{,}000\ \frac{\text{kips}}{\text{in}^2}}{46\ \frac{\text{kips}}{\text{in}^2}}} = 28.12 \quad [< b/t,\ \text{so not compact}]$$

The flanges are not compact. For the webs, using *AISC Specification* Table B4.1b, case 19,

$$\lambda_{pw} = 2.42\sqrt{\frac{E}{F_y}} = 2.42\sqrt{\frac{29{,}000\ \frac{\text{kips}}{\text{in}^2}}{46\ \frac{\text{kips}}{\text{in}^2}}} = 60.76 \quad [< h/t,\ \text{so not compact}]$$

The webs are not compact. Determine whether the member is slender. For the flanges, using *AISC Specification* Table B4.1b, case 17,

$$\lambda_{rf} = 1.40\sqrt{\frac{E}{F_y}} = 1.40\sqrt{\frac{29{,}000\ \frac{\text{kips}}{\text{in}^2}}{46\ \frac{\text{kips}}{\text{in}^2}}} = 35.15 \quad [> b/t,\ \text{so not slender}]$$

The flanges are not slender. For the web, using *AISC Specification* Table B4.1b, case 19,

$$\lambda_{rw} = 5.70\sqrt{\frac{E}{F_y}} = 5.70\sqrt{\frac{29{,}000\ \frac{\text{kips}}{\text{in}^2}}{46\ \frac{\text{kips}}{\text{in}^2}}} = 143.12 \quad [> h/t,\ \text{so not slender}]$$

The webs are not slender. Calculate the nominal moment capacity using Eq. 5.24 based on the limit state for flange local buckling. First, use Eq. 5.23 to calculate M_p.

$$M_p = F_y Z_x = \left(46\ \frac{\text{kips}}{\text{in}^2}\right)\left(\frac{56.4\ \text{in}^3}{12\ \frac{\text{in}}{\text{ft}}}\right) = 216.20\ \text{ft-kips}$$

$$M_n = M_p - (M_p - F_y S_x)\left(3.57\left(\frac{b}{t}\right)\sqrt{\frac{F_y}{E}} - 4.0\right) \le M_p$$

$$= 216.20\ \text{ft-kips} - \left(216.20\ \text{ft-kips} - \left(46\ \frac{\text{kips}}{\text{in}^2}\right)\left(\frac{46.1\ \text{in}^3}{12\ \frac{\text{in}}{\text{ft}}}\right)\right)$$

$$\times\left((3.57)(31.3)\sqrt{\frac{46\ \frac{\text{kips}}{\text{in}^2}}{29{,}000\ \frac{\text{kips}}{\text{in}^2}}} - 4.0\right)$$

$$= 198.42\ \text{ft-kips} \quad [\le M_p = 216.20\ \text{ft-kips}]$$

Use Eq. 5.27 to calculate the nominal moment capacity based on the limit state for web local buckling.

$$M_n = M_p - (M_p - F_yS_x)\left(0.305\left(\frac{h}{t}\right)\sqrt{\frac{F_y}{E}} - 0.738\right) \le M_p$$

$$= 216.20 \text{ ft-kips} - \left(216.20 \text{ ft-kips} - \left(46 \ \frac{\text{kips}}{\text{in}^2}\right)\left(\frac{46.1 \text{ in}^3}{12 \ \frac{\text{in}}{\text{ft}}}\right)\right)$$

$$\times\left((0.305)(65.7)\sqrt{\frac{46 \ \frac{\text{kips}}{\text{in}^2}}{29{,}000 \ \frac{\text{kips}}{\text{in}^2}}} - 0.738\right)$$

$$= 213.83 \text{ ft-kips} \quad [\le M_p = 216.20 \text{ ft-kips}]$$

The nominal moment for the limit state of flange local buckling is less than that for web local buckling and therefore is the controlling value. Calculate the available flexural strength.

LRFD	ASD
$\phi_b M_n = (0.90)(198.42 \text{ ft-kips})$ $= 178.58 \text{ ft-kips}$	$\frac{M_n}{\Omega_b} = \frac{198.42 \text{ ft-kips}}{1.67}$ $= 118.81 \text{ ft-kips}$

AISC Manual Table 3-12 lists $\phi_b M_{nx}$ as 178 ft-kips for LRFD and M_{nx}/Ω_b as 119 ft-kips for ASD.

12. ROUND HSS MEMBERS

This section applies to round hollow structural section (HSS) members having D/t ratios of less than $0.45E/F_y$.

To obtain the nominal flexural strength, M_n, calculate the values obtained from the limit states of yielding and local buckling, and take the lower of the two values. The values of M_n for these states are calculated with the following equations.

For *yielding*, the nominal flexural strength is

$$M_n = M_p = F_yZ \quad \text{[AISC Eq. F8-1]} \qquad 5.28$$

Local buckling does not apply to round compact members. For noncompact sections, slenderness must be checked using *AISC Specification* Table B4.1b, case 20.

If walls are not slender,

$$M_n = \left(\frac{0.021E}{\frac{D}{t}} + F_y \right) S \quad \text{[AISC Eq. F8-2]} \tag{5.29}$$

For sections with slender walls,

$$M_n = F_{\text{cr}} S \quad \text{[AISC Eq. F8-3]} \tag{5.30}$$

For Eq. 5.30, the critical flexural stress is

$$F_{\text{cr}} = \frac{0.33E}{\frac{D}{t}} \quad \text{[AISC Eq. F8-4]} \tag{5.31}$$

Example 5.8
Compact Round HSS Member

Determine the available flexural strength for an HSS14.000 × 0.375 member that has the following properties.

Section properties		Material properties
$A = 15.0 \text{ in}^2$	$S = 49.8 \text{ in}^3$	ASTM A500, grade B steel
$t = 0.349 \text{ in}$	$Z = 65.1 \text{ in}^3$	$F_y = 42 \text{ ksi}$
$I = 349 \text{ in}^4$	$D/t = 40.1$	$F_u = 58 \text{ ksi}$

Solution

Determine whether the member is compact using *AISC Specification* Table B4.1b, case 20.

$$\lambda_p = \frac{0.07E}{F_y} = \frac{(0.07)\left(29{,}000 \ \frac{\text{kips}}{\text{in}^2}\right)}{42 \ \frac{\text{kips}}{\text{in}^2}} = 48.33 \quad [> D/t, \text{ so compact}]$$

The member is compact, so use Eq. 5.28. Calculate the nominal moment capacity.

$$M_n = F_y Z = \left(42 \ \frac{\text{kips}}{\text{in}^2}\right)\left(\frac{65.1 \text{ in}^3}{12 \ \frac{\text{in}}{\text{ft}}}\right) = 227.85 \text{ ft-kips}$$

Calculate the available flexural strength.

LRFD	ASD
$\phi_b M_n = (0.90)(227.85 \text{ ft-kips}) = 205.07 \text{ ft-kips}$	$\dfrac{M_n}{\Omega_b} = \dfrac{227.85 \text{ ft-kips}}{1.67} = 136.44 \text{ ft-kips}$

AISC Manual Table 3-14 lists $\phi_b M_n$ as 205 ft-kips and M_n/Ω_b as 136 ft-kips.

Example 5.9

Noncompact Round HSS Member

Determine the available flexural strength for an HSS14.000 × 0.250 member that has the following properties.

Section properties		Material properties
$A = 10.1 \text{ in}^2$	$S = 34.1 \text{ in}^3$	ASTM A500, grade B steel
$t = 0.233$ in	$Z = 44.2 \text{ in}^3$	$F_y = 42$ ksi
$I = 239 \text{ in}^4$	$D/t = 60.1$	$F_u = 58$ ksi

Solution

Determine whether the member is compact using *AISC Specification* Table B4.1, case 15.

$$\lambda_p = \frac{0.07E}{F_y} = \frac{(0.07)\left(29{,}000 \ \dfrac{\text{kips}}{\text{in}^2}\right)}{42 \ \dfrac{\text{kips}}{\text{in}^2}} = 48.33 \quad [< D/t, \text{ so not compact}]$$

The member is not compact. Use Eq. 5.28 and Eq. 5.29 to determine the nominal flexural strength. For yielding,

$$M_n = F_y Z = \left(42 \ \frac{\text{kips}}{\text{in}^2}\right)\left(\frac{44.2 \text{ in}^3}{12 \ \dfrac{\text{in}}{\text{ft}}}\right) = 154.70 \text{ ft-kips}$$

For local buckling,

$$
\begin{aligned}
M_n &= \left(\frac{0.021E}{\frac{D}{t}} + F_y \right) S \\
&= \left(\frac{(0.021)\left(29{,}000\ \frac{\text{kips}}{\text{in}^2}\right)}{60.1} + 42\ \frac{\text{kips}}{\text{in}^2} \right) \left(\frac{34.1\ \text{in}^3}{12\ \frac{\text{in}}{\text{ft}}} \right) \\
&= 148.14\ \text{ft-kips} \quad [\text{controls}]
\end{aligned}
$$

The limit state of local buckling governs. Calculate the available flexural strength.

LRFD	ASD
$\phi_b M_n = (0.90)(148.14\ \text{ft-kips})$ $= 133.33\ \text{ft-kips}$	$\frac{M_n}{\Omega_b} = \frac{148.14\ \text{ft-kips}}{1.67} = 88.71\ \text{ft-kips}$

AISC Manual Table 3-14 lists $\phi_b M_n$ as 133 ft-kips and M_n/Ω_b as 88.8 ft-kips.

13. TEES AND DOUBLE ANGLES LOADED IN THE PLANE OF SYMMETRY

In the design and analysis of tees and double angles, the first thing to consider is whether the stem of the member is in tension or compression. The nominal strength, M_n, of these members is governed by the lowest value obtained from the limit states of

- yielding of stems (plastic moment)
- lateral-torsional buckling
- flange local buckling
- local buckling of tee stems in flexural compression

For *yielding*, use Eq. 5.32 or Eq. 5.33. For stems in tension,

$$M_n = M_p = F_y Z_x \le 1.6 M_y \quad [\text{AISC Eq. F9-1 and Eq. F9-2}] \qquad 5.32$$

For stems in compression,

$$M_n = M_p = F_y Z_x \le M_y \quad [\text{AISC Eq. F9-1 and Eq. F9-3}] \qquad 5.33$$

For *lateral-torsional buckling*, use Eq. 5.34.

$$M_n = M_{\text{cr}} = \left(\frac{\pi \sqrt{E I_y G J}}{L_b} \right) \left(B + \sqrt{1 + B^2} \right) \quad [\text{AISC Eq. F9-4}] \qquad 5.34$$

In Eq. 5.34, the term B is

$$B = \pm 2.3\left(\frac{d}{L_b}\right)\sqrt{\frac{I_y}{J}} \quad \text{[AISC Eq. F9-5]} \tag{5.35}$$

In Eq. 5.35, use the plus sign when the stem is in tension and the minus sign when the stem is in compression. Use the minus sign if the tip of the stem is in compression anywhere along the unbraced length.

The limit state of *flange local buckling* does not apply for tees with compact sections; otherwise, use Eq. 5.36 or Eq. 5.37. For tees with noncompact sections,

$$M_n = M_p - \left(M_p - 0.7F_yS_{xc}\right)\left(\frac{\lambda - \lambda_{pf}}{\lambda_{rf} - \lambda_{pf}}\right) \le 1.6M_y \quad \text{[AISC Eq. F9-6]} \tag{5.36}$$

S_{xc} is the elastic section modulus referred to the compression flange, and λ is the ratio $b_f/2t_f$.

For tees with slender sections,

$$M_n = \frac{0.70ES_{xc}}{\left(\frac{b_f}{2t_f}\right)^2} \quad \text{[AISC Eq. F9-7]} \tag{5.37}$$

For *local buckling of tee stems in flexural compression*, use Eq. 5.38.

$$M_n = F_{cr}S_x \quad \text{[AISC Eq. F9-8]} \tag{5.38}$$

The critical stress, F_{cr}, to use in Eq. 5.38 depends on the width-to-thickness ratio d/t_w. If $d/t_w \le 0.84\sqrt{E/F_y}$, the critical stress is

$$F_{cr} = F_y \quad \text{[AISC Eq. F9-9]} \tag{5.39}$$

If $0.84\sqrt{E/F_y} < d/t_w \le 1.03\sqrt{E/F_y}$, the critical stress is

$$F_{cr} = \left(2.55 - 1.84\left(\frac{d}{t_w}\right)\sqrt{\frac{F_y}{E}}\right)F_y \quad \text{[AISC Eq. F9-10]} \tag{5.40}$$

If $1.03\sqrt{E/F_y} < d/t_w$, the critical stress is

$$F_{cr} = \frac{0.69E}{\left(\frac{d}{t_w}\right)^2} \quad \text{[AISC Eq. F9-11]} \tag{5.41}$$

Example 5.10

WT Shape Flexural Member

The WT section shown is used as a beam spanning 10 ft with the stem of the tee in tension. The beam is braced continuously and supports a uniform dead load including the beam weight of 0.10 kip/ft and a uniform live load of 0.30 kip/ft.

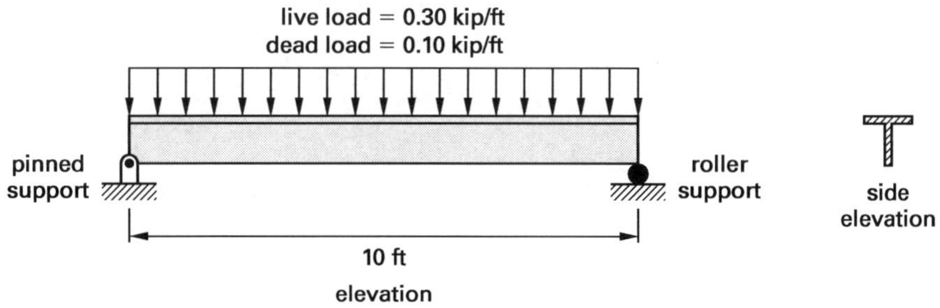

Material properties

ASTM A992 steel

$F_y = 50$ ksi

$F_u = 65$ ksi

There are no deflection requirements. Select a WT section to meet the loading requirements.

Solution

Calculate the required flexural strength. The loading is

LRFD	ASD
$w_u = 1.2w_D + 1.6w_L$ $= (1.2)\left(0.10\ \frac{\text{kip}}{\text{ft}}\right) + (1.6)\left(0.30\ \frac{\text{kip}}{\text{ft}}\right)$ $= 0.60\ \text{kip/ft}$	$w_a = w_D + w_L$ $= 0.10\ \frac{\text{kip}}{\text{ft}} + 0.30\ \frac{\text{kip}}{\text{ft}}$ $= 0.40\ \text{kip/ft}$

The moment is

LRFD	ASD
$M_u = \frac{w_u L^2}{8}$ $= \frac{\left(0.60\ \frac{\text{kip}}{\text{ft}}\right)(10\ \text{ft})^2}{8}$ $= 7.5$ ft-kips	$M_a = \frac{w_a L^2}{8}$ $= \frac{\left(0.40\ \frac{\text{kip}}{\text{ft}}\right)(10\ \text{ft})^2}{8}$ $= 5.0$ ft-kips

Because the stem of the tee is in flexure, make a trial selection of the tee based on the flexural yielding limit state.

LRFD	ASD
$M_u \le \phi_b F_y Z_x$ $Z_{x,\text{req}} = \dfrac{M_u}{\phi_b F_y} = \dfrac{(7.5\text{ ft-kips})\left(12\ \dfrac{\text{in}}{\text{ft}}\right)}{(0.90)\left(50\ \dfrac{\text{kips}}{\text{in}^2}\right)} = 2.0\text{ in}^3$	$M_a \le \dfrac{F_y Z_x}{\Omega_b}$ $Z_{x,\text{req}} = \dfrac{M_a \Omega_b}{F_y} = \dfrac{(5.0\text{ ft-kips})(1.67)\left(12\ \dfrac{\text{in}}{\text{ft}}\right)}{50\ \dfrac{\text{kips}}{\text{in}^2}} = 2.0\text{ in}^3$

Try a WT5 × 7.5 with the following section properties.

$b_f = 4.00$ in　　$I_y = 1.45\text{ in}^4$　　$\bar{y} = 1.37$ in

$t_f = 0.270$ in　　$S_x = 1.50\text{ in}^3$　　$J = 0.0518\text{ in}^4$

$I_x = 5.45\text{ in}^4$　　$Z_x = 2.71\text{ in}^3$　　$G = 11{,}200\text{ kips/in}^2$

$$S_{xc} = \frac{I_x}{\bar{y}} = \frac{5.45\text{ in}^4}{1.37\text{ in}} = 3.98\text{ in}^3$$

From Eq. 5.32, the nominal flexural strength based on yielding is the lower of

$$M_n \le \begin{cases} F_y Z_x = \dfrac{\left(50\ \dfrac{\text{kips}}{\text{in}^2}\right)(2.71\text{ in}^3)}{12\ \dfrac{\text{in}}{\text{ft}}} \\ \quad = 11.29\text{ ft-kips} \\ 1.6M_y = 1.6F_y S_x = \dfrac{(1.6)\left(50\ \dfrac{\text{kips}}{\text{in}^2}\right)(1.50\text{ in}^3)}{12\ \dfrac{\text{in}}{\text{ft}}} \\ \quad = 10.00\text{ ft-kips} \quad [\text{controls}] \end{cases}$$

Check whether the flange is compact using *AISC Specification* Table B4.1b, case 10.

$$\lambda_p = 0.38\sqrt{\frac{E}{F_y}} = 0.38\sqrt{\frac{29{,}000\ \frac{\text{kips}}{\text{in}^2}}{50\ \frac{\text{kips}}{\text{in}^2}}} = 9.15$$

$$\lambda = \frac{b_f}{2t_f} = \frac{4.00\ \text{in}}{(2)(0.270\ \text{in})} = 7.41 \quad [\lambda < \lambda_p, \text{ so compact}]$$

Calculate the lateral-torsional buckling strength. From Eq. 5.35 (using the plus sign because the stem is in tension),

$$\begin{aligned} B &= \pm 2.3\left(\frac{d}{L_b}\right)\sqrt{\frac{I_y}{J}} \\ &= (2.3)\left(\frac{5\ \text{in}}{(10\ \text{ft})\left(12\ \frac{\text{in}}{\text{ft}}\right)}\right)\sqrt{\frac{1.45\ \text{in}^4}{0.0518\ \text{in}^4}} \\ &= 0.507 \end{aligned}$$

From Eq. 5.34, the lateral-torsional buckling strength is

$$\begin{aligned} M_n &= \left(\frac{\pi\sqrt{EI_yGJ}}{L_b}\right)\left(B + \sqrt{1+B^2}\right) \\ &= \left(\frac{\pi\sqrt{\left(29{,}000\ \frac{\text{kips}}{\text{in}^2}\right)(1.45\ \text{in}^4)\left(11{,}200\ \frac{\text{kips}}{\text{in}^2}\right)(0.0518\ \text{in}^4)}}{(10\ \text{ft})\left(12\ \frac{\text{in}}{\text{ft}}\right)^2}\right) \\ &\quad \times\left(0.507 + \sqrt{1+(0.507)^2}\right) \\ &= 17.54\ \text{ft-kips} \end{aligned}$$

The stem is compact, so the limit state of flange local buckling does not apply; the stem is in tension, so local buckling of the stem in compression does not apply. The yielding strength of 10.00 ft-kips is lowest and governs.

Calculate the available flexural strength.

LRFD	ASD
$\phi_b M_n = (0.90)(10.00 \text{ ft-kips}) = 9.00 \text{ ft-kips}$ $[\geq M_u = 7.5 \text{ ft-kips}]$	$\frac{M_n}{\Omega_b} = \frac{10.00 \text{ ft-kips}}{1.67} = 5.99 \text{ ft-kips}$ $[\geq M_a = 5.0 \text{ ft-kips}]$

The trial member meets the requirements for LRFD and ASD. Similar calculations show that the next lighter section, a WT5 × 6, does not have the available flexural strength to meet the requirements.

14. REDUCTION REQUIREMENTS FOR FLANGE HOLES

AISC Specification Sec. F13 covers several issues pertaining to the proportions of beams and girders. Section F13.1 covers hole reductions.

Under some circumstances, a reduction may be required in the nominal flexural strength at holes in the flanges. This is required in order to prevent tensile rupture of the tension flange. The limit state of tensile rupture does not apply if Eq. 5.39 is met.

$$F_u A_{fn} \geq Y_t F_y A_{fg} \quad \text{[AISC Sec. F13.1]} \qquad 5.39$$

A_{fg} and A_{fn} are the gross and net flange areas, respectively. The hole reduction factor, Y_t, is 1.0 if $F_y/F_u \leq 0.8$ and is 1.1 otherwise.

If Eq. 5.39 is not met, then the nominal flexural strength at the holes in the tension flange is limited by Eq. 5.40.

$$M_n \leq \left(\frac{F_u A_{fn}}{A_{fg}}\right) S_x \quad \text{[AISC Eq. F13-1]} \qquad 5.40$$

Example 5.11

Reduction for Holes in Tension Flange for M_n

A W12 × 40 steel beam has holes in the tension flange for $^7/_8$ in diameter bolts. The bolt holes are not staggered, so there are two holes, one in each flange, in a line perpendicular to the beam web.

Section properties		Material properties
$A = 11.7 \text{ in}^2$	$S_x = 51.5 \text{ in}^3$	ASTM A992 steel
$b_f = 8.01 \text{ in}$	$Z_x = 57.0 \text{ in}^3$	$F_y = 50 \text{ ksi}$
$t_f = 0.515 \text{ in}$		$F_u = 65 \text{ ksi}$

Calculate the nominal moment capacity of the section, taking into consideration the holes in the tension flange of the beam.

Solution

The gross area of the tension flange is

$$A_{fg} = b_f t_f = (8.01 \text{ in})(0.515 \text{ in}) = 4.13 \text{ in}^2$$

The total area of the holes is

$$\begin{aligned} A_h &= n_{\text{holes}} t_f d_{\text{hole}} = n_{\text{holes}} t_f (d_{\text{bolt}} + 0.125 \text{ in}) \\ &= (2)(0.515 \text{ in})(0.875 \text{ in} + 0.125 \text{ in}) \\ &= 1.03 \text{ in}^2 \end{aligned}$$

The net area of the tension flange is

$$A_{fn} = A_{fg} - A_h = 4.13 \text{ in}^2 - 1.03 \text{ in}^2 = 3.10 \text{ in}^2$$

Calculate the ratio of yield strength to rupture strength to determine the value of Y_t.

$$\frac{F_y}{F_u} = \frac{50 \ \frac{\text{kips}}{\text{in}}}{65 \ \frac{\text{kips}}{\text{in}}} = 0.77 \quad [< 0.8, \text{ so } Y_t = 1.0]$$

Use Eq. 5.39 to determine whether tensile rupture applies.

$$\begin{gathered} F_u A_{fn} \geq Y_t F_y A_{fg} \\ \left(65 \ \frac{\text{kips}}{\text{in}^2}\right)(3.10 \text{ in}^2) \geq (1.0)\left(50 \ \frac{\text{kips}}{\text{in}^2}\right)(4.13 \text{ in}^2) \\ 201.50 \text{ kips} \geq 206.50 \text{ kips} \quad [\text{no good}] \end{gathered}$$

Equation 5.39 is not satisfied, so a tensile rupture state exists and the nominal moment, M_n, must be reduced. Calculate the reduced available flexural strength using Eq. 5.40.

$$\begin{aligned} M_n &= \left(\frac{F_u A_{fn}}{A_{fg}}\right) S_x \\ &= \frac{\left(\frac{\left(65 \ \frac{\text{kips}}{\text{in}^2}\right)(3.10 \text{ in}^2)}{4.13 \text{ in}^2}\right)(51.5 \text{ in}^3)}{12 \ \frac{\text{in}}{\text{ft}}} \\ &= 209.39 \text{ ft-kips} \end{aligned}$$

The nominal moment capacity of the section is about 209.39 ft-kips.

15. PROPORTIONING LIMITS FOR I-SHAPED MEMBERS

The following proportioning limits apply to plate girders and other fabricated I-shaped beams that are not rolled members.

For singly symmetrical I-shaped members,

$$0.1 \le \frac{I_{yc}}{I_y} \le 0.9 \quad \text{[AISC Eq. F13-2]} \qquad 5.41$$

For I-shaped members with slender webs, the following limits must also be met.

For $a/h \le 1.5$, where a is the clear distance between transverse stiffeners,

$$\left(\frac{h}{t_w}\right)_{\max} = 12.0\sqrt{\frac{E}{F_y}} \quad \text{[AISC Eq. F13-3]} \qquad 5.42$$

For $a/h > 1.5$,

$$\left(\frac{h}{t_w}\right)_{\max} = \frac{0.40E}{F_y} \quad \text{[AISC Eq. F13-4]} \qquad 5.43$$

In unstiffened girders, h/t_w must not be more than 260. The ratio of the web area to the compression flange area must not be more than 10.

16. COVER PLATES

AISC Specification Sec. F13.3 provides the specifications for cover plates and methods of connecting girder flanges to girder webs. Flanges of welded beams and girders can vary in thickness or width through the use of cover plates and by splicing plates together.

For bolted girders, cover plates must not make up more than 70% of the total cross-sectional flange area.

Partial-length cover plates must extend beyond the theoretical cover plate termination by a distance sufficient to develop the cover plate's portion of the strength of the beam or girder at a distance a' from the end of the cover plate.

- When there is a continuous weld greater than or equal to three-fourths of the plate thickness across the end of the plate, $a' = w$.
- When there is a continuous weld smaller than three-fourths of the plate thickness across the end of the plate, $a' = 1.5w$.
- When there is no weld across the end of the plate, $a' = 2.0w$.

17. BEAM SHEAR

Design of members for shear is specified in *AISC Specification* Chap. G, which is divided into the following sections.

G1	General Provisions
G2	Members with Unstiffened or Stiffened Webs
G3	Tension Field Action
G4	Single Angles
G5	Rectangular HSS and Box Members
G6	Round HSS
G7	Weak Axis Shear in Singly and Doubly Symmetric Shapes
G8	Beams and Girders with Web Openings

In rolled W, M, and S shape beams, shear stresses will seldom be the governing design criteria, except for heavily loaded short span beams or heavy concentrated loads near the end of the span. Therefore, it is important to check the beam shear, which can be done quickly, either by calculation or by looking at $\phi_v V_n$ and V_n/Ω_v in the maximum total uniform load tables, *AISC Manual* Table 3-6 through Table 3-9.

The design shear strength, $\phi_v V_n$ (LRFD), and the allowable shear strength, V_n/Ω_v (ASD), are determined as follows.

The nominal shear strength, V_n, of unstiffened or stiffened webs, according to the limit states of shear yielding and shear buckling, is

$$V_n = 0.6F_y A_w C_v \quad [\text{AISC Eq. G2-1}] \qquad 5.44$$

For webs of rolled I-shaped sections that meet the criterion $h/t_w \le 2.24\sqrt{E/F_y}$, the resistance factor for shear, ϕ_v, is 1.00 (LRFD); the safety factor for shear, Ω_v, is 1.50 (ASD); and the web shear coefficient, C_v, is 1.0.

All current ASTM A6 W, S, and HP shapes except W44 × 230, W40 × 149, W36 × 135, W33 × 118, W30 × 90, W24 × 55, W16 × 26, and W12 × 14 meet this criterion for values of F_y up to 50 ksi.

For all other provisions in ASIC *Specification* Chap. G, the resistance factor for shear, ϕ_v, is 0.90 (LRFD), and the safety factor for shear, Ω_v, is 1.67 (ASD). The web shear coefficient, C_v, must be calculated.

For webs of all doubly symmetric shapes and singly symmetric shapes and channels (except for round and HSS) that do not meet the preceding criterion, the web shear coefficient, C_v, is calculated as follows.

If $h/t_w \le 1.10\sqrt{k_v E/F_y}$,

$$C_v = 1.0 \quad [\text{AISC Eq. G2-3}] \qquad 5.45$$

If $1.10\sqrt{k_vE/F_y} < h/t_w \le 1.37\sqrt{k_vE/F_y}$,

$$C_v = \frac{1.10\sqrt{\frac{k_vE}{F_y}}}{\frac{h}{t_w}} \quad \text{[AISC Eq. G2-4]} \qquad 5.46$$

If $h/t_w > 1.37\sqrt{k_vE/F_y}$,

$$C_v = \frac{1.51Ek_v}{\left(\frac{h}{t_w}\right)^2 F_y} \quad \text{[AISC Eq. G2-5]} \qquad 5.47$$

Example 5.12

Rolled Beam Shear Capacity

Using LRFD and ASD, determine the shear strength of a W18 × 50 steel beam that is 30 ft long and that has the following properties.

Section properties		Material properties
$A = 14.7\ \text{in}^2$	$b_f = 7.50$ in	ASTM A992 steel
$d = 18.00$ in	$t_f = 0.57$ in	$F_y = 50$ ksi
$t_w = 0.355$ in		$F_u = 65$ ksi

Solution

Check the h/t_w ratio ($h = d = 18.00$ in).

$$\frac{h}{t_w} \le 2.24\sqrt{\frac{E}{F_y}}$$

$$\frac{18.00\ \text{in}}{0.355\ \text{in}} \le 2.24\sqrt{\frac{29{,}000\ \frac{\text{kips}}{\text{in}^2}}{50\ \frac{\text{kips}}{\text{in}^2}}}$$

$$50.70 \le 53.95 \quad \text{[OK]}$$

Therefore, $\phi_v = 1.00$, $\Omega = 1.50$, and $C_v = 1.0$. (In practice, checking the h/t_w ratio is not necessary. All current rolled I-shaped members except those listed earlier in this section meet the h/t_w ratio requirements.)

From Eq. 5.44,

$$
\begin{aligned}
V_n &= 0.6F_yA_wC_v = 0.6F_y\left(dt_w\right)C_v \\
&= (0.6)\left(50\ \frac{\text{kips}}{\text{in}^2}\right)(18.00\ \text{in})(0.355\ \text{in})(1.0) \\
&= 191.70\ \text{kips}
\end{aligned}
$$

The shear strength is

LRFD	ASD
$\phi_vP_n = (1.0)(191.70\ \text{kips})$ $= 191.70\ \text{kips}$	$\frac{P_n}{\Omega_v} = \frac{191.70\ \text{kips}}{1.50} = 127.80\ \text{kips}$

AISC Manual Table 3-6 provides the following values: ϕ_vP_n = 192 kips, P_n/Ω_v = 128 kips.

18. SHEAR CAPACITY OF RECTANGULAR HSS AND BOX MEMBERS

The nominal shear capacity, V_n, of rectangular and square sections is determined by AISC Sec. G2.1, where the shear area is $A_w = 2ht_w$. In calculating the effective shear area, $t_w = t$ and $k_v = 5$. h is the width resisting the shear force and is taken as the clear distance between the flanges less the inside corner radius on each side. When the corner radius is unknown, h is taken as the corresponding outside dimension less three times the web thickness.

Example 5.13

Shear Capacity of Rectangular HSS

Determine the design and allowable shear capacities of an HSS6 × 4 × ¼ member that has the following properties.

Section properties

$t = 0.233$ in

$A = 4.30\ \text{in}^2$

$b/t = 14.2$

$h/t = 22.8$

$I_x = 20.9\ \text{in}^4$

$S_x = 6.96\ \text{in}^3$

$r_x = 2.20$ in

$Z_x = 8.53\ \text{in}^3$

$I_y = 11.1\ \text{in}^4$

$S_y = 5.56\ \text{in}^3$

$r_y = 1.61$ in

$Z_y = 6.45\ \text{in}^3$

Material properties

ASTM A500, grade B steel

$F_y = 46$ ksi

$F_u = 58$ ksi

Solution

Calculate the effective height, *h*.

$$h = d - 3t_w = 6\ \text{in} - (3)(0.233\ \text{in}) = 5.30\ \text{in}$$

Calculate the shear area.

$$A_w = 2ht_w = (2)(5.30\ \text{in})(0.233\ \text{in}) = 2.47\ \text{in}^2$$

Calculate h/t_w.

$$\frac{h}{t_w} = \frac{5.30\ \text{in}}{0.233\ \text{in}} = 22.75$$

Determine the web shear coefficient, C_v.

$$\frac{h}{t_w} \le 1.10\sqrt{\frac{k_v E}{F_y}}$$

$$22.75 \le 1.10\sqrt{\frac{(5)\left(29{,}000\ \frac{\text{kips}}{\text{in}^2}\right)}{46\ \frac{\text{kips}}{\text{in}^2}}}$$

$$\le 61.76$$

Therefore, use Eq. 5.45.

$$C_v = 1.0$$

(Most standard HSS members listed in the *AISC Manual* have $C_v = 1.0$ at $F_y = 46$ ksi.) Use Eq. 5.44 to calculate the nominal shear.

$$\begin{aligned} V_n &= 0.60F_y A_w C_v \\ &= (0.60)\left(46\ \frac{\text{kips}}{\text{in}^2}\right)(2.47\ \text{in}^2)(1.0) \\ &= 68.17\ \text{kips} \end{aligned}$$

Calculate the design strength and the allowable strength.

LRFD	ASD
$\phi_v V_n = (0.90)(68.17\ \text{kips})$ $= 61.35\ \text{kips}$	$\frac{V_n}{\Omega_v} = \frac{68.17\ \text{kips}}{1.67}$ $= 40.82\ \text{kips}$

19. SHEAR CAPACITY OF ROUND HSS

Equations for the shear capacity of round HSS members are given in *AISC Specification* Sec. G6. The nominal shear strength, V_n, of round HSS members according to the limit states of shear yielding and shear buckling is

$$V_n = \frac{F_{cr}A_g}{2} \quad \text{[AISC Eq. G6-1]} \qquad 5.48$$

F_{cr} is the larger of the values given by Eq. 5.49 and Eq. 5.50, but may not be greater than $0.6F_y$.

$$F_{cr} = \frac{1.60E}{\sqrt{\frac{L_v}{D}}\left(\frac{D}{t}\right)^{5/4}} \quad \text{[AISC Eq. G6-2a]} \qquad 5.49$$

$$F_{cr} = \frac{0.78E}{\left(\frac{D}{t}\right)^{3/2}} \quad \text{[AISC Eq. G6-2b]} \qquad 5.50$$

L_v is the distance from maximum to zero shear force.

Example 5.14

Shear Capacity of Round HSS

Determine the design shear and allowable shear strengths for a round HSS10.000 × 0.25 member with the following properties. The beam is 20 ft long and is subjected to a uniform load.

Section properties		Material properties
$t = 0.233$ in	$r = 3.45$ in	ASTM A500, grade B steel
$A = 7.15$ in^2	$Z_x = 22.2$ in^3	$F_y = 42$ ksi
$I = 85.3$ in^4	$D/t = 42.9$	$F_u = 58$ ksi
$S = 17.1$ in^3		

Solution

Because the beam is uniformly loaded, the distance from the maximum shear force and zero shear force is L_v = 10 ft = 120 in. Determine the critical shear stress, F_{cr}, using Eq. 5.49 and Eq. 5.50.

$$F_{cr} \geq \begin{cases} \dfrac{1.60E}{\sqrt{\dfrac{L_v}{D}}\left(\dfrac{D}{t}\right)^{5/4}} = \dfrac{(1.60)\left(29{,}000\ \dfrac{\text{kips}}{\text{in}^2}\right)}{\sqrt{\dfrac{120\text{ in}}{10\text{ in}}}(42.9)^{5/4}} = 122.00\text{ ksi} \\ \dfrac{0.78E}{\left(\dfrac{D}{t}\right)^{3/2}} = \dfrac{(0.78)\left(29{,}000\ \dfrac{\text{kips}}{\text{in}^2}\right)}{(42.9)^{3/2}} = 80.50\text{ ksi} \end{cases}$$

$$F_{cr} \leq 0.60F_y = (0.60)\left(42\ \frac{\text{kips}}{\text{in}^2}\right) = 25.20\text{ ksi} \quad [\text{controls}]$$

Calculate the nominal shear strength, V_n, using Eq. 5.48.

$$V_n = \frac{F_{cr}A_g}{2} = \frac{\left(25.20\ \dfrac{\text{kips}}{\text{in}^2}\right)(7.15\text{ in}^2)}{2} = 90.09\text{ kips}$$

Calculate the design shear strength and the allowable shear strength.

LRFD	ASD
$\phi_v V_n = (0.90)(90.09\text{ kips}) = 81.08\text{ kips}$	$\dfrac{V_n}{\Omega_v} = \dfrac{90.09\text{ kips}}{1.67} = 53.95\text{ kips}$

6 Flanges and Webs with Concentrated Loads

Nomenclature

A	area	in^2
A_1	loaded area	in^2
A_2	maximum area of supporting surface that is geometrically similar to and concentric with loaded area A_1	in^2
b	width	in
B	width of plate	in
d	depth	in
D	dead load	lbf
E	modulus of elasticity	lbf/in^2
f_c'	specified compressive strength of concrete	lbf/in^2
F	strength or stress	lbf/in^2
h	height	in
I	moment of inertia	in^4
k	distance from outer face of flange to web toe of fillet	in
k_c	coefficient for slender unstiffened elements	–
k_{des}	distance from outer face of flange to web toe of fillet, as a decimal value for design calculations	in
k_{det}	distance from outer face of flange to web toe of fillet, as a fractional value for detailing calculations	in
K	effective length factor	–
KL	effective length	in
KL/r	slenderness ratio	–
l_b	length of bearing	in
L	length	in
L	live load	lbf
M	moment, flexural strength, or moment strength	in-lbf
n	effective cantilever length	in

p	bearing stress	lbf/in^2
P	axial strength	lbf
P	force	lbf
r	radius of gyration	in
R	strength	lbf
R_1, R_3, R_5	beam end bearing constants	lbf
R_2, R_4, R_6	beam end bearing constants	lbf/in
S	elastic section modulus	in^3
t	thickness	in
w	load per unit length	lbf/in
x	distance from end of member	in
Z	plastic section modulus	in^3

Symbols

ϕ	resistance factor (LRFD)	–
Ω	safety factor (ASD)	–

Subscripts

a	required (ASD)
c	compression flange
cr	critical
cross	cross-shaped column
e	elastic critical buckling (Euler)
eff	effective
f	flange
g	gross
max	maximum
min	minimum
n	net or nominal
p	plastic bending
req	required
st	stiffener
u	required (LRFD) or ultimate tensile
w	web
x	about x-axis
y	about y-axis or yield

1. INTRODUCTION

The design requirements for concentrated loads applied to flanges and webs of I-shaped members are contained in Sec. J10 of the *AISC Specification*, which governs design considerations for the following limit states.

- flange local bending (AISC Sec. J10.1)
- web local yielding (AISC Sec. J10.2)
- web local crippling (AISC Sec. J10.3)
- web sidesway buckling (AISC Sec. J10.4)
- web compression buckling (AISC Sec. J10.5)
- web panel zone shear (AISC Sec. J10.6)

The required strength must be less than or equal to the available strength.

$$R_u \le \phi R_n \quad [\text{LRFD}] \qquad 6.1$$

$$R_a \le \frac{R_n}{\Omega} \quad [\text{ASD}] \qquad 6.2$$

When the required strength is greater than the available strength for the applicable limit states, transverse *stiffener plates* (also called *continuity plates*) or *doubler plates* and their attaching welds must be sized for the difference between required and available strengths. Figure 6.1 shows the use of stiffener plates and doubler plates.

Figure 6.1 *Stiffener Plates and Doubler Plates*

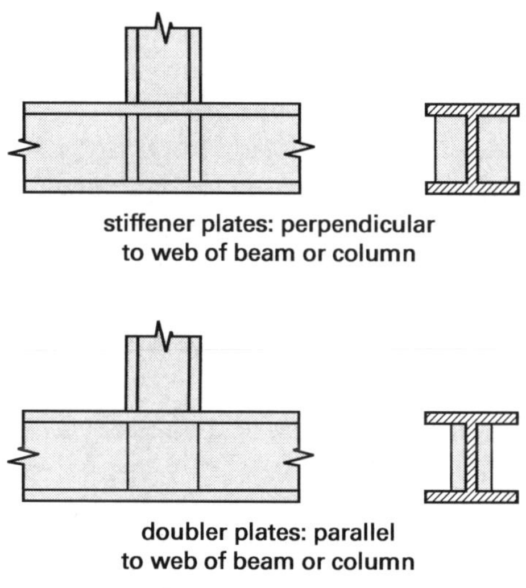

In Fig. 6.2, forces P_1 and P_2 induce single tensile forces on the flange and web. Forces P_3 and P_4 induce single compressive forces on the flange and web. Moments M_1 and M_2 double the tensile forces P_5 and P_7 and double the compressive forces P_6 and P_8 on the flanges and web of the beam.

***Figure 6.2** I-Shaped Beam with Flanges and Webs Subjected to Concentrated Loads*

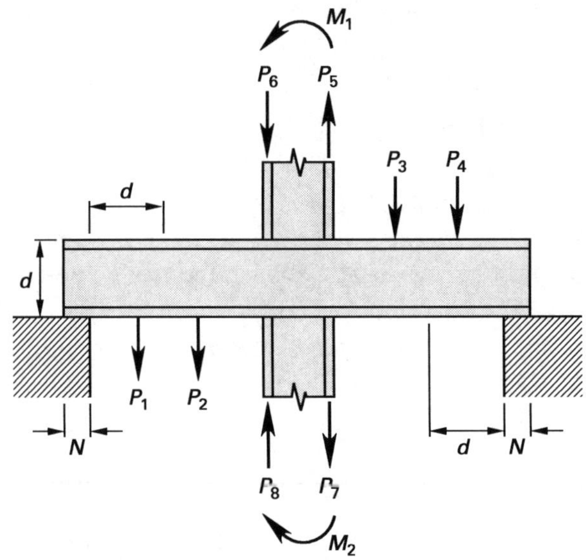

Forces P_1 and P_4 are applied at a distance from the end of the beam that is less than or equal to the beam depth, d. Forces P_2 and P_3 are applied at a distance from the beam end greater than d. This distinction often affects calculations, as shown later in this chapter.

2. FLANGE LOCAL BENDING

The limit state of flange local bending applies to tensile *single-concentrated forces* and the tensile component of *double-concentrated forces.*

For the limit state of flange local bending, the design strength, ϕR_n, and the allowable strength, R_n/Ω, are calculated using Eq. 6.3 for the value of R_n. For LRFD, $\phi = 0.90$, and for ASD, $\Omega = 1.67$.

$$R_n = 6.25F_{yf}t_f^2 \quad \text{[AISC Eq. J10-1]} \qquad 6.3$$

If the length of the loading across the member flange is less than 0.15 times the flange width, $0.15b_f$, Eq. 6.3 need not be checked.

When the concentrated force to be resisted is applied at a distance less than $10t_f$ from the end of the member, then R_n must be reduced by 50%. When required, use a pair of transverse stiffeners.

3. WEB LOCAL YIELDING

This section applies to single-concentrated forces and both components of double-concentrated forces. Figure 6.3 shows the nomenclature used in calculating web yielding and web crippling.

Figure 6.3 *Nomenclature for Web Yielding and Web Crippling*

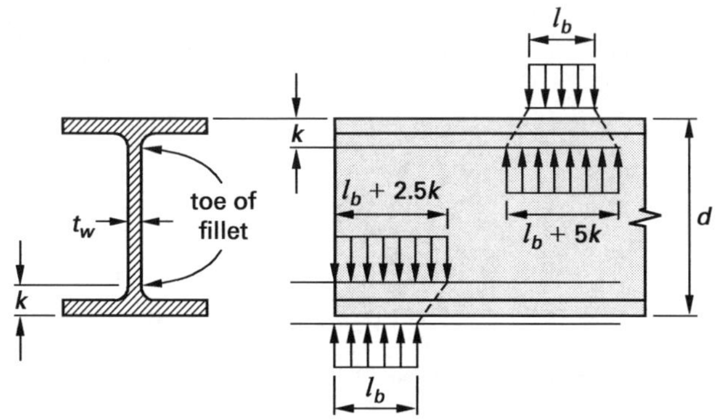

For the limit state of web local yielding, ϕ = 1.00 (LRFD) and Ω = 1.50 (ASD). Available strength is determined as follows.

When the concentrated force to be resisted is applied at a distance greater than d from the end of the member, the nominal strength is

$$R_n = F_{yw}t_w(5k+l_b) \quad \text{[AISC Eq. J10-2]} \qquad 6.4$$

When the concentrated force to be resisted is applied at a distance of d or less from the end of the member, the nominal strength is

$$R_n = F_{yw}t_w(2.5k+l_b) \quad \text{[AISC Eq. J10-3]} \qquad 6.5$$

In these equations, k is the distance from the outer face of the flange to the web toe of the fillet, and l_b is the length of bearing (not less than k for end beam reactions). For W-series beams, use k_{des} and not k_{det} from *AISC Manual* Table 1-1, because these are engineering calculations and not detailing dimensions. When required, a pair of transverse web stiffeners or a doubler plate must be provided.

4. WEB LOCAL CRIPPLING

This section applies to compressive single-concentrated forces or the compressive component of double-concentrated forces. For the limit state of web crippling, ϕ = 0.75 (LRFD) and $\Omega = 2.00$ (ASD). The available strength is determined as follows.

When the concentrated compressive force to be resisted is applied at a distance of $d/2$ or more from the end of the member, then the nominal strength is

$$R_n = 0.80t_w^2\left(1+3\left(\frac{l_b}{d}\right)\left(\frac{t_w}{t_f}\right)^{1.5}\right)\sqrt{\frac{EF_{yw}t_f}{t_w}} \quad \text{[AISC Eq. J10-4]} \qquad 6.6$$

When the concentrated compressive force to be resisted is applied at a distance less than $d/2$ from the end of the member, then the nominal strength is calculated with Eq. 6.7 or Eq. 6.8, depending on the value of l_b/d.

For $l_b/d \le 0.2$,

$$R_n = 0.40t_w^2\left(1+\left(\frac{3l_b}{d}\right)\left(\frac{t_w}{t_f}\right)^{1.5}\right)\sqrt{\frac{EF_{yw}t_f}{t_w}} \quad \text{[AISC Eq. J10-5a]} \qquad 6.7$$

For $l_b/d > 0.2$,

$$R_n = 0.40t_w^2\left(1+\left(\frac{4l_b}{d}-0.2\right)\left(\frac{t_w}{t_f}\right)^{1.5}\right)\sqrt{\frac{EF_{yw}t_f}{t_w}} \quad \text{[AISC Eq. J10-5b]} \qquad 6.8$$

Example 6.1

Concentrated Load on Beam Flange

The simply supported W10 × 30 ASTM A992 steel beam shown has a 1 in thick × 5 in wide steel plate welded to the bottom flange of the beam at midspan. A load is suspended from the plate consisting of 5 kips dead load and 15 kips live load. The plate-to-flange weld is adequate, the beam is laterally braced, and the 5 in dimension of the plate is perpendicular to the beam web.

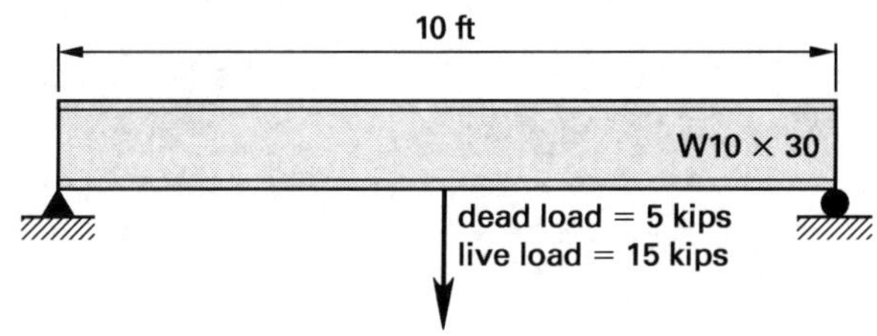

Section properties

$A = 8.84\ \text{in}^2$	$b_f = 5.81$ in	$k_{des} = 0.810$ in
$d = 10.5$ in	$t_f = 0.510$ in	$k_1 = {}^{11}/_{16}$ in
$t_w = 0.300$ in		

Determine whether web stiffener plates must be added to the beam.

Solution

The two limit states that apply to this problem are flange local bending and web local yielding. Calculate the required strengths.

LRFD	ASD
$P_u = 1.2D + 1.6L$ $= (1.2)(5\ \text{kips}) + (1.6)(15\ \text{kips})$ $= 30\ \text{kips}$	$P_a = D + L$ $= 5\ \text{kips} + 15\ \text{kips}$ $= 20\ \text{kips}$

Check the flange local bending using Eq. 6.3.

$$\begin{aligned} R_n &= 6.25F_{yf}t_f^2 \\ &= (6.25)\left(50\ \frac{\text{kips}}{\text{in}^2}\right)(0.510\ \text{in})^2 \\ &= 81.28\ \text{kips} \end{aligned}$$

Calculate the available flange local bending strength.

LRFD	ASD
$\phi R_n = (0.90)(81.28\ \text{kips}) = 73.15\ \text{kips}$	$\frac{R_n}{\Omega} = \frac{81.28\ \text{kips}}{1.67} = 48.67\ \text{kips}$

The available flange local bending strengths are greater than the required strengths, so stiffeners are not required for flange local bending.

Check the web local yielding. The point of load application is greater than *d* distance from the beam end, so use Eq. 6.4. Here, the width of the plate, l_b, is 1 in.

$$\begin{aligned} R_n &= F_{yw}t_w(5k + l_b) \\ &= \left(50\ \frac{\text{kips}}{\text{in}^2}\right)(0.300\ \text{in})\big((5)(0.810\ \text{in}) + 1\ \text{in}\big) \\ &= 75.75\ \text{kips} \end{aligned}$$

Calculate the available web local yielding strength.

LRFD	ASD
$\phi R_n = (1.00)(75.75\ \text{kips}) = 75.75\ \text{kips}$	$\frac{R_n}{\Omega} = \frac{75.75\ \text{kips}}{1.50} = 50.50\ \text{kips}$

The available web local yielding strengths are greater than the required strengths, so stiffeners are not required for web local yielding.

5. BEAM END BEARING REQUIREMENTS

Steel beams often bear on masonry or concrete walls or piers. In these cases, beam bearing plates are frequently used to distribute the beam reaction over a greater area, reducing the stresses imparted to the supporting elements.

Whether or not a bearing plate is used, the beam ends must be checked to ensure that web local yielding and web local crippling do not occur at the ends of the beams. The special formulas used to check these conditions are derived from the more general formulas for web yielding and web crippling described in Sec. 6.3 and Sec. 6.4.

To simplify these calculations, *AISC Manual* Table 9-4 gives beam end bearing constants R_1, R_2, R_3, R_4, R_5, and R_6 for W-series beams.

The formulas used to derive the values for the beam end bearing constants, as given in *AISC Manual* Part 9, are as follows.

$$R_1 = 2.5kF_{yw}t_w \quad \text{[AISC Eq. 9-39]} \tag{6.9}$$

$$R_2 = F_{yw}t_w \quad \text{[AISC Eq. 9-40]} \tag{6.10}$$

$$R_3 = 0.40t_w^2\sqrt{\frac{EF_{yw}t_f}{t_w}} \quad \text{[AISC Eq. 9-41]} \tag{6.11}$$

$$R_4 = 0.40t_w^2\left(\frac{3}{d}\right)\left(\frac{t_w}{t_f}\right)^{1.5}\sqrt{\frac{EF_{yw}t_f}{t_w}} \quad \text{[AISC Eq. 9-42]} \tag{6.12}$$

$$R_5 = 0.40t_w^2\left(1-0.2\left(\frac{t_w}{t_f}\right)^{1.5}\right)\sqrt{\frac{EF_{yw}t_f}{t_w}} \quad \text{[AISC Eq. 9-43]} \tag{6.13}$$

$$R_6 = 0.40t_w^2\left(\frac{4}{d}\right)\left(\frac{t_w}{t_f}\right)^{1.5}\sqrt{\frac{EF_{yw}t_f}{t_w}} \quad \text{[AISC Eq. 9-44]} \tag{6.14}$$

For I-shaped sections not listed in *AISC Manual* Table 9-4, beam end bearing constants are calculated with Eq. 6.9 through Eq. 6.14.

Web Local Yielding at Beam Ends

At beam ends, the formula that is used to determine the available strength for web local yielding, ϕR_n or R_n/Ω, is different depending on where the force is applied. When the compressive force to be resisted is applied at a distance less than d from the end of the member, the formulas are

$$\phi R_n = \phi R_1 + l_b(\phi R_2) \quad \text{[LRFD, AISC Eq. 9-45a]} \tag{6.15}$$

$$l_b = \frac{\phi R_n - \phi R_1}{\phi R_2} \quad \text{[LRFD]} \tag{6.16}$$

$$\frac{R_n}{\Omega} = \frac{R_1}{\Omega} + l_b\left(\frac{R_2}{\Omega}\right) \quad \text{[ASD, AISC Eq. 9-45b]} \tag{6.17}$$

$$l_b = \frac{\frac{R_n}{\Omega} - \frac{R_1}{\Omega}}{\frac{R_2}{\Omega}} \quad \text{[ASD]} \tag{6.18}$$

When the compressive force to be resisted is applied at a distance of d or more from the end of the member, the formulas are

$$\phi R_n = 2\phi R_1 + l_b(\phi R_2) \quad \text{[LRFD, AISC Eq. 9-46a]} \tag{6.19}$$

$$l_b = \frac{\phi R_n - 2\phi R_1}{\phi R_2} \quad \text{[LRFD]} \tag{6.20}$$

$$\frac{R_n}{\Omega} = 2\left(\frac{R_1}{\Omega}\right) + l_b\left(\frac{R_2}{\Omega}\right) \quad \text{[ASD, AISC Eq. 9-46b]} \tag{6.21}$$

$$l_b = \frac{\frac{R_n}{\Omega} - 2\left(\frac{R_1}{\Omega}\right)}{\frac{R_2}{\Omega}} \quad \text{[ASD]} \tag{6.22}$$

In accordance with *AISC Specification* Sec. J10.2, the bearing length l_b must be greater than or equal to k.

Web Local Crippling at Beam Ends

At the ends of beams, the available strength for web local crippling can be determined by the following formulas. When the compressive force to be resisted is applied at a distance less than $d/2$ from the end of the member, then the ratio l_b/d should be checked. For $l_b/d \le 0.2$,

$$\phi R_n = \phi R_3 + l_b(\phi R_4) \quad \text{[LRFD, AISC Eq. 9-47a]} \tag{6.23}$$

$$l_b = \frac{\phi R_n - \phi R_3}{\phi R_4} \quad \text{[LRFD]} \tag{6.24}$$

$$\frac{R_n}{\Omega} = \frac{R_3}{\Omega} + l_b\left(\frac{R_4}{\Omega}\right) \quad \text{[ASD, AISC Eq. 9-47b]} \tag{6.25}$$

$$l_b = \frac{\frac{R_n}{\Omega} - \frac{R_3}{\Omega}}{\frac{R_4}{\Omega}} \quad \text{[ASD]} \tag{6.26}$$

For $l_b/d > 0.2$,

$$\phi R_n = \phi R_5 + l_b(\phi R_6) \quad \text{[LRFD, AISC Eq. 9-48a]} \tag{6.27}$$

$$l_b = \frac{\phi R_n - \phi R_5}{\phi R_6} \quad \text{[LRFD]} \tag{6.28}$$

$$\frac{R_n}{\Omega} = \frac{R_5}{\Omega} + l_b\left(\frac{R_6}{\Omega}\right) \quad \text{[ASD, AISC Eq. 9-48b]} \tag{6.29}$$

$$l_b = \frac{\frac{R_n}{\Omega} - \frac{R_5}{\Omega}}{\frac{R_6}{\Omega}} \quad \text{[ASD]} \tag{6.30}$$

When the compressive force to be resisted is applied at a distance of $d/2$ or more from the end of the member, then

$$\phi R_n = 2\left(\phi R_3 + l_b(\phi R_4)\right) \quad \text{[LRFD, AISC Eq. 9-49a]} \tag{6.31}$$

$$l_b = \frac{\phi R_n - 2(\phi R_3)}{2(\phi R_4)} \quad \text{[LRFD]} \tag{6.32}$$

$$\frac{R_n}{\Omega} = 2\left(\frac{R_3}{\Omega} + l_b\left(\frac{R_4}{\Omega}\right)\right) \quad \text{[ASD, AISC Eq. 9-49b]} \tag{6.33}$$

$$l_b = \frac{\frac{R_n}{\Omega} - 2\left(\frac{R_3}{\Omega}\right)}{2\left(\frac{R_4}{\Omega}\right)} \quad \text{[ASD]} \tag{6.34}$$

Because ϕR_n is greater than or equal to the factored reaction and R_n/Ω is greater than or equal to the service load reaction, substituting the factored reaction for ϕR_n and the service load reaction for R_n/Ω in Eq. 6.15 through Eq. 6.34 will result in a direct solution for the required bearing length.

Example 6.2

Web Yielding and Web Crippling

A W12 × 50 ASTM A992 steel beam bears on a concrete pier and has a dead load end reaction of 15 kips and a live load reaction of 45 kips. Assume $l_b/d \leq 0.2$.

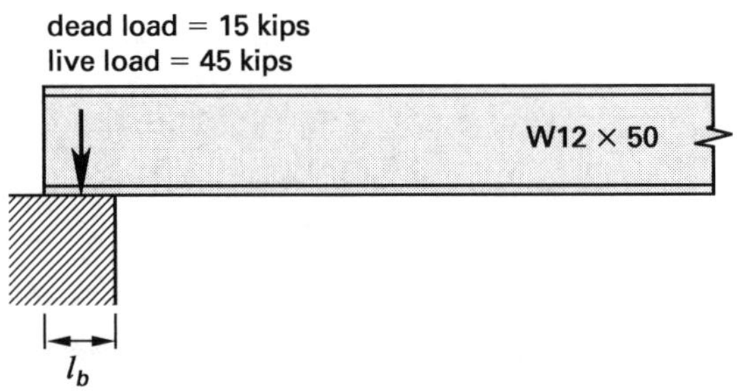

Determine the length of bearing required to prevent web yielding and web crippling.

Solution

Calculate the required strengths.

LRFD	ASD
$P_u = 1.2D + 1.6L$ $= (1.2)(15 \text{ kips}) + (1.6)(45 \text{ kips})$ $= 90 \text{ kips}$	$P_a = D + L$ $= 15 \text{ kips} + 45 \text{ kips}$ $= 60 \text{ kips}$

Calculate the required bearing length, l_b, to prevent web local yielding. The load is at the end of the beam, so the distance from the end of the beam is less than *d*. Therefore, Eq. 6.15 through Eq. 6.18 are used to determine the minimum required bearing length to resist web local yielding.

The beam bearing constants for a W12 × 50 are found in *AISC Manual* Table 9-4.

method	R_1 (kips)	R_2 (kips/in)	R_3 (kips)	R_4 (kips/in)	R_5 (kips)	R_6 (kips/in)
LRFD (ϕR)	52.7	18.5	65.0	7.03	59.3	9.37
ASD (R/Ω)	35.2	12.3	43.4	4.69	39.5	6.25

From Eq. 6.15 through Eq. 6.18,

LRFD	ASD
$\phi R_n = \phi R_1 + l_b(\phi R_2)$ $l_b = \dfrac{R_n - R_1}{R_2}$ $= \dfrac{90 \text{ kips} - 52.7 \text{ kips}}{18.5 \dfrac{\text{kips}}{\text{in}}}$ $= 2.02 \text{ in}$	$\dfrac{R_n}{\Omega} = \dfrac{R_1}{\Omega} + l_b\left(\dfrac{R_2}{\Omega}\right)$ $l_b = \dfrac{R_n - R_1}{R_2}$ $= \dfrac{60 \text{ kips} - 35.2 \text{ kips}}{12.3 \dfrac{\text{kips}}{\text{in}}}$ $= 2.02 \text{ in}$

Calculate the required bearing length, l_b, to prevent web local crippling. Because $l_b/d \leq 0.2$, use Eq. 6.23 through Eq. 6.26.

LRFD	ASD
$\phi R_n = \phi R_3 + l_b(\phi R_4)$ $l_b = \dfrac{R_n - R_3}{R_4}$ $= \dfrac{90 \text{ kips} - 65.0 \text{ kips}}{7.03 \ \dfrac{\text{kips}}{\text{in}}}$ $= 3.56 \text{ in}$	$\dfrac{R_n}{\Omega} = \dfrac{R_3}{\Omega} + l_b\left(\dfrac{R_4}{\Omega}\right)$ $l_b = \dfrac{R_n - R_3}{R_4}$ $= \dfrac{60 \text{ kips} - 43.4 \text{ kips}}{4.69 \ \dfrac{\text{kips}}{\text{in}}}$ $= 3.54 \text{ in}$

The value for l_b for web local crippling is larger than the value for web local yielding, and the larger value governs the bearing length. The minimum required bearing length is 3.54 in. Based on practical considerations, a bearing length of 3.75 in or 4 in would be used.

6. BEAM BEARING PLATES

Bearing plates are primarily used to distribute the reaction of a concentrated load over a greater area, thus reducing the stresses imparted to the supporting element. Even when not needed for this reason, bearing plates are also frequently used to level the bearing surface and bring it to the required elevation. A bed of nonshrink grout is usually placed between the bearing plate and the top of the concrete or masonry element supporting the beam. The design procedures for concrete or masonry supporting elements are identical; however, the load distribution geometry is somewhat different.

Load Distribution Geometry for Bearing on Concrete

Section 10.14 of ACI 318, *Building Code Requirements for Structural Concrete*, specifies the design bearing strength for direct bearing on the loaded area, A_1, and the geometry that permits an increase in the design bearing strength at the lower base area, A_2, as shown in Fig. 6.4. To find A_2, take the loaded area, A_1, as the upper base of a frustum with side slopes of 1:2 (1 vertical to 2 horizontal). The frustum may be that of a pyramid, cone, or tapered wedge, depending on the shape of the loaded area. When this imaginary frustum is extended down as far as is possible while wholly contained within the support, its lower base has the area A_2.

The design bearing strength of concrete must not exceed $\phi(0.85f_c'A_1)$ except when the supporting surface is wider than the loaded area on all sides. In this case, the design strength of the loaded area may be multiplied by $\sqrt{A_2/A_1}$, but not by more than 2.0.

Figure 6.4 Load Bearing Distribution on Concrete

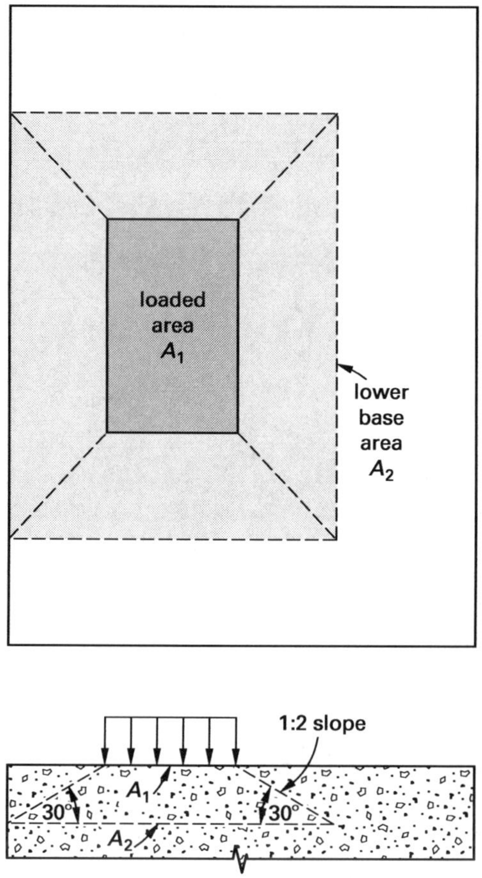

Load Distribution Geometry for Bearing on Masonry

Section 1.9.5 of ACI 530, *Building Code Requirements for Masonry Structures*, specifies the design bearing strength for direct bearing on the loaded area, A_1, and the geometry that permits an increase in the design bearing strength at the lower base area, A_2, as shown in Fig. 6.5. Nomenclature for the beam bearing plate is given in Fig. 6.6. If the loaded area is taken as the upper base of a frustum with side slopes of 1:1, and the largest possible frustum wholly contained within the support is found, then A_2 is the area of the lower base of this frustum.

Bearing plates are designed as cantilevered beam sections with the effective cantilever length, n, being equal to half the width of the plate minus the distance from the bottom of the beam to the web toe of the fillet, or $B/2 - k_{des}$.

$$n = \frac{B}{2} - k_{des} \qquad 6.35$$

The moment of the cantilevered section then becomes

$$M = \frac{wL^2}{2} = \frac{pL^2}{2} \qquad 6.36$$

Figure 6.5 Load Bearing Distribution on Masonry

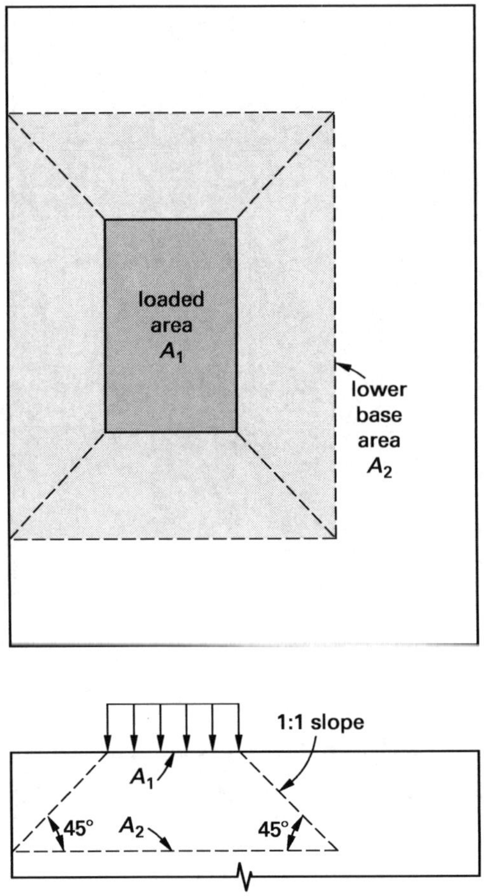

Figure 6.6 Nomenclature for Beam Bearing Plate

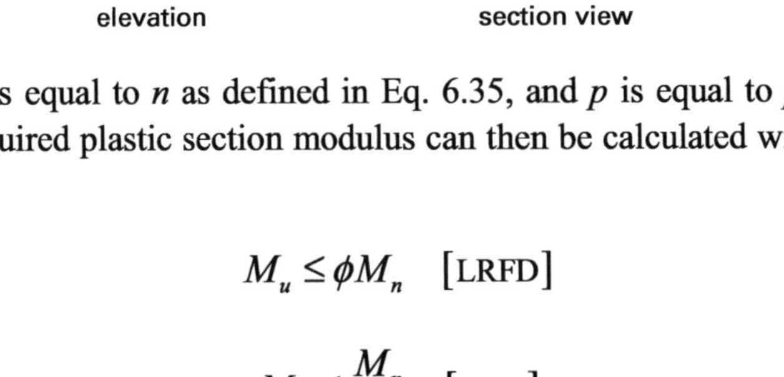

In Eq. 6.36, L is equal to n as defined in Eq. 6.35, and p is equal to p_u (LRFD) or p_a (ASD). The required plastic section modulus can then be calculated with the following formulas.

$$M_u \le \phi M_n \quad \text{[LRFD]} \qquad 6.37$$

$$M_a \le \frac{M_n}{\Omega} \quad \text{[ASD]} \qquad 6.38$$

$$M_n = M_p = F_y Z \le 1.6 M_y \quad \text{[LRFD and ASD]} \qquad 6.39$$

Because lateral bucking will not occur,

$$M_u \le \phi M_n = \phi F_y Z \quad [\text{LRFD}] \tag{6.40}$$

$$M_a \le \frac{F_y Z}{\Omega} \quad [\text{ASD}] \tag{6.41}$$

$$Z_{\text{req}} = \frac{M_u}{\phi F_y} \quad [\text{LRFD}] \tag{6.42}$$

$$Z_{\text{req}} = \frac{\Omega M_a}{F_y} \quad [\text{ASD}] \tag{6.43}$$

Calculate the required bearing plate thickness. From *AISC Manual* Table 17-27, the required plate thickness can be calculated from the plastic section modulus of a rectangular section.

$$Z = \frac{bt^2}{4} \tag{6.44}$$

$$t = \sqrt{\frac{4Z}{b}} \tag{6.45}$$

Combining Eq. 6.35, Eq. 6.36, Eq. 6.42 (LRFD) or Eq. 6.43 (ASD), and Eq. 6.45, the required base plate thickness is

$$t = \sqrt{\frac{2p_u\left(\frac{B}{2} - k_{\text{des}}\right)^2}{\phi F_y}} \quad [\text{LRFD}] \tag{6.46}$$

$$t = \sqrt{\frac{2p_a\left(\frac{B}{2} - k_{\text{des}}\right)^2}{\frac{F_y}{\Omega}}} \quad [\text{ASD}] \tag{6.47}$$

Example 6.3

Beam Bearing Plate

A W12 × 50 ASTM A992 steel beam bears on a concrete pier. The beam has a dead load end reaction of 15 kips and a live load reaction of 45 kips. The length of bearing required is limited to 8 in.

Section properties

$A = 14.6\ \text{in}^2$	$k_{\text{des}} = 1.14\ \text{in}$	$Z_x = 71.9\ \text{in}^3$
$d = 12.2\ \text{in}$	$k_{\text{det}} = 1^1/_2\ \text{in}$	$I_y = 56.3\ \text{in}^4$
$t_w = 0.370\ \text{in}$	$k_1 = {}^{15}/_{16}\ \text{in}$	$S_y = 13.9\ \text{in}^3$
$b_f = 8.08\ \text{in}$	$I_x = 391\ \text{in}^4$	$Z_y = 21.3\ \text{in}^3$
$t_f = 0.640\ \text{in}$	$S_x = 64.2\ \text{in}^3$	

Determine the width and thickness of an ASTM A36 steel bearing plate if the allowable bearing pressure is limited to 0.55 ksi.

Solution

Calculate the required strengths.

LRFD	ASD
$P_u = 1.2D + 1.6L$ $= (1.2)(15\ \text{kips}) + (1.6)(45\ \text{kips})$ $= 90\ \text{kips}$	$P_a = D + L$ $= 15\ \text{kips} + 45\ \text{kips}$ $= 60\ \text{kips}$

Calculate the required width, B, of the bearing plate.

$$A_{\text{req}} = \frac{P_a}{p_a} = \frac{60\ \text{kips}}{0.55\ \frac{\text{kips}}{\text{in}^2}} = 109.09\ \text{in}^2$$

$$B = \frac{A_{\text{req}}}{N} = \frac{109.09\ \text{in}^2}{8\ \text{in}} = 13.64\ \text{in}\quad [\text{use 14 in}]$$

Calculate the bearing stresses.

LRFD	ASD
$p_u = \frac{P_u}{A_{\text{plate}}} = \frac{90\ \text{kips}}{(8\ \text{in})(14\ \text{in})} = 0.80\ \text{ksi}$	$p_a = \frac{P_a}{A_{\text{plate}}} = \frac{60\ \text{kips}}{(8\ \text{in})(14\ \text{in})} = 0.54\ \text{ksi}$

Calculate the effective cantilevered length, n, using Eq. 6.35.

$$n = \frac{B}{2} - k_{\text{des}} = \frac{14\ \text{in}}{2} - 1.14\ \text{in} = 5.86\ \text{in}$$

Calculate the required bearing plate thickness using Eq. 6.46 and Eq. 6.47.

LRFD	ASD
$t=\sqrt{\dfrac{2p_u\left(\dfrac{B}{2}-k_{\text{des}}\right)^2}{\phi F_y}}$ $=\sqrt{\dfrac{(2)\left(0.80\ \dfrac{\text{kip}}{\text{in}^2}\right)(5.86\ \text{in})^2}{(0.90)\left(36\ \dfrac{\text{kips}}{\text{in}^2}\right)}}$ $=1.30\ \text{in}$	$t=\sqrt{\dfrac{2p_a\left(\dfrac{B}{2}-k_{\text{des}}\right)^2}{\dfrac{F_y}{\Omega}}}$ $=\sqrt{\dfrac{(2)\left(0.54\ \dfrac{\text{kip}}{\text{in}^2}\right)(5.86\ \text{in})^2}{\dfrac{36\ \dfrac{\text{kips}}{\text{in}^2}}{1.67}}}$ $=1.31\ \text{in}$

7. STIFFENER AND DOUBLER PLATE REQUIREMENTS

Stiffeners and/or doubler plates must be provided if the required strength exceeds the available strength for the applicable limit states when concentrated loads are applied to the flanges. The stiffening elements (stiffeners or doublers) and the welds connecting them to the member must be sized for the difference between the required strength and the available strength. Installing stiffeners or doublers is time consuming and expensive. It is frequently more economical to increase the weight and/or size of the base member rather than to incorporate stiffening elements.

Each stiffener's width, when added to one-half the thickness of the column web, must be at least one-third the width of the flange or moment connection plate that delivers the force.

The thickness of a stiffener must be at least one-half the thickness of the flange or moment connection plate that delivers the concentrated load, and at least $^{1}/_{16}$ the width of the flange or plate.

Each transverse stiffener must extend at least one-half the depth of the member, except that in the following two cases, the transverse stiffener must extend the full depth of the web.

- If a beam or girder is not otherwise restrained against rotation about its longitudinal axis, a pair of full-depth transverse stiffeners is needed at the member's unframed ends.
- If the available web compression buckling strength is less than the required strength, then one of the following must be provided: a single, full-depth transverse stiffener; a pair of transverse stiffeners; or a doubler plate.

At times, the bearing length required to prevent web yielding or web crippling may be greater than the available bearing length. In these cases, the required bearing length can be reduced by welding full-depth stiffener plates to each side of the beam web. The width of each web stiffener must be at least one-third the width of the flange. For maximum effectiveness, it's preferable that the stiffeners extend from the web to approximately the face of the flange. The minimum thickness of the stiffener is governed by the limiting width-thickness ratios for compression elements, as given in *AISC Specification* Table B4.1a.

The cross section of a pair of web stiffeners is taken to be that of a cross-shaped column composed of the two web stiffeners and a specified length of the web, called the effective web length. For interior stiffeners, the effective web length is

$$L_{w,\text{eff}} = 25t_w \quad [\text{AISC Sec. J10.8}] \tag{6.48}$$

For end stiffeners, it is

$$L_{w,\text{eff}} = 12t_w \quad [\text{AISC Sec. J10.8}] \tag{6.49}$$

The proportioning and design of stiffeners are governed by the following formulas. For a pair of stiffeners, the maximum stiffener width is

$$b_{\text{st,max}} = \frac{b_f - t_w}{2} \tag{6.50}$$

For a pair of stiffeners, the minimum stiffener width is

$$b_{\text{st,min}} = \frac{b_f - t_w}{6} \tag{6.51}$$

The height of each stiffener is

$$h_{\text{st}} = d - 2k_{\text{des}} \tag{6.52}$$

The limiting width-thickness ratio for a projecting compression element is

$$\frac{b}{t} \le 0.64\sqrt{\frac{k_c E}{F_y}} \quad [\text{AISC Table B4.1a, case 2}] \tag{6.53}$$

In Eq. 6.53, the factor k_c is

$$k_c = \frac{4}{\sqrt{\frac{h}{t_w}}} \quad [0.35 \le k \le 0.76] \quad [\text{AISC Table B4.1a, note (a)}] \tag{6.54}$$

The gross area of the cross-shaped column formed by the beam web and stiffeners is calculated from Eq. 6.55 or Eq. 6.56. For end stiffeners,

$$A_{g,\text{cross}} = A_{\text{st}} + 12t_w^2 \tag{6.55}$$

For interior stiffeners,

$$A_{g,\text{cross}} = A_{\text{st}} + 25t_w^2 \qquad 6.56$$

The nominal compressive strength of the stiffener is

$$P_n = F_{\text{cr}} A_{g,\text{cross}} \quad \text{[AISC Eq. E3-1]} \qquad 6.57$$

The critical stress is calculated with Eq. 6.58 or Eq. 6.59.

For $KL/r \le 4.71\sqrt{E/F_y}$,

$$F_{\text{cr}} = 0.658^{F_y/F_e} F_y \quad \text{[AISC Eq. E3-2]} \qquad 6.58$$

For $KL/r > 4.71\sqrt{E/F_y}$,

$$F_{\text{cr}} = 0.877F_e \quad \text{[AISC Eq. E3-3]} \qquad 6.59$$

In Eq. 6.58 and Eq. 6.59, the elastic critical buckling stress is

$$F_e = \frac{\pi^2 E}{\left(\frac{KL}{r}\right)^2} \quad \text{[AISC Eq. E3-4]} \qquad 6.60$$

From *AISC Specification* Sec. J10.8, the effective length factor for a stiffener is $K = 0.75$.

Example 6.4

Bearing Stiffener Design

The W18 × 35 ASTM A992 steel beam shown bears on a concrete pier. The bearing length, l_b, is limited to 12 in. Assume that the concrete pier has adequate bearing strength. The beam end reactions for dead and live loads are 22.5 kips and 67.5 kips, respectively.

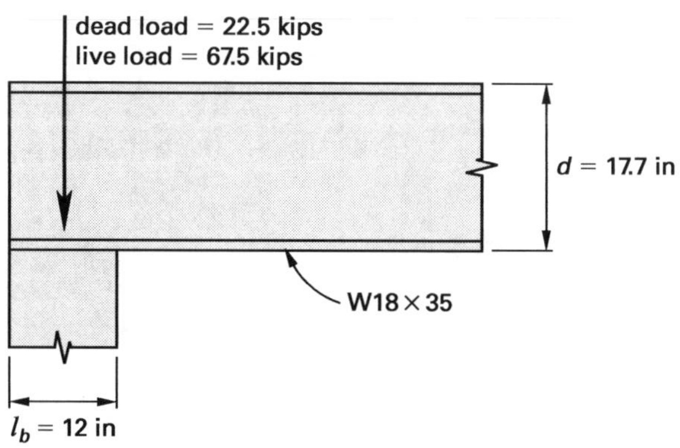

Section properties

$A = 10.3 \text{ in}^2$	$k_{des} = 0.827$ in	$Z_x = 66.5 \text{ in}^3$
$d = 17.7$ in	$k_{det} = 1^1/_8$ in	$I_y = 15.3 \text{ in}^4$
$t_w = 0.300$ in	$k_1 = {}^3/_4$ in	$S_y = 5.12 \text{ in}^3$
$b_f = 6.00$ in	$I_x = 510 \text{ in}^4$	$Z_y = 8.06 \text{ in}^3$
$t_f = 0.425$ in	$S_x = 57.6 \text{ in}^3$	

Beam bearing constants

method	R_1 (kips)	R_2 (kips/in)	R_3 (kips)	R_4 (kips/in)	R_5 (kips)	R_6 (kips/in)
LRFD (ϕR)	31.0	15.0	38.7	3.89	34.1	5.19
ASD (R/Ω)	20.7	10.0	25.8	2.59	22.7	3.46

Determine whether bearing stiffeners are required. If bearing stiffeners are required, design the stiffeners.

Solution

Determine the required strengths.

LRFD	ASD
$R_u = 1.2D + 1.6L$ $= (1.2)(22.5 \text{ kips}) + (1.6)(67.5 \text{ kips})$ $= 135 \text{ kips}$	$R_a = D + L$ $= 22.5 \text{ kips} + 67.5 \text{ kips}$ $= 90 \text{ kips}$

Determine the appropriate formula for resistance to web local yielding.

$l_b/2$ is less than $d/2$ (6 in < 8.85 in), so use Eq. 6.15 (LRFD) or Eq. 6.17 (ASD). The resistance to web local yielding is

LRFD	ASD
$\phi R_n = \phi R_1 + l_b(\phi R_2)$ $= 31 \text{ kips} + (12 \text{ in})\left(15 \ \frac{\text{kips}}{\text{in}}\right)$ $= 211 \text{ kips}$	$\frac{R_n}{\Omega} = \frac{R_1}{\Omega} + l_b\left(\frac{R_2}{\Omega}\right)$ $= 20.7 \text{ kips} + (12 \text{ in})\left(10.0 \ \frac{\text{kips}}{\text{in}}\right)$ $= 140.7 \text{ kips}$

Determine the right formula for resistance to web crippling.

$$\frac{l_b}{d} = \frac{12 \text{ in}}{17.7 \text{ in}} = 0.68 \quad [> 0.2]$$

l_b/d is greater than 0.2, so use Eq. 6.27 (LRFD) or Eq. 6.29 (ASD). The resistance to web crippling is

LRFD	ASD
$\phi R_n = \phi R_5 + l_b(\phi R_6)$ $= 34.1\text{ kips} + (12\text{ in})\left(5.19\ \frac{\text{kips}}{\text{in}}\right)$ $= 96.38\text{ kips}$ $[\text{controls}; < R_u = 135\text{ kips}]$	$\frac{R_n}{\Omega} = \frac{R_5}{\Omega} + l_b\left(\frac{R_6}{\Omega}\right)$ $= 22.7\text{ kips} + (12\text{ in})\left(3.46\ \frac{\text{kips}}{\text{in}}\right)$ $= 64.22\text{ kips}$ $[\text{controls}; < R_a = 90\text{ kips}]$

The value for web local crippling is lower, so it governs. The resistance to web crippling is less than the required strength, so a bearing stiffener is required. Calculate the required strength of the bearing stiffeners.

LRFD	ASD
$R_{st} = R_u - \phi R_n$ $= 135\text{ kips} - 96.38\text{ kips}$ $= 38.62\text{ kips}$	$R_{st} = R_a - \frac{R_n}{\Omega}$ $= 90\text{ kips} - 64.22\text{ kips}$ $= 25.78\text{ kips}$

Fabricate stiffener plates from ASTM A572, grade 50 steel with F_y = 50 ksi and F_u = 65 ksi. *AISC Manual* Table 2-4 indicates that the preferred steel for plates is ASTM A36 with F_y = 36 ksi and F_u = 58 ksi. Determine the maximum and minimum stiffener widths using Eq. 6.50 and Eq. 6.51 (see illustration).

$$\begin{aligned} b_{\text{st,max}} &= \frac{b_f - t_w}{2} = \frac{6.00\text{ in} - 0.300\text{ in}}{2} \\ &= 2.85\text{ in} \end{aligned}$$

$$\begin{aligned} b_{\text{st,min}} &= \frac{b_f - t_w}{6} = \frac{6.00\text{ in} - 0.300\text{ in}}{6} \\ &= 0.95\text{ in} \end{aligned}$$

The cost of the steel in the stiffener is relatively unimportant, so choose a plate size near the maximum. Using a 2.75 in wide plate, however, would place the edge of the stiffener too close to the edge of the flange. Use a 2.50 in wide plate. Determine the minimum stiffener thickness. From Eq. 6.52, the calculated web height is

$$\begin{aligned} h_{\text{st}} &= d - 2k_{\text{des}} \\ &= 17.7\text{ in} - (2)(0.827\text{ in}) \\ &= 16.05\text{ in} \end{aligned}$$

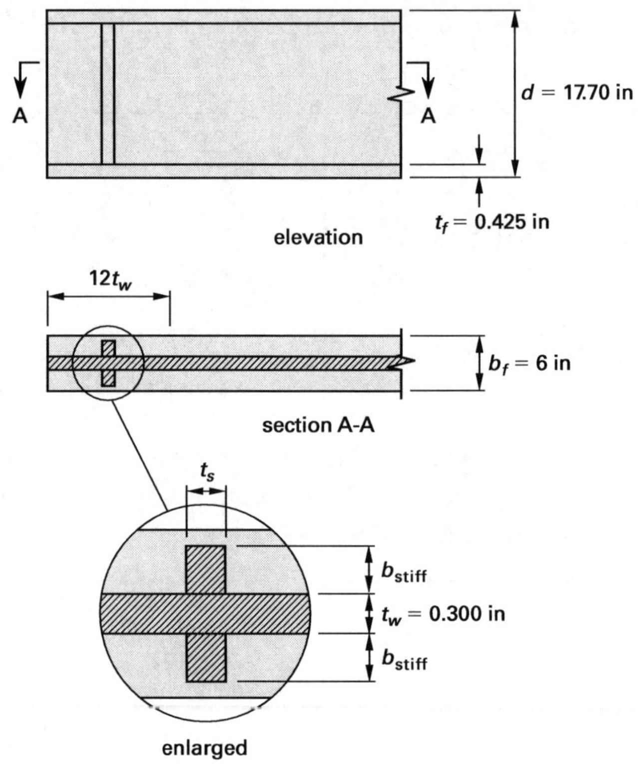

Calculate k_c from Eq. 6.54.

$$\begin{aligned} k_c &= \frac{4}{\sqrt{\frac{h}{t_w}}} \quad [0.35 \geq k_c \geq 0.76] \\ &= \frac{4}{\sqrt{\frac{16.05 \text{ in}}{0.300 \text{ in}}}} \\ &= 0.55 \end{aligned}$$

Calculate the limiting width-thickness ratio for the stiffener using Eq. 6.53.

$$\begin{aligned} \frac{b_{st}}{t_{st}} &\leq 0.64\sqrt{\frac{k_c E}{F_y}} \\ &\leq 0.64\sqrt{\frac{(0.55)\left(29{,}000 \ \frac{\text{kips}}{\text{in}^2}\right)}{50 \ \frac{\text{kips}}{\text{in}^2}}} \\ &\leq 11.43 \end{aligned}$$

Determine the minimum thickness.

$$\frac{b_{\text{st}}}{t_{\text{st}}} \le 11.43$$

$$t_{\text{st}} \ge \frac{b_{\text{st}}}{11.43}$$

$$t_{\text{st,min}} = \frac{2.50 \text{ in}}{11.43} = 0.22 \text{ in} \quad [\text{use } 0.25 \text{ in}]$$

Use a pair of $^1/_4$ in × 2.50 in plate stiffeners. To confirm that these stiffeners will meet the required strength, first determine the gross area of the cross-shaped column using Eq. 6.55.

$$\begin{aligned} A_{\text{st}} &= n_{\text{st}} b_{\text{st}} t_{\text{st}} = (2)(2.50 \text{ in})(0.25 \text{ in}) \\ &= 1.25 \text{ in}^2 \\ A_{g,\text{cross}} &= A_{\text{st}} + 12t_w^2 \\ &= 1.25 \text{ in}^2 + (12)(0.300 \text{ in})^2 \\ &= 2.33 \text{ in}^2 \end{aligned}$$

From Eq. 6.49, the effective web length is

$$L_{w,\text{eff}} = 12t_w = (12)(0.300 \text{ in}) = 3.6 \text{ in}$$

Calculate the moment of inertia of the cross-shaped column about the centerline of the beam web. This is equal to the moment of inertia of the stiffeners through the web plus the moment of inertia of the remainder of the web portion. (The latter term is often neglected in practice because it is relatively insignificant.)

$$\begin{aligned} I_{\text{cross}} &= I_{\text{st}} + I_w = \frac{(bd^3)_{\text{st}}}{12} + \frac{(bd^3)_w}{12} \\ &= \frac{t_{\text{st}}(t_w + 2b_{\text{st}})^3}{12} + \frac{(L_{w,\text{eff}} - t_{\text{st}})t_w^3}{12} \\ &= \frac{(0.25 \text{ in})(0.300 \text{ in} + (2)(2.50 \text{ in}))^3}{12} + \frac{(3.6 \text{ in} - 0.25 \text{ in})(0.300 \text{ in})^3}{12} \\ &= 3.10 \text{ in}^4 + 0.007 \text{ in}^4 \\ &= 3.11 \text{ in}^4 \end{aligned}$$

The radius of gyration for the column is

$$r_{\text{cross}} = \sqrt{\frac{I_{\text{cross}}}{A_{g,\text{cross}}}} = \sqrt{\frac{3.11 \text{ in}^4}{2.33 \text{ in}^2}} = 1.16 \text{ in}$$

Calculate the effective slenderness ratio for the column, using an effective length factor of 0.75.

$$\frac{KL}{r} = \frac{Kh_{\text{cross}}}{r_{\text{cross}}} = \frac{(0.75)(16.05 \text{ in})}{1.16 \text{ in}} = 10.38$$

Determine the correct formula for calculating the critical flexural buckling stress.

$$4.71\sqrt{\frac{E}{F_y}} = 4.71\sqrt{\frac{29{,}000 \ \frac{\text{kips}}{\text{in}^2}}{50 \ \frac{\text{kips}}{\text{in}^2}}} = 113.43$$

This is greater than the slenderness ratio, KL/r, so use Eq. 6.58. From Eq. 6.60, the elastic critical buckling stress is

$$F_e = \frac{\pi^2 E}{\left(\frac{KL}{r}\right)^2} = \frac{\pi^2\left(29{,}000 \ \frac{\text{kips}}{\text{in}^2}\right)}{(10.38)^2} = 2656.46 \text{ ksi}$$

Calculate the critical flexural buckling stress using Eq. 6.58.

$$F_{\text{cr}} = 0.658^{F_y/F_e}F_y = (0.658)^{50 \ \frac{\text{kips}}{\text{in}^2} \Big/ 2656 \ \frac{\text{kips}}{\text{in}^2}}\left(50 \ \frac{\text{kips}}{\text{in}^2}\right) = 49.61 \text{ ksi}$$

From Eq. 6.57, the nominal axial compression load capacity for the cross-shaped stiffener column is

$$P_n = F_{\text{cr}}A_{g,\text{cross}} = \left(49.61 \ \frac{\text{kips}}{\text{in}^2}\right)(2.33 \text{ in}^2) = 115.59 \text{ kips}$$

Calculate the design strength (LRFD) or allowable strength (ASD) for the stiffener column.

LRFD	ASD
$P_{u,\text{st}} = \phi_c P_n$ $= (0.90)(115.59 \text{ kips})$ $= 104.03 \text{ kips}$ $[> R_{\text{st}} = 38.62 \text{ kips, so OK}]$	$P_{a,\text{st}} = \dfrac{P_n}{\Omega_c}$ $= \dfrac{115.59 \text{ kips}}{1.67}$ $= 69.22 \text{ kips}$ $[> R_{\text{st}} = 25.78 \text{ kips, so OK}]$

In both LRFD and ASD solutions, the calculated strength is about 2.6 times greater than the required strength. The minimum width of the stiffeners was calculated as 0.95 in, so in theory the stiffener width of 2.50 in could be decreased, and by doing so it may be possible to decrease the thickness as well. However, most of the cost of installing the stiffeners will be in the fabrication and welding, not the steel. Moreover, as a matter of practice many engineering firms specify a minimum thickness of $^1/_4$ in for stiffeners that are exposed to the elements.

Use a pair of full-height stiffeners: 2.50 in by $^1/_4$ in.

7 Steel Column Design

Nomenclature

A	area	in^2
A_1	loaded area	in^2
A_2	maximum area of supporting surface that is geometrically similar to and concentric with loaded area A_1	in^2
b	width	in
B	width of bearing plate	in
C_w	warping constant	in^6
d	depth	in
D	dead load	lbf
E	modulus of elasticity	lbf/in^2
f	compressive stress	lbf/in^2
f_a	bearing stress of service loads	lbf/in^2
f_u	bearing stress of factored loads	lbf/in^2
f_c'	specified compressive strength of concrete	lbf/in^2
F	strength or stress	lbf/in^2
G	shear modulus of elasticity of steel	lbf/in^2
G_A, G_B	end condition coefficient	–
h	height	in
H	flexural constant	–
I	moment of inertia	in^4
J	torsional constant	in^4
K	effective length factor	–
KL	effective length	in
KL/r	slenderness ratio	–
l	critical base plate cantilever dimension, largest of m, n, and $\lambda n'$	in
L	length	in
L	live load	lbf
m	cantilever dimension for base plate along plate length	in
n	cantilever dimension for base plate along plate width	in
n'	factor used in calculating base plate cantilever dimension	in

l_b	length of bearing plate	in
P	axial strength	lbf
P_p	nominal bearing strength of concrete	lbf
Q	reduction factor	–
Q_a	reduction factor for slender stiffened elements	–
Q_s	reduction factor for slender unstiffened elements	–
r	radius of gyration	in
$\overline{r}_o$	polar radius of gyration about shear center	in
R	strength	lbf
S	elastic section modulus	in^3
t	thickness	in
t_{des}	design wall thickness	in
x_o, y_o	coordinates of shear center with respect to centroid	in
Z	plastic section modulus	in^3

Symbols

λ	factor used in calculating base plate cantilever dimension	–
λ_r	limiting width-thickness ratio for noncompactness	–
$\lambda n'$	base plate cantilever dimension	in
ϕ	resistance factor	–
Ω	safety factor	–

Subscripts

a	required (ASD)
c	compression or compressive
col	column
cr	critical
e	effective or elastic critical buckling (Euler)
f	flange
g	gross
gir	girder
min	minimum
n	nominal
p	base plate
u	required (LRFD) or ultimate tensile
w	web

x about x-axis

y about y-axis or yield

z about z-axis

1. INTRODUCTION

Chapter E of the *AISC Specification* governs the design of columns and other compression members. The chapter is divided into the following sections.

E1	General Provisions
E2	Effective Length
E3	Flexural Buckling of Members Without Slender Elements
E4	Torsional and Flexural-Torsional Buckling of Members Without Slender Elements
E5	Single Angle Compression Members
E6	Built-Up Members
E7	Members with Slender Elements

Tension members and flexural members bent about a single axis can be designed directly using a simple mathematical solution or beam charts or graphs. The design of columns and other compression members is more complex.

The difference is that members subjected to a compressive load have a tendency to buckle even when they are concentrically loaded. Then, as soon as the compression member starts to buckle, it is subjected not only to axial loads but bending loads as well. The lateral displacement and deflection due to combined vertical and lateral loading are known as *P-Δ and P-δ effects*. Making a column stiffer decreases its tendency to buckle, but also decreases its efficiency and cost effectiveness.

Chapter C and App. 7 in the *AISC Specification* give two methods of meeting the stability requirements for structures. Chapter C permits the use of the *direct analysis method* (DM or DAM) for all structures, while App. 7 describes the *effective length method* (ELM), which is permitted as an alternative for structures meeting certain conditions.[1] Table C-C1.1 in the *AISC Commentary on the Specification for Structural Steel Buildings* (hereinafter referred to as the *AISC Commentary*), compares the provisions of those requirements.[2]

The two methods account for P-Δ and P-δ effects in different ways. In the direct analysis method, notional lateral loads are applied in the analysis of the structure. In the effective length method, calculations on each column or other compression member use a modified *effective length* instead of the member's actual length.

[1]The sixth through thirteenth editions of the *AISC Manual* gave the effective length method for providing structural stability and designing compression members. The direct analysis method was added as an alternative in the thirteenth edition, where it appeared in App. 7.
[2]The *AISC Commentary* immediately follows the *AISC Specification* in the *AISC Manual*.

The effective length is the member's actual length multiplied by an *effective length factor*, *K*. This factor varies depending on how the ends of the member are restrained, as well as on whether the member is braced along its length to resist sidesway. In general, the more restraint, the lower the effective length factor.

Chapter E in the *AISC Specification* addresses the design of compression members. The same formulas in Chap. E are used with both the direct analysis method and the effective length method. Which formulas must be used when calculating buckling, however, will depend on each member's *slenderness ratio*, *KL*/*r*. This is the ratio of the member's effective length, *KL*, to its radius of gyration, *r*. When using the direct analysis method, the effective length factor is always taken as 1.0.

The *AISC Manual* once limited the slenderness ratio to a maximum of 200. This is no longer a requirement, but it is still recommended to allow for possible issues during fabrication, handling, and erection.

$$\frac{KL}{r} \le 200 \tag{7.1}$$

The strength of a member with a slenderness ratio of 200 is only approximately 12.5% of what it would be if its slenderness ratio were 1. The least radius of gyration, *r*, for a single angle is about the *Z*-*Z* axis.

The nominal strength, P_n, for a compression member is

$$P_n = F_{\text{cr}} A_g \quad \text{[AISC Eq. E3-1]} \tag{7.2}$$

For LRFD, the design compressive strength, $\phi_c P_n$, can be determined using Eq. 7.3. The compression resistance factor, ϕ_c, is 0.90.

$$\phi_c P_n = \phi_c F_{\text{cr}} A_g \quad \text{[LRFD]} \tag{7.3}$$

For ASD, the allowable compressive strength, P_n/Ω_c, can be determined using Eq. 7.4. The compression safety factor, Ω_c, is 1.67.

$$\frac{P_n}{\Omega_c} = \frac{F_{\text{cr}} A_g}{\Omega_c} \quad \text{[ASD]} \tag{7.4}$$

2. EFFECTIVE LENGTH OF COMPRESSION MEMBERS

The first step in designing a compression member is determining the effective length, *KL*, which is a function of that member's end conditions. If the compression member is braced against sidesway (i.e., sidesway is inhibited), the effective length factor, *K*, will be less than or equal to one. If the compression member is not braced against sidesway (i.e., sidesway is uninhibited), the effective length factor will be greater than one.

Appendix 7 of the *AISC Commentary* gives two methods of determining the effective length factor. The effective length for each axis (*X*-*X*, *Y*-*Y*, and *Z*-*Z*) must be computed, and the one producing the largest KL/r ratio will govern the design of the member. For any axis, the actual length and the effective unbraced length may be different.

Table 7.1 gives theoretical and recommended design K-factors for columns with six different kinds of end conditions. To determine the K-value necessary to calculate the effective length along each axis, use Table 7.1 unless other information is provided.

Table 7.1 *Approximate Values of Effective Length Factor, K*

end 1	end 2	theoretical K-value	recommended design K-value
built-in (rotation fixed, translation fixed)	built-in	0.5	0.65
built-in	pinned (rotation free, translation fixed)	0.7	0.8
built-in	rotation fixed, translation free	1.0	1.2
built-in	free	2.0	2.1
pinned	pinned	1.0	1.0
pinned	rotation fixed, translation free	2.0	2.0

Source: Compiled from *AISC Commentary* Table C-A-7.1.

If information is known, however, about the size and length of members framing into the ends of the column, or (equivalently) about the values for the end condition coefficients G_A and G_B, then determine K from one of the two alignment charts found in App. 7 of the *AISC Commentary*. The first of these, Fig. C-A-7.1, is for frames in which sidesway is inhibited (braced frame, $K \leq 1.0$). The second alignment chart, Fig. C-A-7.2, is for frames in which sidesway is uninhibited (moment frames, $K > 1.0$). These alignment charts are shown in Fig. 7.1.

To use these charts, calculate the values for the end condition coefficients, G_A and G_B, for each of the axes using Eq. 7.5. The modulus of elasticity, E, in Eq. 7.5 is the same for the column and for the girder or beam.

$$G = \frac{\sum \frac{EI_{\text{col}}}{L_{\text{col}}}}{\sum \frac{EI_{\text{gir}}}{L_{\text{gir}}}} = \frac{\sum \frac{I_{\text{col}}}{L_{\text{col}}}}{\sum \frac{I_{\text{gir}}}{L_{\text{gir}}}} \quad \text{[AISC Sec. C-C2.2b]} \qquad 7.5$$

For columns supported by a footing or foundation, but not rigidly connected to it, G can be taken as 10. If the column is rigidly attached to a properly designed footing, G can be taken as 1.0. The effective length of compression members is calculated in the same manner whether using ASD or LRFD.

Figure 7.1 *Alignment Charts for Determining Effective Length Factor, K*

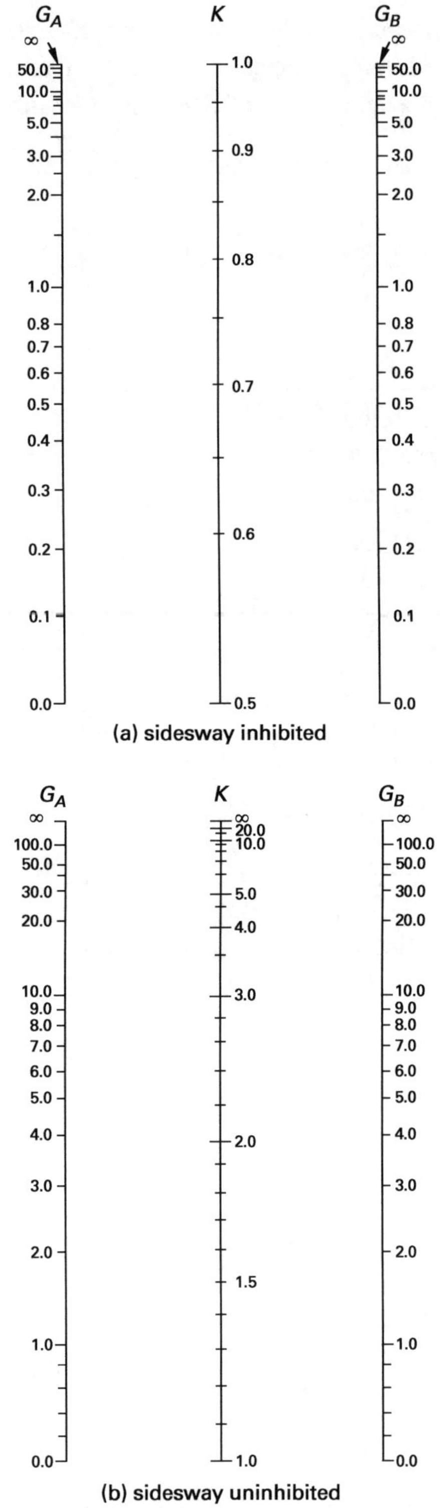

3. COMPRESSIVE STRENGTH FOR FLEXURAL MEMBERS WITHOUT SLENDER ELEMENTS

The information in this section applies to compression members with compact and noncompact sections for uniformly compressed elements. Slender elements can buckle before the overall member buckles. Two types of elements must be considered.

- *Unstiffened elements* are those that are unsupported along only one edge parallel to the direction of the compression force. (See flanges, *AISC Specification* Table B4.1a, case 2.)
- *Stiffened elements* are those that are supported along two edges parallel to the direction of the compression force. (See webs of I-shaped members, *AISC Specification* Table B4.1a, case 5.)

In designing these compression members, the following facts should be considered.

- All W shapes have nonslender flanges for ASTM A992 steel.
- Only one column section has a slender web: W14 × 43.
- Many W shapes meant to be used as beam sections have slender webs for uniform compression.

The critical stress for flexural buckling, F_{cr}, is determined by Eq. 7.6 through Eq. 7.8. When $KL/r \leq 4.71\sqrt{E/F_y}$ (or $F_y/F_e \leq 2.25$),

$$F_{cr} = 0.658^{F_y/F_e} F_y \quad \text{[AISC Eq. E3-2]} \tag{7.6}$$

When $KL/r > 4.71\sqrt{E/F_y}$ (or $F_y/F_e > 2.25$),

$$F_{cr} = 0.877 F_e \quad \text{[AISC Eq. E3-3]} \tag{7.7}$$

In both Eq. 7.6 and Eq. 7.7, the elastic critical buckling stress, F_e, is

$$F_e = \frac{\pi^2 E}{\left(\frac{KL}{r}\right)^2} \quad \text{[AISC Eq. E3-4]} \tag{7.8}$$

Figure 7.2 shows the column curve for the available critical stress, F_{cr}, and Table 7.2 gives the transition point limiting values for KL/r.

Figure 7.2 Column Curve for Available Critical Stress, F_{cr}

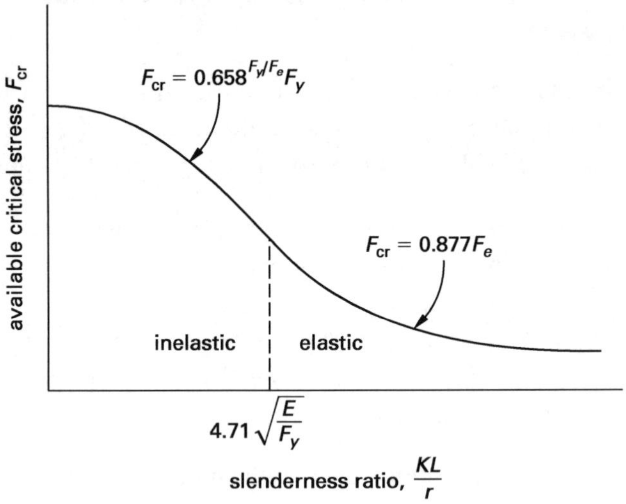

Table 7.2 Transition Point Limiting Values of KL/r

F_y (ksi)	limiting KL/r	F_e (ksi)
36	134	16.0
42	123	18.9
46	118	20.6
50	113	22.2
60	104	26.7
70	96	31.1

Example 7.1

Concentric Axial Loaded Column

A steel column is required to support a concentric axial dead load of 82 kips and concentric axial live load of 246 kips. The actual column height is 12 ft with no intermediate bracing. At the base of the column, the x- and y-axes are fixed against translation and free to rotate; at the top of the column, the x- and y-axes are fixed against rotation and free to translate.

Material properties

ASTM A992

$F_y = 50$ ksi

$F_u = 65$ ksi

Determine the lightest W12 section that will support the load requirements.

Solution

Determine the effective length of the column. From Table 7.1, with end conditions as described, the effective length factor is $K = 2.0$. Therefore,

$$KL = (2)(12 \text{ ft}) = 24 \text{ ft}$$

Determine the required available strength.

LRFD	ASD
$P_u = 1.2D + 1.6L$ $= (1.2)(82 \text{ kips}) + (1.6)(246 \text{ kips})$ $= 492 \text{ kips}$	$P_a = D + L$ $= 82 \text{ kips} + 246 \text{ kips}$ $= 328 \text{ kips}$

From *AISC Manual* Table 4-1, select the lightest W12 member that has at least this available strength. This is a W12 × 72. For LRFD,

$$\phi_c P_n = 493 \text{ kips} \geq P_u = 492 \text{ kips}$$

For ASD,

$$\frac{P_n}{\Omega_c} = 328 \text{ kips} \geq P_a = 328 \text{ kips}$$

The available strengths given in *AISC Manual* Table 4-1 through Table 4-3 apply when the effective length of the y-axis controls, as in this example. To use these tables when the effective length of the x-axis controls, multiply the x-axis effective length by the r_x/r_y ratio given at the bottom of each column.

Example 7.2
Concentric Axial Loaded Column

A W8 × 21 steel column design is controlled by the effective length of the y-axis with $K_yL_y = 12$ ft.

Section properties		Material properties
$A = 6.16 \text{ in}^2$	$S_x = 18.2 \text{ in}^3$	ASTM A992 steel
$d = 8.28$ in	$r_x = 3.49$ in	$F_y = 50$ ksi
$t_w = 0.250$ in	$Z_x = 20.4 \text{ in}^3$	$F_u = 65$ ksi
$b_f = 5.27$ in	$I_y = 9.77 \text{ in}^4$	
$t_f = 0.400$ in	$S_y = 3.71 \text{ in}^3$	
$b_f/2t_f = 6.59$	$r_y = 1.26$ in	
$h/t_w = 27.5$	$Z_y = 5.69 \text{ in}^3$	
$I_x = 75.3 \text{ in}^4$		

Determine the design load capacity and allowable load capacity of the column.

Solution

Determine the slenderness ratio.

$$\frac{K_y L_y}{r_y} = \frac{(1)(12\text{ ft})\left(12\ \frac{\text{in}}{\text{ft}}\right)}{1.26\text{ in}} = 114.29$$

Determine whether the member is in the inelastic or elastic range.

$$4.71\sqrt{\frac{E}{F_y}} = 4.71\sqrt{\frac{29{,}000\ \frac{\text{kips}}{\text{in}^2}}{50\ \frac{\text{kips}}{\text{in}^2}}} = 113.43 \quad [< KL/r]$$

The member is in the elastic range. Calculate the available critical stress using Eq. 7.7 and Eq. 7.8.

$$\begin{aligned} F_e &= \frac{\pi^2 E}{\left(\frac{KL}{r}\right)^2} \\ &= \frac{\pi^2\left(29{,}000\ \frac{\text{kips}}{\text{in}^2}\right)}{(114.29)^2} \\ &= 21.91\text{ ksi} \\ F_{\text{cr}} &= 0.877F_e \\ &= (0.877)\left(21.91\ \frac{\text{kips}}{\text{in}^2}\right) \\ &= 19.22\text{ ksi} \end{aligned}$$

Determine the available strengths using $\phi_c = 0.90$ and $\Omega_c = 1.67$.

LRFD	ASD
$\phi_c P_n = \phi_c F_{\text{cr}} A_g$ $= (0.90)\left(19.22\ \frac{\text{kips}}{\text{in}^2}\right)(6.16\text{ in}^2)$ $= 106.56\text{ kips}$	$\frac{P_n}{\Omega_c} = \frac{F_{\text{cr}} A_g}{\Omega_c}$ $= \frac{\left(19.22\ \frac{\text{kips}}{\text{in}^2}\right)(6.16\text{ in}^2)}{1.67}$ $= 70.90\text{ kips}$

As an alternative to the preceding solution, *AISC Manual* Table 4-22 gives the available critical stress for *KL/r* ratios from 1 to 200 and for yield strengths of 35 ksi, 36 ksi, 42 ksi, 46 ksi, and 50 ksi. For a slenderness ratio of $KL/r = 115$, the values for available critical stress are $\phi_c F_{cr} = 17.1$ ksi and $F_{cr}/\Omega_c = 11.4$ ksi.

LRFD	ASD
$\phi_c P_n = (\phi_c F_{cr}) A_g$ $= \left(17.1 \ \frac{\text{kips}}{\text{in}^2}\right)(6.16 \text{ in}^2)$ $= 105.34 \text{ kips}$	$\frac{P_n}{\Omega_c} = \left(\frac{F_{cr}}{\Omega_c}\right) A_g$ $= \left(11.4 \ \frac{\text{kips}}{\text{in}^2}\right)(6.16 \text{ in}^2)$ $= 70.22 \text{ kips}$

The design load capacity of the column is 105.34 kips and the allowable load capacity is 70.22 kips.

4. TORSIONAL AND FLEXURAL-TORSIONAL BUCKLING OF MEMBERS WITHOUT SLENDER ELEMENTS

This section applies to singly symmetrical and unsymmetrical members, as well as to certain doubly symmetrical members, such as cruciform or built-up columns with compact and noncompact sections for uniformly compressed members. This section does not apply to single angles, which are covered in *AISC Specification* Sec. E5.

The nominal compressive strength, P_n, is determined from the limit states of flexural-torsional buckling and torsional buckling.

$$P_n = F_{cr} A_g \quad \text{[AISC Eq. E4-1]} \tag{7.9}$$

For double angle and T-shaped compression members, use *AISC Specification* Eq. E4-2.

$$F_{cr} = \left(\frac{F_{cr,y} + F_{cr,z}}{2H}\right)\left(1 - \sqrt{1 - \frac{4F_{cr,y}F_{cr,z}H}{(F_{cr,y} + F_{cr,z})^2}}\right) \quad \text{[AISC Eq. E4-2]} \tag{7.10}$$

$F_{cr,y}$ is taken as F_{cr} as determined by Eq. 7.6 and Eq. 7.7, for flexural buckling about the y-axis of symmetry and $KL/r = KL/r_y$. $F_{cr,z}$ is calculated from Eq. 7.11.

$$F_{cr,z} = \frac{GJ}{A_g \bar{r}_o^2} \quad \text{[AISC Eq. E4-3]} \tag{7.11}$$

For all other cases, F_{cr} is determined from Eq. 7.6 and Eq. 7.7 using the torsional or flexural-torsional elastic buckling stress, F_e, as determined by Eq. 7.12 and Eq. 7.13. For doubly symmetrical members,

$$F_e = \left(\frac{\pi^2 E C_w}{(K_z L)^2} + GJ \right) \left(\frac{1}{I_x + I_y} \right) \quad \text{[AISC Eq. E4-4]} \qquad 7.12$$

For singly symmetrical members where y is the axis of symmetry,

$$F_e = \left(\frac{F_{ey} + F_{ez}}{2H} \right) \left(1 - \sqrt{1 - \frac{4F_{ey}F_{ez}H}{(F_{ey} + F_{ez})^2}} \right) \quad \text{[AISC Eq. E4-5]} \qquad 7.13$$

In Eq. 7.13, the flexural constant, H, is

$$H = 1 - \frac{x_o^2 + y_o^2}{\overline{r}_o^2} \quad \text{[AISC Eq. E4-10]} \qquad 7.14$$

The square of the polar radius of gyration about the shear center, $\overline{r}_o^2$, is

$$\overline{r}_o^2 = x_o^2 + y_o^2 + \frac{I_x + I_y}{A_g} \quad \text{[AISC Eq. E4-11]} \qquad 7.15$$

The other values for Eq. 7.13 are

$$F_{ex} = \frac{\pi^2 E}{\left(\frac{K_x L}{r_x} \right)^2} \quad \text{[AISC Eq. E4-7]} \qquad 7.16$$

$$F_{ey} = \frac{\pi^2 E}{\left(\frac{K_y L}{r_y} \right)^2} \quad \text{[AISC Eq. E4-8]} \qquad 7.17$$

$$F_{ez} = \left(\frac{\pi^2 E C_w}{(K_z L)^2} + GJ \right) \left(\frac{1}{A_g \overline{r}_o^2} \right) \quad \text{[AISC Eq. E4-9]} \qquad 7.18$$

Example 7.3

Axial Loaded WT Compression Member

A WT7 × 34 steel member is loaded in compression. K_xL_x = 25 ft and K_yL_y = 25 ft.

Section properties		Material properties
$A = 10.0\text{ in}^2$	$H = 0.916$	ASTM A992 steel
$r_x = 1.81\text{ in}$	$d = 7.02\text{ in}$	$F_y = 50\text{ ksi}$
$r_y = 2.46\text{ in}$	$t_w = 0.415\text{ in}$	$F_u = 65\text{ ksi}$
$\bar{r}_o = 3.19\text{ in}$	$b_f = 10.0\text{ in}$	
$J = 1.50\text{ in}^4$	$t_f = 0.72\text{ in}$	

Determine the design strength (LRFD) and the allowable strength (ASD) of the member.[3]

Solution

Check for slender elements with *AISC Specification* Table B4.1a. For the web (case 4),

$$\lambda_{rw} = 0.75\sqrt{\frac{E}{F_y}} = 0.75\sqrt{\frac{29{,}000\ \frac{\text{kips}}{\text{in}^2}}{50\ \frac{\text{kips}}{\text{in}^2}}} = 18.1$$

$$\frac{d}{t_w} = \frac{7.02\text{ in}}{0.415\text{ in}} = 16.9\quad [< \lambda_{rw},\ \text{so not slender}]$$

For the flanges (case 1),

$$\lambda_{rf} = 0.56\sqrt{\frac{E}{F_y}} = 0.56\sqrt{\frac{29{,}000\ \frac{\text{kips}}{\text{in}^2}}{50\ \frac{\text{kips}}{\text{in}^2}}} = 13.5$$

$$\frac{b_f}{2t_f} = \frac{10\text{ in}}{(2)(0.720\text{ in})} = 6.94\quad \left[< \lambda_{rf},\ \text{so not slender}\right]$$

Neither the web nor the flanges are slender; therefore, because there are no slender elements, *AISC Specification* Sec. E3 and Sec. E4 will apply.

[3] The values of $\bar{r}_o$ and H for this member are provided in Table 1-32 of the *AISC Manual: LRFD*, third edition. Unfortunately they are not provided in the *AISC Manual*, fourteenth edition.

Check the flexural buckling about the x- and y-axes using Eq. 7.6 through Eq. 7.8.

$$\frac{KL}{r_x} = \frac{(25\ \text{ft})\left(12\ \frac{\text{in}}{\text{ft}}\right)}{1.81\ \text{in}} = 165.75$$

$$\frac{KL}{r_y} = \frac{(25\ \text{ft})\left(12\ \frac{\text{in}}{\text{ft}}\right)}{2.46\ \text{in}} = 121.95$$

$$4.71\sqrt{\frac{E}{F_y}} = 4.71\sqrt{\frac{29{,}000\ \frac{\text{kips}}{\text{in}^2}}{50\ \frac{\text{kips}}{\text{in}^2}}} = 113.43 \quad \left[< KL/r_x, < KL/r_y\right]$$

KL/r_x is greater than both KL/r_y and $4.71\sqrt{\frac{E}{F_y}}$. Therefore, for flexural buckling about the y-axis of symmetry, $F_{\text{cr},x}$ is taken as F_{cr} as determined by Eq. 7.7 and Eq. 7.8.

$$F_{ex} = \frac{\pi^2 E}{\left(\frac{KL}{r_x}\right)^2} = \frac{\pi^2\left(29{,}000\ \frac{\text{kips}}{\text{in}^2}\right)}{(165.75)^2} = 10.42\ \text{ksi}$$

$$F_{\text{cr},x} = 0.877F_{ex} = (0.877)\left(10.42\ \frac{\text{kips}}{\text{in}^2}\right) = 9.14\ \text{ksi}$$

Check compressive strength for torsional and flexural-torsional buckling using Eq. 7.10 and Eq. 7.11 first calculating $F_{\text{cr},y}$. The shear modulus of elasticity for steel is 11,200 ksi.

$$F_{ey} = \frac{\pi^2 E}{\left(\frac{KL}{r_y}\right)^2} = \frac{\pi^2\left(29{,}000\ \frac{\text{kips}}{\text{in}^2}\right)}{(121.95)^2} = 19.25\ \text{ksi}$$

$$F_{\text{cr},y} = 0.877F_{ey} = (0.877)\left(19.25\ \frac{\text{kips}}{\text{in}^2}\right) = 16.88\ \text{ksi}$$

$$F_{\text{cr},z} = \frac{GJ}{A_g\bar{r}_o^2} = \frac{\left(11{,}200\ \frac{\text{kips}}{\text{in}^2}\right)(1.50\ \text{in}^4)}{(10.0\ \text{in}^2)(3.19\ \text{in})^2} = 165.09\ \text{ksi}$$

$$F_{cr} = \left(\frac{F_{cr,y} + F_{cr,z}}{2H}\right)\left(1 - \sqrt{1 - \frac{4F_{cr,y}F_{cr,z}H}{\left(F_{cr,y} + F_{cr,z}\right)^2}}\right)$$

$$= \left(\frac{16.88\ \frac{\text{kips}}{\text{in}^2} + 165.09\ \frac{\text{kips}}{\text{in}^2}}{(2)(0.916)}\right)$$

$$\times\left(1 - \sqrt{1 - \frac{(4)\left(16.88\ \frac{\text{kips}}{\text{in}^2}\right)\left(165.09\ \frac{\text{kips}}{\text{in}^2}\right)(0.916)}{\left(16.88\ \frac{\text{kips}}{\text{in}^2} + 165.09\ \frac{\text{kips}}{\text{in}^2}\right)^2}}\right)$$

$$= 16.84\ \text{ksi} \quad [> F_{cr,x} = 9.14\ \text{ksi}]$$

Therefore, the controlling factor is compressive strength for flexural buckling about the x-axis.

Calculate the nominal strength, P_n, of the member.

$$P_n = F_{cr}A_g = \left(9.14\ \frac{\text{kips}}{\text{in}^2}\right)\left(10.0\ \text{in}^2\right) = 91.4\ \text{kips}$$

Determine the design (LRFD) and allowable (ASD) compressive strengths of the member.

LRFD	ASD
$\phi_c P_n = (0.90)(91.4\ \text{kips})$ $= 82.26\ \text{kips}$	$\frac{P_n}{\Omega_c} = \frac{91.4\ \text{kips}}{1.67}$ $= 54.73\ \text{kips}$

These numbers can also be found or double-checked in *AISC Manual* Table 4-7. By interpolation, the design strength is 82.35 kips and the allowable strength is 54.80 kips.

5. MEMBERS WITH SLENDER ELEMENTS

When a member contains slender elements, to prevent local buckling, the gross area of that member is modified in calculations by the reduction factors Q_s and Q_a. *AISC Specification* Sec. E7 gives the requirements for the design of members with unstiffened or stiffened slender elements. The nominal compressive strength, P_n, is determined from the limit states of flexural, torsional, and flexural-torsional buckling. For each limit state,

$$P_n = F_{cr}A_g \quad \text{[AISC Eq. E7-1]} \qquad 7.19$$

The flexural buckling stress, F_{cr}, is determined by Eq. 7.20 and Eq. 7.21. When $KL/r \leq 4.71\sqrt{E/QF_y}$ (or $QF_y/F_e \leq 2.25$),

$$F_{cr} = Q\left(0.658^{QF_y/F_e}\right)F_y \quad \text{[AISC Eq. E7-2]} \tag{7.20}$$

When $KL/r > 4.71\sqrt{E/QF_y}$ (or $QF_y/F_e > 2.25$),

$$F_{cr} = 0.877F_e \quad \text{[AISC Eq. E7-3]} \tag{7.21}$$

In Eq. 7.20 and Eq. 7.21, the reduction factor, Q, is equal to 1.0 for members with compact and noncompact sections and is equal to Q_sQ_a for members with slender-element sections. (Compact, noncompact, and slender-element sections are defined in *AISC Specification* Sec. B4.)

Stiffened Slender Elements

The reduction factor, Q_a, for slender stiffened elements is

$$Q_a = \frac{A_e}{A} \quad \text{[AISC Eq. E7-16]} \tag{7.22}$$

The reduced effective width, b_e, of a slender element is determined from Eq. 7.23 or Eq. 7.24. For uniformly compressed slender elements, with $b/t \geq 1.49\sqrt{E/f}$, except for flanges of square and rectangular sections of uniform thickness,

$$b_e = 1.92t\sqrt{\frac{E}{f}}\left(1 - \frac{0.34}{\frac{b}{t}}\sqrt{\frac{E}{f}}\right) \leq b \quad \text{[AISC Eq. E7-17]} \tag{7.23}$$

In Eq. 7.23, f is taken as F_{cr} as calculated by Eq. 7.20 and Eq. 7.21 with $Q = 1.0$. For flanges of square and rectangular slender-element sections of uniform thickness, with $b/t \geq 1.40\sqrt{E/f}$,

$$b_e = 1.92t\sqrt{\frac{E}{f}}\left(1 - \frac{0.38}{\frac{b}{t}}\sqrt{\frac{E}{f}}\right) \leq b \quad \text{[AISC Eq. E7-18]} \tag{7.24}$$

In Eq. 7.24, f is taken as P_n/A_e. Calculating this requires iteration. For simplicity, f may also be taken as equal to F_y. This will give a slightly conservative estimate of the compression capacity of the member.

Example 7.4

Axial Loaded HSS Compression Member with Slender Elements

An HSS12 × 8 × $^3/_{16}$ steel column has pinned connections top and bottom. The length of the column is 30 ft, and there are no intermediate braces.

Section properties		Material properties
$A_g = 6.76\ \text{in}^2$	$b/t = 43.0$	ASTM A500, grade B steel
$r_x = 4.56$ in	$h/t = 66.0$	$F_y = 46$ ksi
$r_y = 3.35$ in	$t_{\text{des}} = 0.174$ in	$F_u = 58$ ksi

Determine the nominal load capacity, the design strength (LRFD), and the allowable strength (ASD).

Solution

Determine the slenderness ratios.

$$\frac{KL}{r_x} = \frac{(1)(30\ \text{ft})\left(12\ \frac{\text{in}}{\text{ft}}\right)}{4.56\ \text{in}} = 78.95$$

$$\frac{KL}{r_y} = \frac{(1)(30\ \text{ft})\left(12\ \frac{\text{in}}{\text{ft}}\right)}{3.35\ \text{in}} = 107.46 \quad [\text{controls}]$$

Calculate the limiting width-thickness ratios using *AISC Specification* Table B4.1a, case 6.

$$\lambda_r = 1.40\sqrt{\frac{E}{F_y}} = 1.40\sqrt{\frac{29{,}000\ \frac{\text{kips}}{\text{in}^2}}{46\ \frac{\text{kips}}{\text{in}^2}}} = 35.15$$

$$\frac{b}{t} = 43.0 \quad [> \lambda_r,\ \text{so slender}]$$

Therefore, the width is a slender element.

$$\frac{h}{t} = 66.0 \quad [> \lambda_r,\ \text{so slender}]$$

Therefore, the height is a slender element. For determining the width-thickness ratio, b and h are each taken as the corresponding outside dimension minus three times the design wall thickness, per *AISC Specification* Sec. B4.1b(d). Calculate b and h.

$$\begin{aligned} b &= \text{outside dimension} - 3t_{\text{des}} \\ &= 8.00\text{ in} - (3)(0.174\text{ in}) \\ &= 7.48\text{ in} \\ h &= \text{outside dimension} - 3t_{\text{des}} \\ &= 12.00\text{ in} - (3)(0.174\text{ in}) \\ &= 11.5\text{ in} \end{aligned}$$

Calculate the reduction factor, Q_a. To use Eq. 7.22, first the effective area must be determined. The easiest way is to calculate the area of the "unused" portions of the walls of the HSS—that is, the portions of the walls that are in excess of the effective length—and subtract them from the gross area, which is given. To calculate the effective length of the flanges of a square or rectangular slender element section of uniform thickness, use Eq. 7.24, taking f conservatively as F_y.

For the 8 in walls,

$$b_e \le \begin{cases} 1.92t\sqrt{\dfrac{E}{f}}\left(1-\dfrac{0.38}{\dfrac{b}{t}}\sqrt{\dfrac{E}{f}}\right) \\ = (1.92)(0.174\text{ in})\sqrt{\dfrac{29{,}000\ \dfrac{\text{kips}}{\text{in}^2}}{46\ \dfrac{\text{kips}}{\text{in}^2}}}\left(1-\dfrac{0.38}{43.0}\sqrt{\dfrac{29{,}000\ \dfrac{\text{kips}}{\text{in}^2}}{46\ \dfrac{\text{kips}}{\text{in}^2}}}\right) \\ = 6.53\text{ in} \quad [\text{controls}] \\ b = 7.48\text{ in} \end{cases}$$

The length that cannot be used in this direction is $b - b_e = 7.48\text{ in} - 6.53\text{ in} = 0.950\text{ in}$.

For the 12 in walls,

$$b_e \le \begin{cases} 1.92t\sqrt{\dfrac{E}{f}}\left(1-\dfrac{0.38}{\dfrac{b}{t}}\sqrt{\dfrac{E}{f}}\right) \\ =(1.92)(0.174 \text{ in})\sqrt{\dfrac{29{,}000 \ \dfrac{\text{kips}}{\text{in}^2}}{46 \ \dfrac{\text{kips}}{\text{in}^2}}}\left(1-\dfrac{0.38}{66.0}\sqrt{\dfrac{29{,}000 \ \dfrac{\text{kips}}{\text{in}^2}}{46 \ \dfrac{\text{kips}}{\text{in}^2}}}\right) \\ =7.18 \text{ in} \quad [\text{controls}] \\ b=11.5 \text{ in} \end{cases}$$

The length that cannot be used in this direction is $b - b_e$ = 11.5 in – 7.18 in = 4.32 in. Subtract the unused areas from the gross area, given as 6.76 in^2, to get the effective area.

$$\begin{aligned} A_e &= 6.76 \text{ in}^2 - (2)(0.174 \text{ in})(0.950 \text{ in}) - (2)(0.174 \text{ in})(4.32 \text{ in}) \\ &= 4.93 \text{ in}^2 \end{aligned}$$

Use Eq. 7.22 to determine the reduction factor.

$$Q = Q_a = \frac{A_e}{A} = \frac{4.93 \text{ in}^2}{6.76 \text{ in}^2} = 0.729$$

Determine the appropriate equation for F_{cr}.

$$4.71\sqrt{\frac{E}{QF_y}} = 4.71\sqrt{\frac{29{,}000 \ \dfrac{\text{kips}}{\text{in}^2}}{(0.729)\left(46 \ \dfrac{\text{kips}}{\text{in}^2}\right)}} = 139 \quad [> KL/r_y = 107.46]$$

Because 139 is greater than KL/r_y, use Eq. 7.20.

Using Eq. 7.8, the elastic critical buckling stress is

$$F_e = \frac{\pi^2 E}{\left(\dfrac{KL}{r}\right)^2} = \frac{\pi^2\left(29{,}000 \ \dfrac{\text{kips}}{\text{in}^2}\right)}{(107.46)^2} = 24.79 \text{ ksi}$$

Use Eq. 7.20 to find critical flexural buckling stress.

$$\begin{aligned} F_{cr} &= 0.658^{QF_y/F_e} QF_y \\ &= (0.658)^{(0.729)\left(46\ \frac{\text{kips}}{\text{in}^2}\right)/24.79\ \frac{\text{kips}}{\text{in}^2}} (0.729)\left(46\ \frac{\text{kips}}{\text{in}^2}\right) \\ &= 19.04\ \text{ksi} \end{aligned}$$

Calculate the nominal load capacity, P_n.

$$\begin{aligned} P_n = F_{cr}A_g &= \left(19.04\ \frac{\text{kips}}{\text{in}^2}\right)(6.76\ \text{in}^2) \\ &= 128.71\ \text{kips} \end{aligned}$$

Calculate the design strength, $\phi_c P_n$, and the allowable strength, P_n/Ω_c.

LRFD	ASD
$\phi_c P_n = (0.90)(128.71\ \text{kips}) = 115.84\ \text{kips}$	$\frac{P_n}{\Omega_c} = \frac{128.71\ \text{kips}}{1.67} = 77.07\ \text{kips}$

The compressive stress, f, was conservatively taken as F_y, so the calculated capacities should be slightly less than the tabulated loads in *AISC Manual* Table 4-3. For a value of 30 ft for KL with respect to r_y, those loads are as follows.

LRFD	ASD
$\phi_c P_n = 125\ \text{kips}$	$\frac{P_n}{\Omega_c} = 83.2\ \text{kips}$

6. SINGLE ANGLE COMPRESSION MEMBERS

The nominal compressive strength, P_n, of single angle compression members is calculated in accordance with *AISC Specification* Sec. E3 or Sec. E7 as appropriate, using a slenderness ratio as determined by that section. Provisions in *AISC Specification* Sec. E4 apply to angles with $b/t > 20$.

The effects of eccentricity on single angle members may be neglected when the following conditions are met.

- Members are loaded at the ends in compression through the same leg.
- Members are attached by welding or by at least two bolts per connection.
- There are no intermediate transverse loads.

Members are evaluated as axially loaded compression members using the appropriate effective slenderness ratios, as follows.

case 1

This applies to

- individual members
- web members of planar trusses with adjacent web members attached to the same side of the gusset plate or chord

For equal leg angles, and for unequal leg angles connected through the longer leg, start by calculating L/r_x. If $0 \le L/r_x \le 80$,

$$\frac{KL}{r} = 72 + 0.75\left(\frac{L}{r_x}\right) \quad \text{[AISC Eq. E5-1]} \qquad 7.25$$

If $L/r_x > 80$,

$$\frac{KL}{r} = 32 + 1.25\left(\frac{L}{r_x}\right) \le 200 \quad \text{[AISC Eq. E5-2]} \qquad 7.26$$

For unequal leg angles with leg length ratios less than 1.7 and connected through the shorter leg, calculate L/r_x. If $0 \le L/r_x \le 80$,

$$\frac{KL}{r} = 72 + 0.75\left(\frac{L}{r_x}\right) + 4\left(\left(\frac{b_{\text{long}}}{b_{\text{short}}}\right)^2 - 1\right) \le 0.95\left(\frac{L}{r_z}\right) \qquad 7.27$$

If $L/r_x > 80$,

$$\frac{KL}{r} = 32 + 1.25\left(\frac{L}{r_x}\right) + 4\left(\left(\frac{b_{\text{long}}}{b_{\text{short}}}\right)^2 - 1\right) \le 0.95\left(\frac{L}{r_z}\right) \qquad 7.28$$

If none of the preceding conditions apply, consult *AISC Specification* Sec. E5(c).

case 2

This applies to web members of box or space trusses with adjacent web members attached to the same side of the gusset plate or chord.

For equal leg angles, and for unequal leg angles connected through the longer leg, calculate L/r_x. If $0 \le L/r_x \le 75$,

$$\frac{KL}{r} = 60 + 0.8\left(\frac{L}{r_x}\right) \quad \text{[AISC Eq. E5-3]} \qquad 7.29$$

If $L/r_x > 75$,

$$\frac{KL}{r} = 45 + \frac{L}{r_x} \le 200 \quad \text{[AISC Eq. E5-4]} \qquad 7.30$$

For unequal leg angles with leg length ratios less than 1.7 and connected through the shorter leg, calculate L/r_x. If $0 \le L/r_x \le 80$,

$$\frac{KL}{r} = 60 + 0.8\left(\frac{L}{r_x}\right) + 6\left(\left(\frac{b_{\text{long}}}{b_{\text{short}}}\right)^2 - 1\right) \le 0.82\left(\frac{L}{r_z}\right) \quad 7.31$$

If $L/r_x > 80$,

$$\frac{KL}{r} = 45 + \frac{L}{r_x} + 6\left(\left(\frac{b_{\text{long}}}{b_{\text{short}}}\right)^2 - 1\right) \le 0.82\left(\frac{L}{r_z}\right) \quad 7.32$$

If none of the preceding conditions apply, consult *AISC Specification* Sec. E5(c).

Example 7.5

Single Angle Compression Member

A single angle compression member is 12 ft long and is attached with two bolts at each end through the same leg.

Section properties		Material properties
$L6 \times 6 \times {}^5/_8$ in	$S_x = S_y = 5.64\ \text{in}^3$	ASTM A36 steel
$A = 7.13\ \text{in}^2$	$r_x = r_y = 1.84$ in	$F_y = 36$ ksi
$I_x = I_y = 24.1\ \text{in}^4$	$r_z = 1.17$ in	$F_u = 58$ ksi

Determine the nominal strength, the design strength (LRFD), and the allowable strength (ASD).

Solution

Determine the effective slenderness ratio. For an individual member with equal legs, either Eq. 7.25 or Eq. 7.26 will be used.

$$\frac{L}{r_x} = \frac{(12\ \text{ft})\left(12\ \frac{\text{in}}{\text{ft}}\right)}{1.84\ \text{in}} = 78.26 \quad [\le 80]$$

$L/r_x < 80$, so use Eq. 7.25.

$$\begin{aligned}\frac{KL}{r} &= 72 + 0.75\left(\frac{L}{r_x}\right)\\ &= 72 + (0.75)(78.26)\\ &= 130.70\end{aligned}$$

Determine whether to use Eq. 7.6 or Eq. 7.7 to find the critical stress.

$$4.71\sqrt{\frac{E}{F_y}} = 4.71\sqrt{\frac{29{,}000\ \frac{\text{kips}}{\text{in}^2}}{36\ \frac{\text{kips}}{\text{in}^2}}}$$
$$= 134 \quad [\geq KL/r = 130.70]$$

This is greater than *KL*/*r*, so use Eq. 7.6 to find the critical stress. First, use Eq. 7.8 to find the elastic critical buckling stress.

$$F_e = \frac{\pi^2 E}{\left(\frac{KL}{r}\right)^2} = \frac{\pi^2\left(29{,}000\ \frac{\text{kips}}{\text{in}^2}\right)}{(130.70)^2} = 16.76\ \text{ksi}$$

Use Eq. 7.6 to find the critical stress.

$$F_{\text{cr}} = 0.658^{F_y/F_e} F_y$$
$$= (0.658)^{36\ \frac{\text{kips}}{\text{in}^2}/16.76\ \frac{\text{kips}}{\text{in}^2}}\left(36\ \frac{\text{kips}}{\text{in}^2}\right)$$
$$= 14.65\ \text{ksi}$$

Determine the nominal strength.

$$P_n = F_{\text{cr}} A_g = \left(14.65\ \frac{\text{kips}}{\text{in}^2}\right)(7.13\ \text{in}^2)$$
$$= 104.45\ \text{kips}$$

Determine the design strength and the allowable strength.

LRFD	ASD
$\phi_c P_n = (0.90)(104.45\ \text{kips}) = 94.01\ \text{kips}$	$\frac{P_n}{\Omega_c} = \frac{104.45\ \text{kips}}{1.67} = 62.54\ \text{kips}$

An alternative way to determine the design strength and allowable strength is to use *AISC Manual* Table 4-11. To get an effective length, KL, with respect to the z-axis, multiply KL/r by r_z.

$$KL = \left(\frac{KL}{r}\right) r_z = \frac{(130.70)(1.17 \text{ in})}{12 \ \frac{\text{in}}{\text{ft}}} = 12.74 \text{ ft}$$

Interpolating between the tabulated values for 12 ft and 13 ft gives

$$\phi_c P_n = 94.08 \text{ kips} \quad [\text{versus 94.01 kips calculated}]$$

$$\frac{P_n}{\Omega_c} = 62.61 \text{ kips} \quad [\text{versus 62.54 kips calculated}]$$

7. COLUMN BASE PLATE DESIGN

Designing column base plates for concentric axial loads is relatively easy. Depending on the magnitude of the loads, the size of the base plate can be determined by the location of the anchor rods rather than by the bearing capacity of the supporting element. The holes for the anchor rods will be either punched or drilled in the plate. In either case, minimum edge distances and minimum clearances are required between the column steel and the anchor rod for the washer, nut, and wrenches. Virtually all columns must have a minimum of four anchor rods to meet Occupational Safety and Health Administration (OSHA) requirements.

Even though the holes in the base plate for the anchor rods are oversized to accommodate misplaced rods, the design assumptions are based on the gross area of the base plate bearing on the supporting element. Hole sizes in the base plates for anchor rods are shown in *AISC Commentary* Table C-J9.1, and are larger than the standard oversized holes used for fitting up parts of the superstructure. The edge distances for the holes should be based on the hole diameter rather than the rod diameter. Heavy plate washers ($^5/_{16}$ in to $^1/_2$ in thick) should be used in lieu of standard washers to prevent deformation of the washers.

As with beam-bearing plates, the length and width of the column base plates should preferably be in full inches. The thickness should be in increments of $^1/_8$ in up to a thickness of 1.25 in and in increments of $^1/_4$ in when the thickness exceeds 1.25 in. Additional information for column base plate design can be found in Part 14 of the *AISC Manual*.

The size of the base plate is a function of the bearing capacity of the concrete supporting element. This must be calculated in accordance with ACI 318, so factored loads (LRFD) must be used, not service loads.

The design bearing strength of concrete is $\phi(0.85f_c'A_1)$ where $\phi = 0.65$. When the area of the concrete supporting element is larger than that of the base plate, A_1, the design bearing strength of the loaded area can be increased. The maximum increase occurs when $\sqrt{A_2/A_1} = 2$. (A_2 is the area of the base of a frustum whose sides have a downward slope of 2 horizontal to 1 vertical, with all sides equidistant from the bearing area of A_1. See Sec. 6.6 and Fig. 6.4.)

The base plate is assumed to be a cantilevered beam bending about a critical section near the edges of the column section. For W, S, M, and HP shapes, the critical sections are defined as $0.95d$ and $0.80b_f$. For rectangular and square HSS shapes, the critical sections are defined as $0.95d$ and $0.95b$. For round HSS and pipe shapes, the critical section is defined as 0.80 times the diameter of the member.

The base plate cantilever distances, as illustrated in Fig. 7.3, are calculated with the following formulas from *AISC Manual* Part 14.

$$m = \frac{N - 0.95d}{2} \quad \text{[AISC Eq. 14-2]} \qquad 7.33$$

$$n = \frac{B - 0.80b_f}{2} \quad \text{[AISC Eq. 14-3]} \qquad 7.34$$

$$n' = \frac{\sqrt{db_f}}{4} \quad \text{[AISC Eq. 14-4]} \qquad 7.35$$

$$\lambda n' = \tfrac{1}{4}\lambda\sqrt{db_f} \qquad 7.36$$

For closely cropped base plates, the factor λ can be conservatively taken as 1.0. Otherwise, calculate λ as

$$\lambda = \frac{2\sqrt{X}}{1+\sqrt{1-X}} \le 1 \quad \text{[AISC Eq. 14-5]} \qquad 7.37$$

Figure 7.3 *Base Plate Critical Bending Planes*

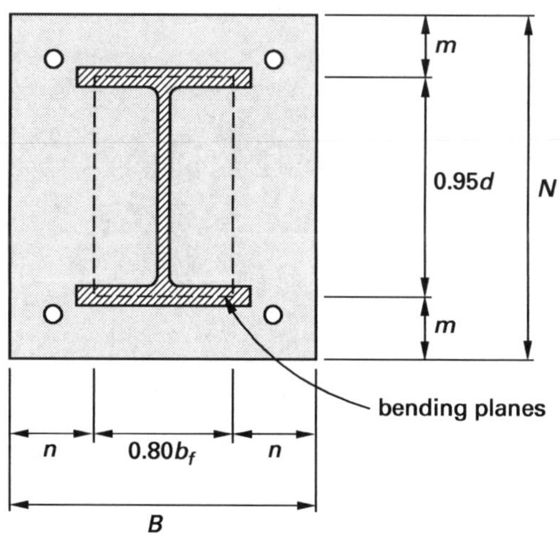

X is calculated as follows.

$$X = \frac{4db_fP_u}{(d+b_f)^2\,\phi_cP_p} \quad \text{[LRFD, AISC Eq. 14-6a]} \tag{7.38}$$

$$X = \frac{4db_fP_a\Omega_c}{(d+b_f)^2\,P_p} \quad \text{[ASD, AISC Eq. 14-6b]} \tag{7.39}$$

The following formulas are used to design the base plate thickness. l is the largest of the values m, n, and $\lambda n'$. The minimum base plate thickness is

$$t_{p,\text{min}} = l\sqrt{\frac{2P_u}{0.9F_yBN}} \quad \text{[LRFD, AISC Eq. 14-7a]} \tag{7.40}$$

$$t_{p,\text{min}} = l\sqrt{\frac{3.33P_a}{F_yBN}} \quad \text{[ASD, AISC Eq. 14-7b]} \tag{7.41}$$

ACI 318 does not have provisions for using unfactored loads. Where a total load is known but the percentages allocated to live and dead loads are unknown, it is acceptable practice to multiply the total load by an average load factor of 1.5 and proceed with the design. The average load factor is based on a live load equaling three times a dead load, which is consistent with the tables in the *AISC Manual*.

Example 7.6

Concentrically Axially Loaded Base Plate

A W12 × 72 steel column supports a dead load of 165 kips and a live load of 165 kips. The column bears on a base plate the same size as a concrete pier ($A_1 = A_2$). The design compressive strength of the concrete is $f_c' = 5$ ksi.

Section properties	Material properties	
$d = 12.3$ in	ASTM A992 steel column	ASTM A36 steel plate
$b_f = 12.0$ in	$F_y = 50$ ksi	$F_y = 36$ ksi
	$F_u = 65$ ksi	$F_u = 58$ ksi

Determine the size and thickness of a square base plate.

Solution

Calculate the required design strengths.

LRFD	ASD
$P_u = 1.2D + 1.6L$ $= (1.2)(165 \text{ kips}) + (1.6)(165 \text{ kips})$ $= 462 \text{ kips}$	$P_a = D + L$ $= 165 \text{ kips} + 165 \text{ kips}$ $= 330 \text{ kips}$

Calculate the required bearing area. Use the factored load because ACI 318 is based on factored loads, not service loads.

$$\begin{aligned} A_1 &\geq \frac{P_u}{\phi(0.85 f_c')} \\ &\geq \frac{(1.2)(165 \text{ kips}) + (1.6)(165 \text{ kips})}{(0.65)(0.85)\left(5 \ \dfrac{\text{kips}}{\text{in}^2}\right)} \\ &\geq 167.24 \text{ in}^2 \quad [= 12.93 \text{ in} \times 12.93 \text{ in}] \end{aligned}$$

Use a base plate that is 14 in × 14 in (having an area of 196 in^2, which is greater than 167.24 in^2). Use Eq. 7.33 through Eq. 7.35 to calculate cantilever projection lengths; the greatest value governs the design.

$$\begin{aligned} m &= \frac{N - 0.95d}{2} \\ &= \frac{14 \text{ in} - (0.95)(12.3 \text{ in})}{2} \\ &= 1.16 \text{ in} \end{aligned}$$

$$\begin{aligned} n &= \frac{B - 0.80b_f}{2} \\ &= \frac{14 \text{ in} - (0.80)(12.0 \text{ in})}{2} \\ &= 2.20 \text{ in} \end{aligned}$$

$$\begin{aligned} \lambda n' &= \tfrac{1}{4}\lambda\sqrt{db_f} \\ &= \left(\frac{1}{4}\right)(1)\sqrt{(12.3 \text{ in})(12.0 \text{ in})} \\ &= 3.04 \text{ in} \quad [\text{controls}] \end{aligned}$$

Calculate the bearing stresses.

LRFD	ASD
$f_u = \dfrac{P_u}{BN} = \dfrac{462 \text{ kips}}{(14 \text{ in})(14 \text{ in})}$ $= 2.36 \text{ ksi}$	$f_a = \dfrac{P_a}{BN} = \dfrac{330 \text{ kips}}{(14 \text{ in})(14 \text{ in})}$ $= 1.68 \text{ ksi}$

Use Eq. 7.40 and Eq. 7.41 to calculate the required thickness of the base plate.

$$l = \lambda n' = 3.04 \text{ in} \quad [\text{controlling value}]$$

LRFD	ASD
$t_{p,\min} = l\sqrt{\dfrac{2P_u}{0.9F_y BN}}$ $= (3.04 \text{ in}) \times \sqrt{\dfrac{(2)(462 \text{ kips})}{(0.9)\left(36 \dfrac{\text{kips}}{\text{in}^2}\right) \times (14 \text{ in})(14 \text{ in})}}$ $= 1.16 \text{ in}$	$t_p = l\sqrt{\dfrac{3.33P_a}{F_y BN}}$ $= (3.04 \text{ in}) \times \sqrt{\dfrac{(3.33)(330 \text{ kips})}{\left(36 \dfrac{\text{kips}}{\text{in}^2}\right)(14 \text{ in})(14 \text{ in})}}$ $= 1.19 \text{ in}$

Use a plate that is 14 in × $1^1/_4$ in × 1 ft 2 in.

Determine whether the base plate thickness is excessive due to taking λ conservatively as 1.0. Verify concrete bearing strength.

LRFD	ASD
$\phi_c P_p = \phi_c(0.85 f'_c A_1)\sqrt{\dfrac{A_2}{A_1}}$ $= (0.65)(0.85)\left(5 \dfrac{\text{kips}}{\text{in}^2}\right) \times (196 \text{ in}^2)\sqrt{\dfrac{196 \text{ in}^2}{196 \text{ in}^2}}$ $= 541.45 \text{ kips} \quad [> 462 \text{ kips, so OK}]$	$\dfrac{P_p}{\Omega_c} = \dfrac{0.85 f'_c A_1}{\Omega_c}\sqrt{\dfrac{A_2}{A_1}}$ $= \dfrac{(0.85)\left(5 \dfrac{\text{kips}}{\text{in}^2}\right)(196 \text{ in}^2)}{2.31} \times \sqrt{\dfrac{196 \text{ in}^2}{196 \text{ in}^2}}$ $= 360.61 \text{ kips} \quad [> 330 \text{ kips, so OK}]$

In some AISC publications, the reduction factor, ϕ, is given as 0.60 rather than 0.65 as specified. A reduction factor of 0.65 results in a more conservative design with a design strength approximately 8% greater than a reduction factor of 0.60 gives.

To calculate λ, start by using Eq. 7.38 and Eq. 7.39 to calculate X.

LRFD	ASD
$X = \dfrac{4db_f P_u}{(d+b_f)^2 \phi_c P_p}$ $= \dfrac{(4)(12.3\text{ in})(12.0\text{ in})(462\text{ kips})}{(12.3\text{ in}+12.0\text{ in})^2(541.45\text{ kips})}$ $= 0.85$	$X = \dfrac{4db_f P_a \Omega_c}{(d+b_f)^2 P_p}$ $= \dfrac{(4)(12.3\text{ in})(12.0\text{ in})(330\text{ kips})}{(12.3\text{ in}+12.0\text{ in})^2(360.61\text{ kips})}$ $= 0.91$

Calculate λ.

LRFD	ASD
$\lambda \le \begin{cases} \dfrac{2\sqrt{X}}{1+\sqrt{1-X}} = \dfrac{2\sqrt{0.85}}{1+\sqrt{1-0.85}} = 1.33 \\ 1.0 \quad [\text{controls}] \end{cases}$	$\lambda \le \begin{cases} \dfrac{2\sqrt{X}}{1+\sqrt{1-X}} = \dfrac{2\sqrt{0.91}}{1+\sqrt{1-0.91}} = 1.47 \\ 1.0 \quad [\text{controls}] \end{cases}$

In this case, the calculated value of λ is 1.0; therefore, taking λ conservatively as 1.0 has resulted in no reduction in the thickness of the base plate. If the calculated value of λ had been less than 1.0, it would have resulted in a thinner calculated thickness for the base plate.

8 Combined Stress Members

Nomenclature

A	cross-sectional area	in^2
b_f	flange width	in
b_x, b_y	coefficient for bending about strong or weak axis	$(ft\text{-}kips)^{-1}$
B	overall width of rectangular HSS member measured perpendicular to plane of connection	in
C	torsional constant	in^3
C_b	lateral-torsional buckling modification factor (beam buckling coefficient)	–
C_v	web shear coefficient defined in *AISC Specification* Sec. G2.1	–
D	dead load	lbf
D	outside diameter	ft
E	modulus of elasticity	lbf/in^2
f_a	required axial stress	lbf/in^2
f_b	required flexural stress	lbf/in^2
F_a	available axial stress	lbf/in^2
F_b	available flexural stress	lbf/in^2
F_{cr}	critical flexural buckling stress	lbf/in^2
F_e	elastic critical buckling stress	lbf/in^2
F_u	specified minimum tensile strength	lbf/in^2
F_y	specified minimum yield stress	lbf/in^2
h	height of wall or web	in
h_e	effective web height	in
H	overall height of rectangular HSS member measured in plane of connection	in
I	moment of inertia	in^4
J	torsional constant	in^4
k_v	web plate buckling coefficient	–
K	effective length factor	–

KL	effective length	ft
L	length	ft
L	live load	lbf
L_b	length between braces or braced points	in
L_p	limiting unbraced length for full plastic moment	in
L_r	limiting unbraced length for inelastic lateral-torsional buckling	in
M_c	available flexural strength	ft-lbf
M_D	moment due to dead load	ft-lbf
M_L	moment due to live load	ft-lbf
M_n	nominal flexural strength	ft-lbf
M_p	plastic bending moment	ft-lbf
M_r	required flexural strength	ft-lbf
p	coefficient for axial compression	lbf^{-1}
P_a	required force (ASD)	lbf
P_c	available axial compressive strength or available tensile strength	lbf
P_D	axial dead load	lbf
P_e	elastic buckling load	lbf
P_L	axial live load	lbf
P_n	nominal tensile strength	lbf
P_r	required axial compressive strength	lbf
P_{tab}	equivalent required tabular load	lbf
P_u	required force (LRFD)	lbf
r	radius of gyration	ft
S	elastic section modulus	in^3
t	thickness	in
t_f	flange thickness	in
t_r	coefficient for tension rupture	lbf^{-1}
t_w	web thickness	in
t_y	coefficient for tension yielding	lbf^{-1}
T_c	available torsional strength	lbf/in^2
T_D	torsional dead load	ft-lbf
T_L	torsional live load	ft-lbf
T_n	nominal torsional strength	lbf/in^2
T_r	required torsional strength	lbf/in^2

V_c	available shear strength	lbf/in^2
V_n	nominal shear strength	lbf/in^2
V_r	required shear strength	lbf/in^2
w_a	required strength per unit length (ASD)	lbf/ft
w_D	dead load per unit length	lbf/ft
w_L	live load per unit length	lbf/ft
w_u	required strength per unit length (LRFD)	lbf/ft
Z	plastic section modulus	in^3

Symbols

λ_p	limiting width-thickness ratio for compactness	–
ϕ	resistance factor (LRFD)	–
Ω	safety factor (ASD)	–

Subscripts

b	bending or flexure
c	compression
g	gross
t	tensile or tension
T	torsional
w	major axis, web, or wall
x	x-axis or strong axis
y	y-axis or weak axis
z	minor axis

1. GENERAL

AISC Specification Chap. H is the primary source for information pertaining to members subjected to combined stresses. Chapter H addresses members subject to axial force and flexure force about one or both axes, with or without torsion, and members subject to torsion only. That chapter is divided as follows.

H1	Doubly and Singly Symmetrical Members Subject to Flexure and Axial Force
H2	Unsymmetrical and Other Members Subject to Flexure and Axial Force
H3	Members Under Torsion and Combined Torsion, Flexure, Shear, and/or Axial Force
H4	Rupture of Flanges with Holes Subject to Tension

The previous chapters of this book have discussed the design of members that are subjected to forces along a single axis such as axial tension, axial compression, and bending about either the *X-X* axis (strong) or the *Y-Y* axis (weak). Most structural members, however, are subjected to loading conditions that will produce combined stresses on the member. As a member is loaded in a secondary or tertiary axis, the force on the primary axis must be reduced so that the effects of the combined loads will not exceed the design strength (LRFD) or the allowable strength (ASD) of the member.

The reduction of loads on one axis to accommodate loads along another axis makes combined stress members more difficult to design. Fortunately, the *AISC Manual* provides some help in simplifying design and analysis.

2. DOUBLY AND SINGLY SYMMETRICAL MEMBERS SUBJECT TO FLEXURE AND AXIAL FORCE

Of the various kinds of members under combined stresses, the easiest to design and/or analyze is a beam subjected to biaxial bending. Typical examples include roof purlins and wall girts that are subjected to bending about their *x*- and *y*-axes.

When an axial load is added, the solution becomes a little more complex. The equations used to check the combined stresses or combined load capacities are often called *interaction formulas* or *unity check formulas*.

Design for Compression and Flexure

The two basic formulas for designing or analyzing doubly and singly symmetrical members subject to flexure and axial force are as follows. In these formulas, P_r is required axial compressive strength, P_c is available axial compressive strength, M_r is required flexural strength, and M_c is available flexural strength. Subscripts x and y pertain to the strong and weak axes, respectively.

For $P_r/P_c \geq 0.2$,

$$\frac{P_r}{P_c} + \left(\frac{8}{9}\right)\left(\frac{M_{rx}}{M_{cx}} + \frac{M_{ry}}{M_{cy}}\right) \leq 1.0 \quad \text{[AISC Eq. H1-1a]} \qquad 8.1$$

For $P_r/P_c < 0.2$,

$$\frac{P_r}{2P_c} + \left(\frac{M_{rx}}{M_{cx}} + \frac{M_{ry}}{M_{cy}}\right) \leq 1.0 \quad \text{[AISC Eq. H1-1b]} \qquad 8.2$$

When there is no axial force, Eq. 8.1 is not applicable. Use Eq. 8.2 with $P_r/2P_c = 0$.

The values for available axial strength, P_e, and available flexural strength, M_{cx} and M_{cy}, must be multiplied by the resistance factor, ϕ, for LRFD solutions and divided by the safety factor, Ω, for ASD solutions.

Design for Tension and Flexure

The interaction of tension and flexure in doubly symmetric members and singly symmetric members constrained to bend about a geometric axis (x and/or y) is limited by Eq. 8.1 and Eq. 8.2. As before, when there is no axial force, Eq. 8.1 is not applicable, and Eq. 8.2 is used with $P_r/2P_c = 0$.

The lateral-torsional buckling modification factor, C_b, is discussed in Chap. 5. For doubly symmetric members, C_b may be increased by $\sqrt{1+(\alpha P_r/P_{ey})}$ for axial tension acting concurrently with flexure. α is 1.0 for LRFD and 1.6 for ASD, and P_{ey} is

$$P_{ey} = \frac{\pi^2 EI_y}{L_b^2} \quad \text{[AISC Sec. H1.2]} \qquad 8.3$$

3. DOUBLY SYMMETRIC MEMBERS IN SINGLE AXIS FLEXURE AND COMPRESSION

For doubly symmetric members in flexure and compression, with moments primarily about their major axes, the combined approach given in *AISC Specification* Sec. H1.1 need not be followed. Instead, two independent limit states—in-plane instability and out-of-plane buckling—may be considered separately.

For the limit state of *in-plane instability*, use Eq. 8.1 and Eq. 8.2 as applicable. Determine P_c, M_r, and M_c in the plane of bending.

For the limit state of *out-of-plane buckling*, use

$$\frac{P_r}{P_{cy}} + \left(1.5 - 0.5\frac{P_r}{P_{ey}}\right) + \left(\frac{M_{rx}}{C_b M_{cx}}\right)^2 \le 1.0 \quad \text{[AISC Eq. H1-2]} \qquad 8.4$$

Example 8.1

I-Shaped Beam with Biaxial Bending

The steel beam shown is laterally supported along the full length of its compression flange.

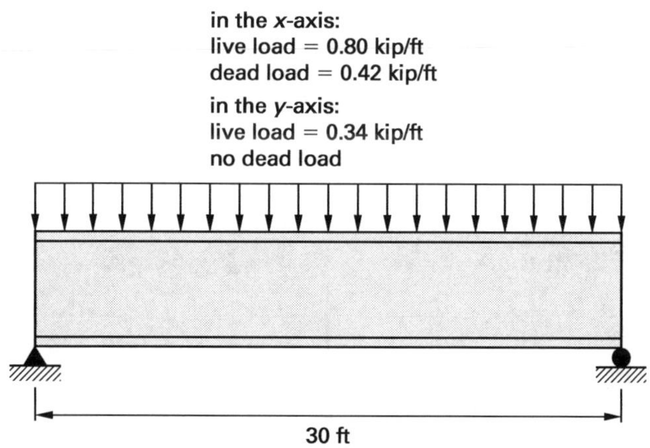

Section properties		Material properties
W16 × 67	$S_y = 23.2\ \text{in}^3$	ASTM A992 steel
$Z_x = 130\ \text{in}^3$	$b_f/2t_f = 7.70$	$F_y = 50$ ksi
$Z_y = 35.5\ \text{in}^3$	$h/t_w = 35.9$	$F_u = 65$ ksi

Determine whether the beam is satisfactory for the applied loads.

Solution

Calculate the required strength for LRFD and ASD.

LRFD	ASD
For the *x*-axis, $$w_u = 1.2w_D + 1.6w_L = (1.2)\left(0.42\ \frac{\text{kip}}{\text{ft}}\right) + (1.6)\left(0.80\ \frac{\text{kip}}{\text{ft}}\right) = 1.78\ \text{kips/ft}$$ $$M_{rx} = \frac{w_uL^2}{8} = \frac{\left(1.78\ \frac{\text{kips}}{\text{ft}}\right)(30\ \text{ft})^2}{8} = 200.25\ \text{ft-kips}$$	For the *x*-axis, $$w_a = w_D + w_L = 0.42\ \frac{\text{kip}}{\text{ft}} + 0.80\ \frac{\text{kip}}{\text{ft}} = 1.22\ \text{kips/ft}$$ $$M_{rx} = \frac{w_aL^2}{8} = \frac{\left(1.22\ \frac{\text{kips}}{\text{ft}}\right)(30\ \text{ft})^2}{8} = 137.25\ \text{ft-kips}$$
For the *y*-axis, $$w_u = 1.2w_D + 1.6w_L = (1.2)\left(0\ \frac{\text{kip}}{\text{ft}}\right) + (1.6)\left(0.34\ \frac{\text{kip}}{\text{ft}}\right) = 0.54\ \text{kip/ft}$$ $$M_{ry} = \frac{w_uL^2}{8} = \frac{\left(0.54\ \frac{\text{kip}}{\text{ft}}\right)(30\ \text{ft})^2}{8} = 60.75\ \text{ft-kips}$$	For the *y*-axis, $$w_a = w_D + w_L = 0\ \frac{\text{kip}}{\text{ft}} + 0.34\ \frac{\text{kip}}{\text{ft}} = 0.34\ \text{kip/ft}$$ $$M_{ry} = \frac{w_aL^2}{8} = \frac{\left(0.34\ \frac{\text{kip}}{\text{ft}}\right)(30\ \text{ft})^2}{8} = 38.25\ \text{ft-kips}$$

Check for the limiting width thickness ratios of the flanges and web. For the flanges, see case 10 from *AISC Specification* Table B4.1b.

$$\begin{aligned}\lambda_p &= 0.38\sqrt{\frac{E}{F_y}} \\ &= 0.38\sqrt{\frac{29{,}000\ \frac{\text{kips}}{\text{in}^2}}{50\ \frac{\text{kips}}{\text{in}^2}}} \\ &= 9.15\quad \left[> b_f/2t_f = 7.70,\ \text{so compact}\right]\end{aligned}$$

Therefore, the flanges are compact. (All but 10 wide-flange shapes have compact flanges for ASTM A992 steel, and all but one have compact flanges for ASTM A36 steel.)

For the web, see case 15 from Table B4.1b.

$$\begin{aligned}\lambda_p &= 3.76\sqrt{\frac{E}{F_y}} \\ &= 3.76\sqrt{\frac{29{,}000\ \frac{\text{kips}}{\text{in}^2}}{50\ \frac{\text{kips}}{\text{in}^2}}} \\ &= 90.55\quad \left[> h/t_w = 35.9,\ \text{so compact}\right]\end{aligned}$$

Therefore, the web is compact. (All wide-flange shapes have compact webs for ASTM A992 steel, so this calculation could be omitted.)

Calculate the nominal flexural strength for the *x*-axis. Because a compression flange is laterally braced for its entire length, the nominal moment capacity is calculated as follows, using Eq. 5.6.

$$\begin{aligned}M_{nx} &= M_p = F_y Z_x \\ &= \frac{\left(50\ \frac{\text{kips}}{\text{in}^2}\right)\left(130\ \text{in}^3\right)}{12\ \frac{\text{in}}{\text{ft}}} \\ &= 541.67\ \text{ft-kips}\end{aligned}$$

Calculate the nominal flexural strength for the y-axis using Eq. 5.19.

$$M_{ny} = M_p \leq \begin{cases} F_y Z_y = \dfrac{\left(50\ \dfrac{\text{kips}}{\text{in}^2}\right)\left(35.5\ \text{in}^3\right)}{12\ \dfrac{\text{in}}{\text{ft}}} \\ \quad = 147.92\ \text{ft-kips} \quad [\text{controls}] \\ 1.6F_y S_y = \dfrac{(1.6)\left(50\ \dfrac{\text{kips}}{\text{in}^2}\right)\left(23.2\ \text{in}^3\right)}{12\ \dfrac{\text{in}}{\text{ft}}} \\ \quad = 154.67\ \text{ft-kips} \end{cases}$$

Calculate the design strength (LRFD) and the allowable strength (ASD) for the member.

LRFD	ASD
For the x-axis, $M_{cx} = \phi_b M_{px}$ $= (0.90)(541.67\ \text{ft-kips})$ $= 487.50\ \text{ft-kips}$	For the x-axis, $M_{cx} = \dfrac{M_{px}}{\Omega_b}$ $= \dfrac{541.67\ \text{ft-kips}}{1.67}$ $= 324.35\ \text{ft-kips}$
For the y-axis, $M_{cy} = \phi_b M_{py} = 0.90M_n$ $= (0.90)(147.92\ \text{ft-kips})$ $= 133.13\ \text{ft-kips}$	For the y-axis, $M_{cy} = \dfrac{M_{py}}{\Omega_b}$ $= \dfrac{147.92\ \text{ft-kips}}{1.67}$ $= 88.57\ \text{ft-kips}$

Alternatively, from *AISC Manual* Table 3-2 and Table 3-4,

LRFD	ASD
For the x-axis, $M_{cx} = \phi_b M_{px} = 488\ \text{ft-kips}$	For the x-axis, $M_{cx} = \dfrac{M_{px}}{\Omega_b} = 324\ \text{ft-kips}$
For the y-axis, $M_{cy} = \phi_b M_{py} = 133\ \text{ft-kips}$	For the y-axis, $M_{cy} = \dfrac{M_{py}}{\Omega_b} = 88.6\ \text{ft-kips}$

Perform the unity check to determine whether the beam is satisfactory. Because there is no axial load, $P_r/P_c = 0 < 0.2$. Therefore, use Eq. 8.2. With no axial load, the first term of Eq. 8.2 is zero.

$$\frac{P_r}{2P_c} + \frac{M_{rx}}{M_{cx}} + \frac{M_{ry}}{M_{cy}} \le 1.0$$

$$0 + \frac{M_{rx}}{M_{cx}} + \frac{M_{ry}}{M_{cy}} \le 1.0$$

LRFD	ASD
$\frac{M_{rx}}{M_{cx}} + \frac{M_{ry}}{M_{cy}} \le 1.0$	$\frac{M_{rx}}{M_{cx}} + \frac{M_{ry}}{M_{cy}} \le 1.0$
$\frac{200.25 \text{ ft-kips}}{487.50 \text{ ft-kips}} + \frac{60.75 \text{ ft-kips}}{133.13 \text{ ft-kips}} = 0.87$	$\frac{137.25 \text{ ft-kips}}{324.35 \text{ ft-kips}} + \frac{38.25 \text{ ft-kips}}{88.57 \text{ ft-kips}} = 0.86$
$[\le 1.0, \text{ so OK}]$	$[\le 1.0, \text{ so OK}]$

The beam is satisfactory for the design loads because the results of the interaction equations are less than or equal to 1.0.

4. COMBINED TENSION AND BENDING

Hangers and vertical and horizontal bracing members are typical members that are subject to combined tension and bending.

There are a number of advantages to using hangers to support loads. Hangers use less steel and do not take up valuable floor space as columns do. They make optimal use of the strength of the material. Unless braced in some manner, a hanger can also be subjected to lateral loads in one or both axes.

When a tension member is subjected to bending loads in either or both of its axes, the design tensile strength (LRFD) or allowable tensile strength (ASD) must be reduced. The same interaction equations, Eq. 8.1 and Eq. 8.2, are used for hangers as for combined compression and bending members. However, C_b may be increased in accordance with *AISC Specification* Sec. H1.2 when calculating the nominal flexural strength.

Example 8.2

Combined Tension and Bending

A hanger is loaded as shown. The member is laterally unbraced except for the rigid connection at its top.

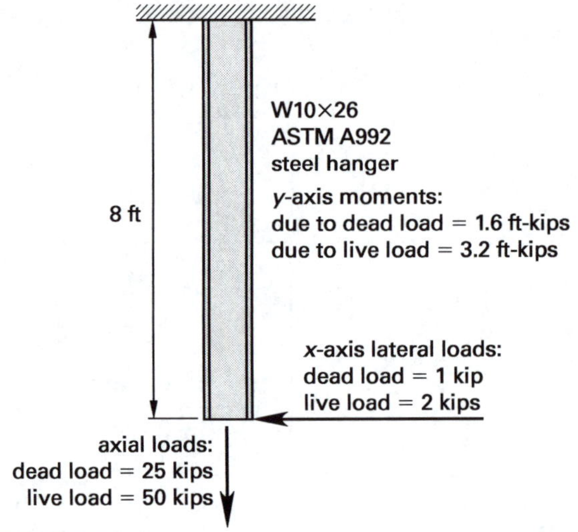

Section properties		Material properties
$A = 7.61\text{ in}^2$	$L_p = 4.80\text{ ft}$	ASTM A992 steel
$S_x = 27.9\text{ in}^3$	$L_r = 14.9\text{ ft}$	$F_y = 50\text{ ksi}$
$S_y = 4.89\text{ in}^3$	$b_f/2t_f = 6.56$	$F_u = 65\text{ ksi}$
$Z_x = 31.3\text{ in}^3$	$h/t_w = 34.0$	
$Z_y = 7.50\text{ in}^3$	$I_y = 14.1\text{ in}^4$	

Determine whether the hanger satisfies the *AISC Manual*'s strength requirements.

Solution

Calculate the required strengths for LRFD and ASD. The axial load is

LRFD	ASD
$\begin{aligned} P_r &= 1.2D + 1.6L \\ &= (1.2)(25\text{ kips}) + (1.6)(50\text{ kips}) \\ &= 110\text{ kips} \end{aligned}$	$\begin{aligned} P_r &= D + L \\ &= 25\text{ kips} + 50\text{ kips} \\ &= 75\text{ kips} \end{aligned}$

The *x*- and *y*-axis bending are

LRFD	ASD
$M_{rx} = 1.2P_{Dx}h + 1.6P_{Lx}h$ $= (1.2)(1 \text{ kip})(8 \text{ ft}) + (1.6)(2 \text{ kips})(8 \text{ ft})$ $= 35.20 \text{ ft-kips}$	$M_{rx} = P_{Dx}h + P_{Lx}h$ $= (1 \text{ kip})(8 \text{ ft}) + (2 \text{ kips})(8 \text{ ft})$ $= 24 \text{ ft-kips}$
$M_{ry} = 1.2M_{Dy} + 1.6M_{Ly}$ $= (1.2)(1.6 \text{ ft-kips}) + (1.6)(3.2 \text{ ft-kips})$ $= 7.04 \text{ ft-kips}$	$M_{ry} = M_{Dy} + M_{Ly}$ $= 1.6 \text{ ft-kips} + 3.2 \text{ ft-kips}$ $= 4.8 \text{ ft-kips}$

Check for the limiting width thickness ratios of the flanges and the web. For the flanges, see *AISC Specification* Table B4.1b, case 10.

$$\lambda_p = 0.38\sqrt{\frac{E}{F_y}} = 0.38\sqrt{\frac{29{,}000 \ \frac{\text{kips}}{\text{in}^2}}{50 \ \frac{\text{kips}}{\text{in}^2}}} = 9.15 \quad [> b_f/2t_f = 6.56, \text{ so compact}]$$

The flanges are compact. (All but 10 wide-flange shapes have compact flanges for ASTM A992 steel, and all but one have compact flanges for ASTM A36 steel.) For the web, see *AISC Specification* Table B4.1b, case 10.

$$\lambda_p = 3.76\sqrt{\frac{E}{F_y}} = 3.76\sqrt{\frac{29{,}000 \ \frac{\text{kips}}{\text{in}^2}}{50 \ \frac{\text{kips}}{\text{in}^2}}} = 90.55 \quad [> h/t_w = 34.0, \text{ so compact}]$$

The web is compact. (This step could be eliminated because all wide-flange shapes have compact webs for ASTM A992 steel.) The nominal tensile strength is

$$P_n = F_y A_g = \left(50 \ \frac{\text{kips}}{\text{in}^2}\right)(7.61 \text{ in}^2) = 380.50 \text{ kips}$$

Calculate the design tensile strength (LRFD) and the allowable tensile strength (ASD).

LRFD	ASD
$P_c = \phi_t P_n = (0.90)(380.50 \text{ kips}) = 342.45 \text{ kips}$	$P_c = \frac{P_n}{\Omega_t} = \frac{380.50 \text{ kips}}{1.67} = 227.84 \text{ kips}$

Calculate the nominal flexural strength about the *x*-axis. Because $h = L_b$ and $L_p < L_b < L_r$, bending about the strong axis falls into zone 2 bending, the inelastic buckling limit, and the nominal design strength will be less than M_p. Therefore, use Eq. 5.9. For a cantilevered beam, C_b is 1.0 (*AISC Specification* Sec. F1); however, it is possible to increase C_b in accordance with Sec. 8.2 (from *AISC Specification* Sec. H1.2) as follows. From Eq. 8.3,

$$P_{ey} = \frac{\pi^2 EI_y}{L_b^2} = \frac{\pi^2\left(29{,}000 \frac{\text{kips}}{\text{in}^2}\right)(14.1 \text{ in}^4)}{(8 \text{ ft})^2\left(12 \frac{\text{in}}{\text{ft}}\right)^2} = 437.90 \text{ kips}$$

LRFD	ASD
$P_u = P_r = 110 \text{ kips}$	$P_a = P_r = 75 \text{ kips}$
$\sqrt{1+\frac{\alpha P_r}{P_{ey}}} = \sqrt{1+\frac{(1)(110 \text{ kips})}{437.90 \text{ kips}}} = 1.12$	$\sqrt{1+\frac{\alpha P_r}{P_{ey}}} = \sqrt{1+\frac{(1.6)(75 \text{ kips})}{437.90 \text{ kips}}} = 1.13$

Therefore, C_b could be taken as (1.12)(1.0) = 1.12. For this problem, use a conservative approach and assume C_b is 1.0. From Eq. 5.6,

$$M_p = F_y Z_x = \frac{\left(50 \frac{\text{kips}}{\text{in}^2}\right)(31.3 \text{ in}^3)}{12 \frac{\text{in}}{\text{ft}}} = 130 \text{ ft-kips}$$

From Eq. 5.9,

$$M_{nx} = C_b\left(M_p - \left(M_p - 0.7F_yS_x\right)\left(\frac{L_b - L_p}{L_r - L_p}\right)\right) \le M_p$$

$$= (1.0)\left(130 \text{ ft-kips} - \left(130 \text{ ft-kips} - \frac{(0.7)\left(50 \ \frac{\text{kips}}{\text{in}^2}\right) \times \left(27.9 \text{ in}^3\right)}{12 \ \frac{\text{in}}{\text{ft}}}\right)\left(\frac{8 \text{ ft} - 4.80 \text{ ft}}{14.90 \text{ ft} - 4.80 \text{ ft}}\right)\right)$$

$$= 114.59 \text{ ft-kips} \quad \left[\le M_p, \text{ so controls}\right]$$

Calculate the nominal flexural strength about the y-axis using Eq. 5.17.

$$M_{ny} = M_p \le \begin{cases} F_yZ_y = \dfrac{\left(50 \ \frac{\text{kips}}{\text{in}^2}\right)\left(7.50 \text{ in}^3\right)}{12 \ \frac{\text{in}}{\text{ft}}} = 31.25 \text{ ft-kips} \quad [\text{controls}] \\ 1.6F_yS_y = \dfrac{(1.6)\left(50 \ \frac{\text{kips}}{\text{in}^2}\right)\left(4.89 \text{ in}^3\right)}{12 \ \frac{\text{in}}{\text{ft}}} = 32.60 \text{ ft-kips} \end{cases}$$

Calculate the design flexural strength and the allowable flexural strength. The nominal strengths for the x- and y-axes are

$$M_{nx} = 114.59 \text{ ft-kips}$$
$$M_{ny} = 31.25 \text{ ft-kips}$$

LRFD	ASD
$M_{cx} = \phi_b M_{nx} = (0.90)(114.59 \text{ ft-kips}) = 103.13 \text{ ft-kips}$	$M_{cx} = \dfrac{M_{nx}}{\Omega_b} = \dfrac{114.59 \text{ ft-kips}}{1.67} = 68.62 \text{ ft-kips}$
$M_{cy} = \phi_b M_{ny} = (0.90)(31.25 \text{ ft-kips}) = 28.13 \text{ ft-kips}$	$M_{cy} = \dfrac{M_{ny}}{\Omega_b} = \dfrac{31.25 \text{ ft-kips}}{1.67} = 18.71 \text{ ft-kips}$

Determine whether Eq. 8.1 or Eq. 8.2 is the correct interaction equation to use.

$$\frac{P_r}{P_c} = \frac{110 \text{ kips}}{342.45 \text{ kips}} = 0.32 \quad [\geq 0.2, \text{ so use Eq. 8.1}]$$

From Eq. 8.1,

LRFD	ASD
$\frac{P_r}{P_c} + \left(\frac{8}{9}\right)\left(\frac{M_{rx}}{M_{cx}} + \frac{M_{ry}}{M_{cy}}\right) \leq 1.0$	$\frac{P_r}{P_c} + \left(\frac{8}{9}\right)\left(\frac{M_{rx}}{M_{cx}} + \frac{M_{ry}}{M_{cy}}\right) \leq 1.0$
$\frac{110 \text{ kips}}{342.45 \text{ kips}} + \left(\frac{8}{9}\right) \times \left(\frac{35.20 \text{ ft-kips}}{103.13 \text{ ft-kips}} + \frac{7.04 \text{ ft-kips}}{28.13 \text{ ft-kips}}\right) = 0.85$	$\frac{75 \text{ kips}}{227.84 \text{ kips}} + \left(\frac{8}{9}\right) \times \left(\frac{24 \text{ ft-kips}}{68.62 \text{ ft-kips}} + \frac{4.8 \text{ ft-kips}}{18.71 \text{ ft-kips}}\right) = 0.87$
$[\leq 1.0, \text{ so OK}]$	$[\leq 1.0, \text{ so OK}]$

Therefore, the W10 × 26 hanger section is satisfactory to resist the imparted loads. It satisfies the requirements of the *AISC Manual*.

5. COMBINED COMPRESSION AND BENDING

Members having combined compression and bending stresses occur frequently in building structures. Columns in moment-resisting frames are a typical example. Beam columns are usually beams that have axial loads due to wind, seismic, or other lateral loads.

Designing these members can be tedious because it is an iterative process. Fortunately, Part 6 of the *AISC Manual* provides guidance and many constants that help in the design and analysis of these members, keeping the number of iterations to a minimum.

Members subjected to combined compression and bending forces are designed to satisfy the requirements of Eq. 8.1 (for $P_r/P_c \geq 0.2$) and Eq. 8.2 (for $P_r/P_c < 0.2$). The tables in Part 6 of the *AISC Manual* can be used in designing and analyzing W shape members subjected to combined axial and bending loads. These tables contain values for five variables that can be used to resolve Eq. 8.1 and Eq. 8.2 more quickly. These five variables are defined in Table 8.1.

***Table 8.1** Definitions of p, b_x, b_y, t_r, and t_y*

	LRFD	ASD
axial compression (kips^{-1})	$p=\dfrac{1}{\phi_c P_n}$	$p=\dfrac{\Omega_c}{P_n}$
strong axis bending (ft-kips)$^{-1}$	$b_x=\dfrac{8}{9\phi_b M_{nx}}$	$b_x=\dfrac{8\Omega_b}{9M_{nx}}$
weak axis bending (ft-kips)$^{-1}$	$b_y=\dfrac{8}{9\phi_b M_{ny}}$	$b_y=\dfrac{8\Omega_b}{9M_{ny}}$
tension rupture (kips^{-1})	$t_r=\dfrac{1}{\phi_t F_u\left(0.75A_g\right)}$	$t_r=\dfrac{\Omega_t}{F_u\left(0.75A_g\right)}$
tension yielding (kips^{-1})	$t_y=\dfrac{1}{\phi_c F_y A_g}$	$t_y=\dfrac{\Omega}{F_y A_g}$

For this use, Eq. 8.1 can be rewritten as

$$pP_r+b_xM_{rx}+b_yM_{ry}\le 1.0 \qquad 8.5$$

Equation 8.2 can be rewritten as

$$0.5pP_r+\tfrac{9}{8}\left(b_xM_{rx}+b_yM_{ry}\right)\le 1.0 \qquad 8.6$$

Example 8.3

Combined Compression and Bending on W Shape Member

A steel column supports the loads shown and has the following properties. Loads are based on the direct analysis method.

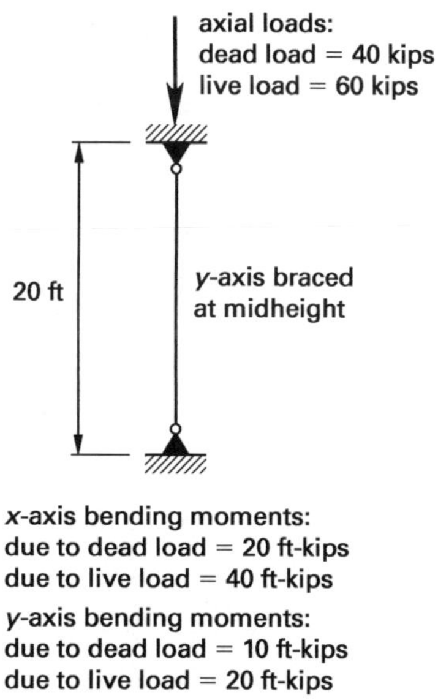

End conditions

top of column, both axes: rotation free, translation fixed

bottom of column, both axes: rotation free, translation fixed

Bracing

x-axis: ends only

y-axis: both ends and midheight

Material properties

ASTM A992 steel

$F_y = 50$ ksi

$F_u = 65$ ksi

Determine the lightest W12 steel section that will support the loads with the given end conditions.

Solution

Loads are based on the direct analysis method, so $K = 1.0$.

$$KL_y = (1.0)(10 \text{ ft}) = 10 \text{ ft}$$

The effective length for the *x*-axis is

$$KL_x = (1.0)(20 \text{ ft}) = 20 \text{ ft}$$

Calculate the required strengths.

LRFD	ASD
The axial load is $P_r = 1.2D + 1.6L$ $= (1.2)(40 \text{ kips}) + (1.6)(60 \text{ kips})$ $= 144 \text{ kips}$	The axial load is $P_r = D + L$ $= 40 \text{ kips} + 60 \text{ kips}$ $= 100 \text{ kips}$
The *x*-axis bending is $M_{rx} = 1.2M_{Dx} + 1.6M_{Lx}$ $= (1.2)(20 \text{ ft-kips})$ $+ (1.6)(40 \text{ ft-kips})$ $= 88 \text{ ft-kips}$	The *x*-axis bending is $M_{rx} = M_{Dx} + M_{Lx}$ $= 20 \text{ ft-kips} + 40 \text{ ft-kips}$ $= 60 \text{ ft-kips}$

LRFD	ASD
The y-axis bending is $M_{ry} = 1.2M_{Dy} + 1.6M_{Ly}$ $= (1.2)(10 \text{ ft-kips}) + (1.6)(20 \text{ ft-kips})$ $= 44 \text{ ft-kips}$	The y-axis bending is $M_{ry} = M_{Dy} + M_{Ly}$ $= 10 \text{ ft-kips} + 20 \text{ ft-kips}$ $= 30 \text{ ft-kips}$

In Table 4-1 of the *AISC Manual*, the lightest W12 column section listed is a W12 × 40. Lighter W12 sections are available, which are used primarily as beam sections. Calculating the eccentricities reveals that they are relatively large, 0.60 ft on the x-axis and 0.30 ft on the y-axis. Based on the relatively large eccentricities, assume that the ratio of the required axial strength to the nominal strength will be approximately 0.25 for the first trial selection. Calculate the approximate equivalent required tabular load based on a ratio of 0.25.

LRFD	ASD
$P_{\text{tab}} = \dfrac{P_r}{\text{ratio}}$ $= \dfrac{144 \text{ kips}}{0.25}$ $= 576 \text{ kips}$	$P_{\text{tab}} = \dfrac{P_r}{\text{ratio}}$ $= \dfrac{100 \text{ kips}}{0.25}$ $= 400 \text{ kips}$

Select tentative column sections from *AISC Manual* Table 4-1 using an effective length with respect to the y-axis of 10 ft.

section	LRFD, $\phi_c P_n$ (kips)	ASD, P_n/Ω_c (kips)
W12 × 45, $r_x/r_y = 2.64$	447	294
W12 × 50, $r_x/r_y = 2.64$	500	332
W12 × 53, $r_x/r_y = 2.11$	592	394
W12 × 58, $r_x/r_y = 2.10$	647	431

All the r_x/r_y ratios for these beams are greater than 2.0, and the unbraced length ratio between the x- and y-axes is 2.0, so the effective unbraced length with respect to the y-axis is 10 ft.

The ratio of required strength to nominal strength was assumed for this trial to be 0.25, so Eq. 8.1 is the interaction formula that applies, unless subsequent calculations indicate that the ratio is actually less than 0.20, in which case Eq. 8.2 is the applicable formula.

Using tables in Part 6 of the *AISC Manual*, select the lightest W12 member that will develop the required strengths. If the assumed ratio of 0.25 is correct, then the

W12 × 53 or the W12 × 58 will be the correct selection. However, check the W12 × 50 first in case it proves to be acceptable.

Perform a unity check for the W12 × 50 member with Eq. 8.5, the modified interaction formula. From *AISC Manual* Table 6-1,

LRFD	ASD
$p\times 10^3 = 2.0 \text{ kips}^{-1}$ $b_x\times 10^3 = 3.54 \;(\text{ft-kips})^{-1}$ $b_y\times 10^3 = 11.1 \;(\text{ft-kips})^{-1}$	$p\times 10^3 = 3.01 \text{ kips}^{-1}$ $b_x\times 10^3 = 5.32 \;(\text{ft-kips})^{-1}$ $b_y\times 10^3 = 16.7 \;(\text{ft-kips})^{-1}$

From Eq. 8.5,

LRFD	ASD
$pP_r + b_xM_{rx} + b_yM_{ry} \le 1.0$ $\left(\frac{2.00}{10^3 \text{ kips}}\right)(144 \text{ kips}) + \left(\frac{3.54}{10^3 \text{ ft-kips}}\right)(88 \text{ ft-kips}) + \left(\frac{11.1}{10^3 \text{ ft-kips}}\right)(44 \text{ ft-kips}) = 1.09$ $[> 1.0, \text{not OK}]$	$pP_r + b_xM_{rx} + b_yM_{ry} \le 1.0$ $\left(\frac{3.01}{10^3 \text{ kips}}\right)(100 \text{ kips}) + \left(\frac{5.32}{10^3 \text{ ft-kips}}\right)(60 \text{ ft-kips}) + \left(\frac{16.7}{10^3 \text{ ft-kips}}\right)(30 \text{ ft-kips}) = 1.12$ $[> 1.0, \text{not OK}]$

This is not good, so the W12 × 50 is unsatisfactory.

Because the first term in the solution of the interaction exceeds 0.20, Eq. 8.1 is the right equation to use. Perform a unity check for the W12 × 53 member with the modified interaction formula and values from Table 6-1. From *AISC Manual* Table 6-1,

LRFD	ASD
$p\times 10^3 = 1.69 \text{ kips}^{-1}$ $b_x\times 10^3 = 3.12 \;(\text{ft-kips})^{-1}$ $b_y\times 10^3 = 8.15 \;(\text{ft-kips})^{-1}$	$p\times 10^3 = 2.54 \text{ kips}^{-1}$ $b_x\times 10^3 = 4.68 \;(\text{ft-kips})^{-1}$ $b_y\times 10^3 = 12.2 \;(\text{ft-kips})^{-1}$

From Eq. 8.5,

LRFD	ASD
$pP_r + b_xM_{rx} + b_yM_{ry} \le 1.0$	$pP_r + b_xM_{rx} + b_yM_{ry} \le 1.0$
$\left(\dfrac{1.69}{10^3 \text{ kips}}\right)(144 \text{ kips})$	$\left(\dfrac{2.54}{10^3 \text{ kips}}\right)(100 \text{ kips})$
$+\left(\dfrac{3.12}{10^3 \text{ ft-kips}}\right)(88 \text{ ft-kips})$	$+\left(\dfrac{4.68}{10^3 \text{ ft-kips}}\right)(60 \text{ ft-kips})$
$+\left(\dfrac{8.15}{10^3 \text{ ft-kips}}\right)(44 \text{ ft-kips})$	$+\left(\dfrac{12.2}{10^3 \text{ ft-kips}}\right)(30 \text{ ft-kips})$
$= 0.88$	$= 0.90$
$[\le 1.0, \text{ so OK}]$	$[\le 1.0, \text{ so OK}]$

This is OK, so the W12 × 53 is satisfactory. Because the first term in the solution of the interaction exceeds 0.20, Eq. 8.1 is used. Therefore the W12 × 53 is the lightest W12 section capable of resisting the required design loads.

Example 8.4

Combined Compression and Bending on HSS Member

An HSS10 × 6 × $^3/_8$ member supports the load shown and has the following properties. Loads shown are based on the direct analysis method.

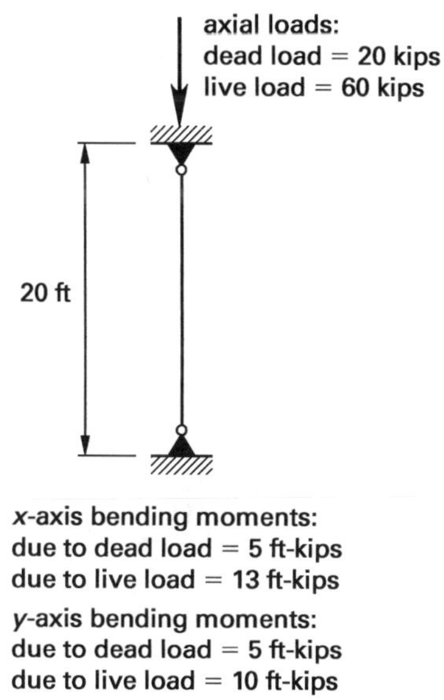

Bracing

x-axis: ends only

y-axis: ends only

Section properties

$t_w = 0.349$ in

$A = 10.4$ in^2

$b/t = 14.2$

$h/t = 25.7$

$I_x = 137$ in^4

$S_x = 27.4$ in^3

$r_x = 3.63$ in

$Z_x = 33.8$ in^3

$I_y = 61.8$ in^4

$S_y = 20.6$ in^3

$r_y = 2.44$ in

$Z_y = 23.7$ in^3

Material properties

ASTM A500, grade B steel

$F_y = 46$ ksi

$F_u = 58$ ksi

Determine whether the column is adequate to support the applied loads.

Solution

Loads are based on the direct analysis method, so $K = 1.0$.

Calculate the required strengths.

LRFD	ASD
The axial load is $P_r = 1.2D + 1.6L$ $= (1.2)(20 \text{ kips}) + (1.6)(60 \text{ kips})$ $= 120 \text{ kips}$	The axial load is $P_r = D + L$ $= 20 \text{ kips} + 60 \text{ kips}$ $= 80 \text{ kips}$
The x-axis bending is $M_{rx} = 1.2M_{Dx} + 1.6M_{Lx}$ $= (1.2)(5 \text{ ft-kips})$ $+ (1.6)(13 \text{ ft-kips})$ $= 26.8 \text{ ft-kips}$	The x-axis bending is $M_{rx} = M_{Dx} + M_{Lx}$ $= 5 \text{ ft-kips} + 13 \text{ ft-kips}$ $= 18 \text{ ft-kips}$
The y-axis bending is $M_{ry} = 1.2M_{Dy} + 1.6M_{Ly}$ $= (1.2)(5 \text{ ft-kips})$ $+ (1.6)(10 \text{ ft-kips})$ $= 22 \text{ ft-kips}$	The y-axis bending is $M_{ry} = M_{Dy} + M_{Ly}$ $= 5 \text{ ft-kips} + 10 \text{ ft-kips}$ $= 15 \text{ ft-kips}$

Determine whether the flanges and webs are compact. For the flanges, use *AISC Specification* Table B4.1b, case 17.

$$1.12\sqrt{\frac{E}{F_y}} = 1.12\sqrt{\frac{29{,}000\ \frac{\text{kips}}{\text{in}^2}}{46\ \frac{\text{kips}}{\text{in}^2}}} = 28.12 \quad [> b/t = 14.2\text{, so compact}]$$

For the webs, use *AISC Specification* Table B4.1b, case 19.

$$2.42\sqrt{\frac{E}{F_y}} = 2.42\sqrt{\frac{29{,}000\ \frac{\text{kips}}{\text{in}^2}}{46\ \frac{\text{kips}}{\text{in}^2}}} = 60.76 \quad [> h/t = 25.7\text{, so compact}]$$

From *AISC Specification* Table B4.1a, case 6, the limiting width-to-thickness ratio for an HSS section subject to axial compression is

$$\frac{b}{t} \le 1.40\sqrt{\frac{E}{F_y}} = 1.40\sqrt{\frac{29{,}000\ \frac{\text{kips}}{\text{in}^2}}{46\ \frac{\text{kips}}{\text{in}^2}}} = 35.15$$

Both b/t and h/t are less than 35.15, so the section meets this requirement.

Calculate the critical slenderness ratio. The y-axis governs for the given conditions.

$$\frac{K_yL_y}{r_y} = \frac{(1)(20\ \text{ft})\left(12\ \frac{\text{in}}{\text{ft}}\right)}{2.44\ \text{in}} = 98.36$$

Determine whether the compression is in the inelastic or the elastic range.

$$4.71\sqrt{\frac{E}{F_y}} = 4.71\sqrt{\frac{29{,}000\ \frac{\text{kips}}{\text{in}^2}}{46\ \frac{\text{kips}}{\text{in}^2}}} = 118.26 \quad [> KL/r\text{, so use Eq. 7.6}]$$

The member is in the inelastic range, and Eq. 7.6 applies. Calculate the elastic critical buckling stress using Eq. 7.8.

$$F_e = \frac{\pi^2 E}{\left(\frac{K_yL_y}{r_y}\right)^2} = \frac{\pi^2\left(29{,}000\ \frac{\text{kips}}{\text{in}^2}\right)}{(98.36)^2} = 29.58\ \text{ksi}$$

Calculate the critical flexural buckling stress using Eq. 7.6.

$$F_{cr} = 0.658^{F_y/F_e} F_y = (0.658)^{46 \frac{\text{kips}}{\text{in}^2} / 29.58 \frac{\text{kips}}{\text{in}^2}} \left(46 \ \frac{\text{kips}}{\text{in}^2} \right) = 23.99 \text{ ksi}$$

Calculate the nominal axial strength using Eq. 7.2.

$$P_n = F_{cr} A_g = \left(23.99 \ \frac{\text{kips}}{\text{in}^2} \right) (10.4 \text{ in}^2) = 249.52 \text{ kips}$$

Calculate the design strength (LRFD) and allowable strength (ASD).

LRFD	ASD
$P_c = \phi_c P_n = (0.90)(249.52 \text{ kips})$ $= 224.57 \text{ kips}$	$P_c = \frac{P_n}{\Omega_c} = \frac{249.52 \text{ kips}}{1.67} = 149.41 \text{ kips}$

Calculate the flexural design strength for the x-axis using Eq. 5.23. The section is compact.

$$M_{nx} = M_{px} = F_y Z_x = \frac{\left(46 \ \frac{\text{kips}}{\text{in}^2} \right)(33.8 \text{ in}^3)}{12 \ \frac{\text{in}}{\text{ft}}}$$

$$= 129.57 \text{ ft-kips}$$

LRFD	ASD
$M_{cx} = \phi_b M_{nx} = (0.90)(129.57 \text{ ft-kips})$ $= 116.61 \text{ ft-kips}$	$M_{cx} = \frac{M_{nx}}{\Omega_c} = \frac{129.57 \text{ ft-kips}}{1.67}$ $= 77.59 \text{ ft-kips}$

Calculate the flexural design strength for the y-axis using Eq. 5.23. The section is compact.

$$M_{ny} = M_{py} = F_y Z_y = \frac{\left(46 \ \frac{\text{kips}}{\text{in}^2} \right)(23.7 \text{ in}^3)}{12 \ \frac{\text{in}}{\text{ft}}}$$

$$= 90.85 \text{ ft-kips}$$

LRFD	ASD
$M_{cy} = \phi_b M_{ny} = (0.90)(90.85 \text{ ft-kips})$ $= 81.77 \text{ ft-kips}$	$M_{cy} = \frac{M_{ny}}{\Omega_c} = \frac{90.85 \text{ ft-kips}}{1.67}$ $= 54.40 \text{ ft-kips}$

Determine which interaction formula is applicable.

$$\frac{P_r}{P_c} = \frac{120 \text{ kips}}{224 \text{ kips}} = 0.54 \quad [\geq 0.2, \text{ so use Eq. 8.1}]$$

Perform the unity check using Eq. 8.1 as the interaction formula.

LRFD	ASD
$\frac{P_r}{P_c} + \left(\frac{8}{9}\right)\left(\frac{M_{rx}}{M_{cx}} + \frac{M_{ry}}{M_{cy}}\right) \leq 1.0$	$\frac{P_r}{P_c} + \left(\frac{8}{9}\right)\left(\frac{M_{rx}}{M_{cx}} + \frac{M_{ry}}{M_{cy}}\right) \leq 1.0$
$\frac{120 \text{ kips}}{224.57 \text{ kips}} + \left(\frac{8}{9}\right) \times \left(\frac{26.8 \text{ ft-kips}}{116.61 \text{ ft-kips}} + \frac{22 \text{ ft-kips}}{81.77 \text{ ft-kips}}\right) = 0.98$	$\frac{80 \text{ kips}}{149.41 \text{ kips}} + \left(\frac{8}{9}\right) \times \left(\frac{18 \text{ ft-kips}}{77.59 \text{ ft-kips}} + \frac{15 \text{ ft-kips}}{54.40 \text{ ft-kips}}\right) = 0.99$
$[\leq 1.0, \text{ so OK}]$	$[\leq 1.0, \text{ so OK}]$

The column is adequate to support the applied loads.

The procedures used earlier in solving Ex. 8.3 are generic and can be used to solve all doubly and singly symmetrical members subject to flexural and axial forces. However, the example problem could also have been solved more quickly using the design aids in the *AISC Manual* as follows.

LRFD	ASD
From *AISC Manual* Table 4-3,	From *AISC Manual* Table 4-3,
$P_c = \phi_c P_n = 225 \text{ kips}$	$P_c = \frac{P_n}{\Omega_c} = 149 \text{ kips}$
From *AISC Manual* Table 3-12,	From *AISC Manual* Table 3-12,
$M_{cx} = \phi_b M_{nx} = 116 \text{ ft-kips}$	$M_{cx} = \frac{M_{nx}}{\Omega_b} = 77.5 \text{ ft-kips}$
From *AISC Manual* Table 3-12,	From *AISC Manual* Table 3-12,
$M_{cy} = \phi_b M_{ny} = 81.8 \text{ ft-kips}$	$M_{cy} = \frac{M_{ny}}{\Omega_b} = 54.4 \text{ ft-kips}$

6. UNSYMMETRICAL AND OTHER MEMBERS SUBJECT TO FLEXURE AND AXIAL FORCE

For members that are subject to flexure and axial stress but that do not conform to the member descriptions in *AISC Specification* Sec. H1, use the interaction formula in Eq. 8.7 to determine whether they are satisfactory.

$$\left|\frac{f_a}{F_a}+\frac{f_{bw}}{F_{bw}}+\frac{f_{bz}}{F_{bz}}\right|\le 1.0 \quad \text{[AISC Eq. H2-1]} \qquad 8.7$$

The subscripts w and z stand for the major and minor principal axes, respectively.

7. MEMBERS UNDER TORSION AND COMBINED TORSION, FLEXURE, SHEAR, AND/OR AXIAL FORCES

Torsional Strength of Round and Rectangular HSS Members

The closed shapes of round and rectangular HSS members make them extremely efficient in resisting torsional forces. The design torsional strength, $\phi_T T_n$ (LRFD, $\phi_T = 0.90$), and the allowable torsional strength, T_n/Ω_T (ASD, $\Omega_T = 1.67$), of these members are determined as follows.

The limit states for deriving the nominal torsional strength are torsional yielding and torsional buckling. The nominal torsional strength is

$$T_n = F_{cr}C \quad \text{[AISC Eq. H3-1]} \qquad 8.8$$

The values of the critical stress, F_{cr}, and the HSS torsional constant, C, are calculated differently for round and rectangular HSS.

Round HSS Members

For round HSS, the critical stress, F_{cr}, is the larger of the values given by Eq. 8.9 and Eq. 8.10, but no larger than $0.60F_y$.

$$F_{cr} = \frac{1.23E}{\sqrt{\frac{L}{D}}\left(\frac{D}{t}\right)^{5/4}} \quad \text{[AISC Eq. H3-2a]} \qquad 8.9$$

$$F_{cr} = \frac{0.60E}{\left(\frac{D}{t}\right)^{3/2}} \quad \text{[AISC Eq. H3-2b]} \qquad 8.10$$

L is the length of the member and D is its outside diameter. The torsional constant, C, can be conservatively taken as

$$C = \frac{\pi(D-t)^2 t}{2} \qquad 8.11$$

Rectangular HSS Members

To calculate the critical stress for rectangular HSS, start by comparing h/t to $2.45\sqrt{E/F_y}$. For $h/t \leq 2.45\sqrt{E/F_y}$,

$$F_{\text{cr}} = 0.6F_y \quad [\text{AISC Eq. H3-3}] \qquad 8.12$$

For $2.45\sqrt{E/F_y} < h/t \leq 3.07\sqrt{E/F_y}$,

$$F_{\text{cr}} = \frac{0.6F_y\left(2.45\sqrt{\frac{E}{F_y}}\right)}{\frac{h}{t}} \quad [\text{AISC Eq. H3-4}] \qquad 8.13$$

For $3.07\sqrt{E/F_y} < h/t \leq 260$,

$$F_{\text{cr}} = \frac{0.458\pi^2 E}{\left(\frac{h}{t}\right)^2} \quad [\text{AISC Eq. H3-5}] \qquad 8.14$$

The torsional constant, C, can be conservatively taken as

$$C = 2(B-t)(H-t)t - (4.5)(4-\pi)t^3 \quad [\text{AISC Sec. H3.1}] \qquad 8.15$$

Example 8.5

Torsional Strength of HSS Member

An HSS12 × 6 × $^3/_{16}$ member is subjected to torsional forces only.

Section properties		Material properties
$t = 0.174$ in	$Z_x = 23.7$ in^3	ASTM A500, grade B steel
$A = 6.06$ in^2	$I_y = 40.0$ in^4	$F_y = 46$ ksi
$b/t = 31.5$	$S_y = 13.3$ in^3	$F_u = 58$ ksi
$h/t = 66.0$	$r_x = 2.57$ in	
$I_x = 116$ in^4	$Z_y = 14.7$ in^3	
$S_x = 19.4$ in^3	$J = 94.6$ in^4	
$r_x = 4.38$ in	$C = 24.0$ in^3	

Determine the nominal torsional strength, the design strength (LRFD), and the allowable strength (ASD).

Solution

Compare h/t to $2.45\sqrt{E/F_y}$ to determine whether Eq. 8.12, Eq. 8.13, or Eq. 8.14 is the appropriate formula for calculating the critical stress, F_{cr}.

$$2.45\sqrt{\frac{E}{F_y}} = 2.45\sqrt{\frac{29{,}000\ \frac{\text{kips}}{\text{in}^2}}{46\ \frac{\text{kips}}{\text{in}^2}}} = 61.52 \quad [< h/t = 66,\ \text{so cannot use Eq. 8.12}]$$

$$3.07\sqrt{\frac{E}{F_y}} = 3.07\sqrt{\frac{29{,}000\ \frac{\text{kips}}{\text{in}^2}}{46\ \frac{\text{kips}}{\text{in}^2}}} = 77.08 \quad [> h/t = 66,\ \text{so use Eq. 8.13}]$$

The torsional constant, C, is given, so it does not have to be calculated. Use Eq. 8.13 to calculate the critical stress, F_{cr}.

$$\begin{aligned} F_{cr} &= \frac{0.6F_y\left(2.45\sqrt{\frac{E}{F_y}}\right)}{\frac{h}{t}} \\ &= \frac{(0.60)\left(46\ \frac{\text{kips}}{\text{in}^2}\right)(2.45)\sqrt{\frac{29{,}000\ \frac{\text{kips}}{\text{in}^2}}{46\ \frac{\text{kips}}{\text{in}^2}}}}{66.0} \\ &= 25.72\ \text{ksi} \end{aligned}$$

Use Eq. 8.8 to determine the nominal torsional resistance.

$$\begin{aligned} T_n = F_{cr}C &= \frac{\left(25.72\ \frac{\text{kips}}{\text{in}^2}\right)(24.0\ \text{in}^3)}{12\ \frac{\text{in}}{\text{ft}}} \\ &= 51.44\ \text{ft-kips} \end{aligned}$$

Determine the design and allowable torsional strengths.

LRFD	ASD
$T_c = \phi_T T_n$ $= (0.90)(51.44\ \text{ft-kips})$ $= 46.30\ \text{ft-kips}$	$T_c = \frac{T_n}{\Omega_T} = \frac{51.44\ \text{ft-kips}}{1.67}$ $= 30.80\ \text{ft-kips}$

HSS Members Subject to Combined Torsion, Shear, Flexure, and Axial Force

When the required torsional strength, T_r, is less than or equal to 20% of the available torsional strength, T_c, then the interaction of torsion, shear, flexure, and/or axial force for HSS members is determined in accordance with *AISC Specification* Sec. H1. Torsional effects are neglected. When $P_r/P_c \geq 0.2$, Eq. 8.1 applies; when $P_r/P_c < 0.2$, Eq. 8.2 applies.

When the required torsional strength, T_r, exceeds 20% of the available torsional strength, T_c, the interaction of torsion, shear, flexure, and/or axial load is limited by Eq. 8.16.

$$\left(\frac{P_r}{P_c}+\frac{M_r}{M_c}\right)+\left(\frac{V_r}{V_c}+\frac{T_r}{T_c}\right)^2 \leq 1.0 \quad \text{[AISC Eq. H3-6]} \qquad 8.16$$

Where there is no axial load, $P_r/P_c = 0$.

Example 8.6

Combined Flexure, Shear, and Torsion

An HSS12 × 6 × $^3/_{16}$ member is loaded as shown.

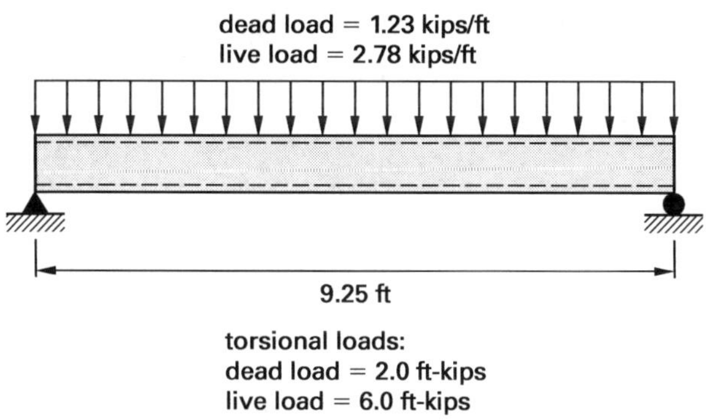

Section properties		Material properties
$t = 0.174$ in	$Z_x = 23.7$ in^3	ASTM A500, grade B steel
$A = 6.06$ in^2	$I_y = 40.0$ in^4	$F_y = 46$ ksi
$b/t = 31.5$	$S_y = 13.3$ in^3	$F_u = 58$ ksi
$h/t = 66.0$	$r_y = 2.57$ in	
$I_x = 116$ in^4	$Z_y = 14.7$ in^3	
$S_x = 19.4$ in^3	$J = 94.6$ in^4	
$r_x = 4.38$ in	$C = 24.0$ in^3	

Determine whether the member is satisfactory for the imparted loads.

Solution

Calculate the required strengths.

LRFD	ASD
$V_r = (1.2w_D + 1.6w_L)\left(\frac{L}{2}\right)$ $= \left((1.2)\left(1.23\ \frac{\text{kips}}{\text{ft}}\right) + (1.6)\left(2.78\ \frac{\text{kips}}{\text{ft}}\right)\right) \times \left(\frac{9.25\text{ ft}}{2}\right)$ $= 27.40\text{ kips}$	$V_r = (w_D + w_L)\left(\frac{L}{2}\right)$ $= \left(1.23\ \frac{\text{kips}}{\text{ft}} + 2.78\ \frac{\text{kips}}{\text{ft}}\right) \times \left(\frac{9.25\text{ ft}}{2}\right)$ $= 18.55\text{ kips}$
$w_u = w_D + w_L$ $= (1.2)\left(1.23\ \frac{\text{kips}}{\text{ft}}\right) + (1.6)\left(2.78\ \frac{\text{kips}}{\text{ft}}\right)$ $= 5.92\text{ kips/ft}$	$w_a = w_D + w_L$ $= 1.23\ \frac{\text{kips}}{\text{ft}} + 2.78\ \frac{\text{kips}}{\text{ft}}$ $= 4.01\text{ kips/ft}$
$M_r = \frac{w_u L^2}{8}$ $= \frac{\left(5.92\ \frac{\text{kips}}{\text{ft}}\right)(9.25\text{ ft})^2}{8}$ $= 63.32\text{ ft-kips}$	$M_r = \frac{w_a L^2}{8}$ $= \frac{\left(4.01\ \frac{\text{kips}}{\text{ft}}\right)(9.25\text{ ft})^2}{8}$ $= 42.89\text{ ft-kips}$
$T_r = 1.2T_D + 1.6T_L$ $= (1.2)(2.0\text{ ft-kips}) + (1.6)(6.0\text{ ft-kips})$ $= 12.0\text{ ft-kips}$	$T_r = T_D + T_L$ $= 2.0\text{ ft-kips} + 6.0\text{ ft-kips}$ $= 8.0\text{ ft-kips}$

Calculate the shear capacity of the section. From Sec. 5.17, the effective web height for shear is taken as the height less three times the wall thickness ($h_e = h - 3t$).

$$\begin{aligned} A_w &= 2h_e t = 2(h - 3t)t \\ &= (2)\big(12\text{ in} - (3)(0.174\text{ in})\big)(0.174\text{ in}) \\ &= 3.99\text{ in}^2 \end{aligned}$$

Use Eq. 5.44.

$$V_n = 0.60F_yA_wC_v$$

For $h/t_w < 260$, the web plate buckling coefficient, k_v, is 5.0 (per *AISC Specification* Sec. G2.1b). Determine the equation to use to calculate C_v.

$$\begin{aligned} 1.10\sqrt{\frac{k_vE}{F_y}} &= 1.10\sqrt{\frac{(5)\left(29{,}000\ \frac{\text{kips}}{\text{in}^2}\right)}{46\ \frac{\text{kips}}{\text{in}^2}}} \\ &= 61.76\quad [< h/t = 66] \end{aligned}$$

Therefore, use Eq. 5.46.

$$\begin{aligned} C_v &= \frac{1.10\sqrt{\frac{k_vE}{F_y}}}{\frac{h}{t}} \\ &= \frac{1.10\sqrt{\frac{(5)\left(29{,}000\ \frac{\text{kips}}{\text{in}^2}\right)}{46\ \frac{\text{kips}}{\text{in}^2}}}}{66} \\ &= 0.94 \end{aligned}$$

Calculate the nominal shear capacity, V_n, using Eq. 5.44.

$$\begin{aligned} V_n &= 0.6F_yA_wC_v \\ &= (0.6)\left(46\ \frac{\text{kips}}{\text{in}^2}\right)(3.99\ \text{in}^2)(0.94) \\ &= 103.52\ \text{kips} \end{aligned}$$

Calculate the design shear strength and the allowable shear strength.

LRFD	ASD
$V_c = \phi_vV_n$ $= (0.90)(103.52\ \text{kips})$ $= 93.17\ \text{kips}$	$V_c = \frac{V_n}{\Omega_v} = \frac{103.52\ \text{kips}}{1.67}$ $= 61.99\ \text{kips}$

Calculate the design and the allowable torsional strengths. (The torsional strengths of the HSS12 × 6 × $^3/_{16}$ member will be the same as calculated in Ex. 8.5, as the members are the same size.)

LRFD	ASD
$T_c = \phi_T T_n$ $= (0.90)(51.44 \text{ ft-kips})$ $= 46.30 \text{ ft-kips}$	$T_c = \frac{T_n}{\Omega_T} = \frac{51.44 \text{ ft-kips}}{1.67}$ $= 30.80 \text{ ft-kips}$

Determine the appropriate formulas for calculating the allowable flexural strength of the tubular section. Check the flange slenderness ratio in accordance with *AISC Specification* Table B4.1, case 12.

$$\lambda_p = 1.12\sqrt{\frac{E}{F_y}} = 1.12\sqrt{\frac{29{,}000 \frac{\text{kips}}{\text{in}^2}}{46 \frac{\text{kips}}{\text{in}^2}}} = 28.12$$

$$\lambda_r = 1.40\sqrt{\frac{E}{F_y}} = 1.40\sqrt{\frac{29{,}000 \frac{\text{kips}}{\text{in}^2}}{46 \frac{\text{kips}}{\text{in}^2}}} = 35.15$$

$$\frac{b}{t} = 31.5 \quad \left[\lambda_p < b/t < \lambda_r, \text{ so noncompact}\right]$$

The section flange is noncompact.

Check the web slenderness ratio in accordance with *AISC Specification* Table B4.1, case 13.

$$\lambda_p = 2.42\sqrt{\frac{E}{F_y}} = 2.42\sqrt{\frac{29{,}000 \frac{\text{kips}}{\text{in}^2}}{46 \frac{\text{kips}}{\text{in}^2}}} = 60.76$$

$$\lambda_r = 5.70\sqrt{\frac{E}{F_y}} = 5.70\sqrt{\frac{29{,}000 \frac{\text{kips}}{\text{in}^2}}{46 \frac{\text{kips}}{\text{in}^2}}} = 143.12$$

$$\frac{h}{t} = 66.0 \quad \left[\lambda_p < h/t < \lambda_r, \text{ so noncompact}\right]$$

The section web is noncompact.

Calculate the plastic moment capacity using Eq. 5.6.

$$M_p = F_y Z = \frac{\left(46\ \frac{\text{kips}}{\text{in}^2}\right)\left(23.7\ \text{in}^3\right)}{12\ \frac{\text{in}}{\text{ft}}} = 90.85\ \text{ft-kips}$$

Calculate the moment capacity based on the limit state of flange local buckling, using Eq. 5.24 because the flanges are noncompact.

$$M_n = M_p - \left(M_p - F_y S_x\right)\left(3.57\left(\frac{b}{t}\right)\sqrt{\frac{F_y}{E}} - 4.0\right) \le M_p$$

$$= 90.85\ \text{ft-kips} - \left(90.85\ \text{ft-kips} - \left(46\ \frac{\text{kips}}{\text{in}^2}\right)\left(\frac{19.4\ \text{in}^3}{12\ \frac{\text{in}}{\text{ft}}}\right)\right)$$

$$\times\left((3.57)(31.5)\sqrt{\frac{46\ \frac{\text{kips}}{\text{in}^2}}{29{,}000\ \frac{\text{kips}}{\text{in}^2}}} - 4.0\right)$$

$$= 82.95\ \text{ft-kips}\quad \left[\le M_p = 90.85\ \text{ft-kips}\right]$$

Calculate the moment capacity based on the limit state of web local buckling, using Eq. 5.27 because the web is noncompact.

$$M_n = M_p - \left(M_p - F_y S_x\right)\left(0.305\left(\frac{h}{t}\right)\sqrt{\frac{F_y}{E}} - 0.738\right) \le M_p$$

$$= 90.85\ \text{ft-kips} - \left(90.85\ \text{ft-kips} - \left(46\ \frac{\text{kips}}{\text{in}^2}\right)\left(\frac{19.4\ \text{in}^3}{12\ \frac{\text{in}}{\text{ft}}}\right)\right)$$

$$\times\left((0.305)(66.0)\sqrt{\frac{46\ \frac{\text{kips}}{\text{in}^2}}{29{,}000\ \frac{\text{kips}}{\text{in}^2}}} - 0.738\right)$$

$$= 89.72\ \text{ft-kips}\quad \left[\le M_p = 90.85\ \text{ft-kips}\right]$$

The smaller value for moment capacity, $M_n = 82.95$ ft-kips, controls.

LRFD	ASD
$M_c = \phi_b M_n = (0.90)(82.95\text{ ft-kips}) = 74.66\text{ ft-kips}$	$M_c = \dfrac{M_n}{\Omega_b} = \dfrac{82.95\text{ ft-kips}}{1.67} = 49.67\text{ kips}$

AISC Manual Table 3-12 lists the design strength as $\phi_b M_n = 74.6$ ft-kips and the allowable strength as $M_n/\Omega_b = 49.6$ ft-kips.

Because T_r is more than 20% of T_c, use Eq. 8.16 to determine whether the member is satisfactory. With no axial loading, $P_r/P_c = 0$. For LRFD,

$$\frac{M_r}{M_c} + \left(\frac{V_r}{V_c} + \frac{T_r}{T_c}\right)^2 = \frac{63.32\text{ ft-kips}}{74.66\text{ ft-kips}} + \left(\frac{27.40\text{ kips}}{93.17\text{ kips}} + \frac{12.0\text{ ft-kips}}{46.30\text{ ft-kips}}\right)^2$$
$$= 1.15 \quad [> 1.00\text{, so no good}]$$

For ASD,

$$\frac{M_r}{M_c} + \left(\frac{V_r}{V_c} + \frac{T_r}{T_c}\right)^2 = \frac{42.89\text{ ft-kips}}{49.67\text{ ft-kips}} + \left(\frac{18.55\text{ kips}}{61.99\text{ kips}} + \frac{8.0\text{ ft-kips}}{30.80\text{ ft-kips}}\right)^2$$
$$= 1.18 \quad [> 1.00\text{, so no good}]$$

Because the member is overstressed approximately 15% to 18%, it is not satisfactory. A greater wall thickness could be used to obtain a satisfactory design while maintaining the same overall member width and depth.

9 Bolted Connections

Nomenclature

A	cross-sectional area	in^2
A_b	nominal unthreaded body area of bolt	in^2
b	width	in
C	coefficient	–
d	distance of fastener from center of gravity of fastener group	in
D	dead load	lbf
e	eccentricity	in
e_x	horizontal component of eccentricity	in
f	stress	lbf/in^2
f_v	required shear stress	lbf/in^2
F	strength or stress	lbf/in^2
F_{nt}	nominal tensile stress from *AISC Specification* Table J3.2	lbf/in^2
F'_{nt}	nominal tensile stress modified to include effects of shearing stress	lbf/in^2
F_{nv}	nominal shear stress from *AISC Specification* Table J3.2	lbf/in^2
F_u	specified minimum tensile strength	lbf/in^2
F_y	specified minimum yield stress	lbf/in^2
I	moment of inertia	in^4
L	live load	lbf
L_e	edge distance (distance from center of hole to edge of material)	in
$L_{e,\text{full}}$	minimum edge distance for full bearing strength	in
M	moment	in-lbf
n	number of bolts	–
P	force or load	lbf
r	radius of gyration	in
r_n	nominal strength per bolt	lbf
R	resultant force	lbf
R	strength	lbf
s	bolt spacing	in
S	elastic section modulus	in^3
t	thickness	in

T	torsional strength	lbf/in^2
U	reduction factor	–
U_{bs}	reduction coefficient for block shear rupture strength	–
x	horizontal component of distance	in
$\overline{x}$	connection eccentricity	in
y	vertical component of distance	in
Z	plastic section modulus	in^3

Symbols

ϕ	resistance factor (LRFD)	–
Ω	safety factor (ASD)	–

Subscripts

a	required (ASD)
e	effective
g	gross
h	holes
min	minimum
n	net or nominal
t	tensile or tension
u	required (LRFD)
v	shear
x	x-axis, strong axis, or horizontal component
y	y-axis, weak axis, or vertical component
z	z-axis

1. GENERAL

AISC Specification Chap. J governs the design and use of bolted and welded connections, joints, and fasteners. Chapter K provides the specifications for connections to HSS members. The following sections of the *AISC Manual* are also used in the analysis and design of connections.

Part 7	Design Considerations for Bolts
Part 8	Design Considerations for Welds
Part 9	Design of Connecting Elements
Part 10	Design of Simple Shear Connections
Part 11	Design of Partially Restrained Moment Connections
Part 12	Design of Fully Restrained (FR) Moment Connections
Part 13	Design of Bracing Connections and Truss Connections

Up to the end of World War II, rivets were commonly used to join structural steel members. After the war, bolted and welded connections replaced rivets. Bolts are preferred in field connections because they make erection easier and faster, are less susceptible to environmental conditions, and present fewer quality control issues.

2. BOLT TYPES AND DESIGNATIONS

The *AISC Manual* divides high-strength structural bolts into two groups on the basis of their material strength. Bolts in Group A have a specified minimum tensile strength, F_u, of 105 ksi and a nominal tensile strength, F_{nt}, of 90 ksi. Bolts in Group A include A325, A325M, F1852, A354 Grade BC, and A449. For bolts in Group B, $F_u = 150$ ksi and $F_{nt} = 90$ ksi. Bolts in Group B include A490, A490M, F2280, and A354 Grade BD. The F1852 and F2280 are twist-off tension-control bolts designed to make it easier to check after installation that the bolt is properly pretensioned.

ASTM A307 bolts, frequently referred to as *unfinished bolts*, are not high-strength bolts. They are used when a higher strength bolt is not required. The allowable tensile strength of an A307 bolt is 20 ksi.

The suffix letters after an ASTM bolt designation are not part of the designation itself. These letters indicate the parameters used in the design of the connection. The suffix N (as in A325-N) indicates that the threads are included in the shear plane. (Figure 9.1 illustrates connections with single and double shear planes.) When the threads of a bolt are included in a shear plane, the bolt load capacity is reduced, because the thread valley reduces the net area of the bolt. The suffix X (as in A490-X) indicates that bolt threads are excluded from the shear plane. Both the N and X suffixes indicate that the connection is a bearing type connection. A conservative engineer or designer always assumes that the threads will be included in the shear plane. The suffix SC (A325-SC, A490-SC) indicates a *slip-critical* connection. Other suffixes that are occasionally used are ST for snug-tight and PT for pretensioned, both of which are for bearing connections.

Figure 9.1 *Diagram of Single and Double Shear Planes*

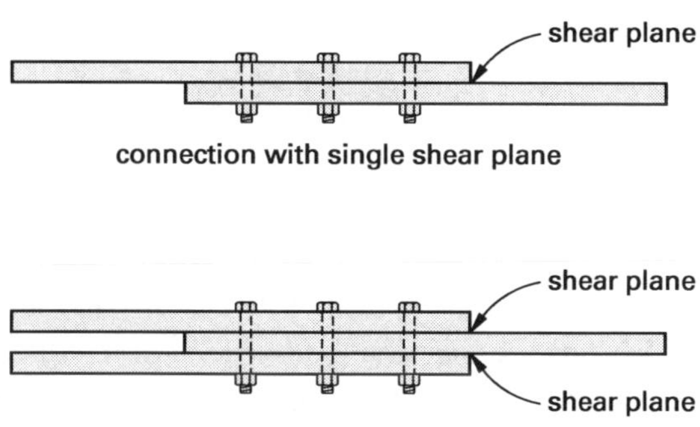

The preferred normal bolt sizes used in structural connections have diameters of $^3/_4$ in, $^7/_8$ in, 1 in, and $1^1/_8$ in. Using the same bolt type and diameter throughout a project simplifies the inventory and quality control procedures. Bolt lengths are determined by the thickness of the plies being joined and whether washers or tension indicators are required. Bolt lengths vary by $^1/_4$ in increments up to a 5 in length and by $^1/_2$ in increments above 5 in.

3. BEARING CONNECTIONS

Bearing connections are the most common type of bolted connections. A bearing connection is generally used wherever a slip-critical or moment connection is not required. Simple shear connections used to connect beam to beam, beam to girder, beam to column, or girder to column are generally bearing connections.

Bearing connections are the easiest to analyze and design. The *AISC Manual* contains numerous tables illustrating standard shear connections and their load capacities. In designing or analyzing bolted connections, the following items must be checked.

- *available shear strength of bolts:* single or double shear planes (*AISC Manual* Table 7-1)
- *available tensile strength of bolts* (*AISC Manual* Table 7-2)
- *slip-critical connections:* available shear strength, when slip is a serviceability limit state (*AISC Manual* Table 7-3)
- *available bearing strength at bolt holes:* bearing for supporting and supported elements, based on bolt spacing (*AISC Manual* Table 7-4)
- *available bearing strength at bolt holes:* bearing for supporting and supported elements, based on edge distance (*AISC Manual* Table 7-5)

4. SLIP-CRITICAL CONNECTIONS

Fully tensioned high-strength bolts (A325, F1852, and A490 bolts) are used to make slip-critical connections. Section 16.2 of the *AISC Manual, Specification for Structural Joints Using High-Strength Bolts,* requires slip-critical connections when bolts and welds are used in the same element of a connection and the load is to be distributed among the bolts and welds. Slip-critical connections are also required for the supports of running machinery and other live loads that produce impact loads or reversal of stresses, as well as for all members carrying cranes with a capacity of at least 5 tons. Section 4.2 of the *Specification for Structural Joints* specifies and provides guidance concerning where slip-critical joints should be used. Generally, these are in tiered structures that are at least 100 ft high.

The engineering design values for bolts in a slip-critical connection are less than the values for bolts in a bearing connection with the bolt threads either included or excluded from the shear plane. As a result, there will be more bolts in a slip-critical connection than in a bearing connection with the same design load. If a slip-critical connection fails, it reverts to being a bearing connection and has a higher overall load capacity. Finger shims with a total thickness less than or equal to $^1/_4$ in may be inserted into a slip-critical connection without any detrimental effect.

5. BOLT HOLES

The diameter of a standard bolt hole is $^1/_{16}$ in larger than the diameter of the bolt. Oversize, short-slotted, and long-slotted holes are used to assist in the fit-up of steel or to permit field adjustment of shelf angles and other similar secondary elements. Refer to *AISC Specification* Table J3.3 for specific details regarding standard, oversized, short-slotted, and long-slotted holes. Short- and long-slotted holes should only be used under the conditions specified in *AISC Specification* Sec. J3.2.

The distance between centers of standard and oversized holes should be no less than $2^2/_3$ times the nominal diameter; a minimum of three times the nominal diameter is preferred. Normal practice is to use 3 in centers for bolts with diameters up to 1 in.

The codes also specify minimum and maximum *edge distances* (distances from the center of the hole to the edge of the material). The minimum edge distance is a function of nominal bolt diameter and whether the material edge is sheared or rolled. Minimum edge distances are given in *AISC Specification* Table J3.4 and Table J3.5.

The maximum distance to the nearest edge of parts in contact is 12 times the thickness of the connected part but not more than 6 in. The maximum longitudinal spacing between connectors is as follows.

- for painted or unpainted members not subject to corrosion, 24 times the thickness of the thinner element but not more than 12 in
- for unpainted members of weathering steel subject to atmospheric corrosion, 14 times the thickness of the thinner element but not more than 7 in

6. BLOCK SHEAR RUPTURE

One limit state for bolted connections is *block shear rupture*, in which shear rupture occurs along a path of connection holes and a tension failure path occurs perpendicular to the shear rupture path. Figure 9.2 shows examples of block shear rupture and tension failure.

Figure 9.2 *Examples of Block Shear Rupture and Tension Failure*

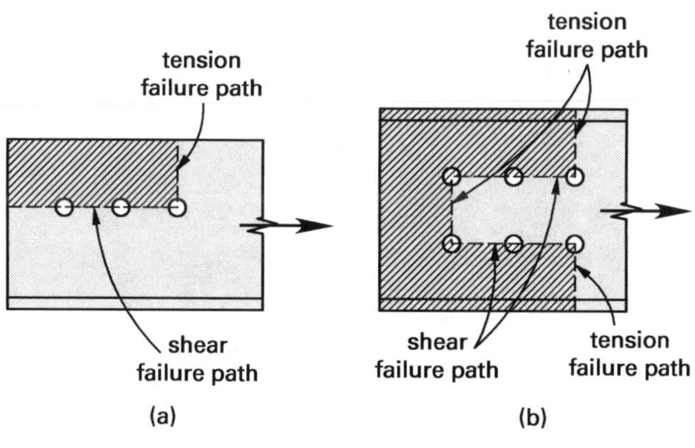

To calculate block shear rupture, use Eq. 9.1 ($\phi = 0.75$ for LRFD, $\Omega = 2.00$ for ASD).

$$R_n = 0.60F_uA_{nv} + U_{bs}F_uA_{nt} \le 0.60F_yA_{gv} + U_{bs}F_uA_{nt} \quad \text{[AISC Eq. J4-5]} \qquad 9.1$$

When tension stress is uniform, $U_{bs} = 1.0$, and when tension stress is nonuniform, $U_{bs} = 0.5$. (See *AISC Commentary* Fig. C-J4.2 for examples.)

7. LAP SPLICE CONNECTIONS

Lap splice connections are the easiest connections to design, whether the connection is in tension or compression. The design principles are identical for a two-member lap splice (single shear plane) and a three-member lap splice (double shear plane). The same principles are also used in designing other types of connections.

Example 9.1

Lap Splice Connection of Two Plates

Two steel plates are connected as shown by a lap splice with six $^3/_4$ in diameter Group A bolts with threads included in the shear plane.

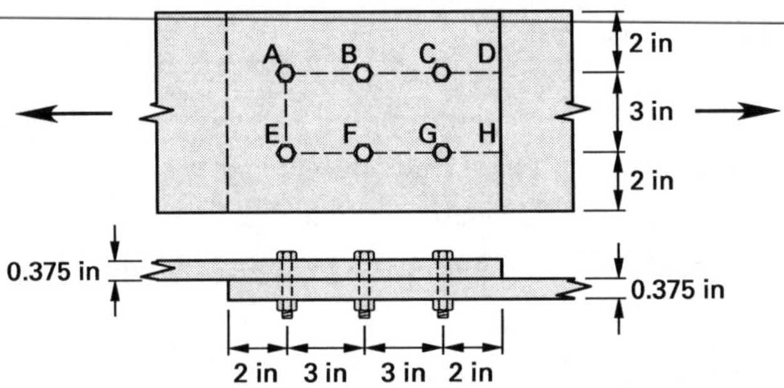

(not to scale)

Section properties	Material properties
plate width = 7 in	ASTM A36 steel plates
plate thickness = $^3/_8$ in	$F_y = 36$ ksi
standard hole size	$F_u = 58$ ksi

Determine the design and allowable strengths of the assembly. Determine the governing limit state for the connection.

Solution

Calculate the gross cross-sectional area of each plate.

$$A_g = bt = (7 \text{ in})(0.375 \text{ in}) = 2.63 \text{ in}^2$$

Use Eq. 4.7 to calculate the area of the holes.

$$A_h = n_{\text{holes}}t(d_{\text{bolt}} + 0.125 \text{ in}) = (2)(0.375 \text{ in})(0.75 \text{ in} + 0.125 \text{ in}) = 0.66 \text{ in}^2$$

Use Eq. 4.6 to calculate the net cross-sectional area of each plate.

$$\begin{aligned} A_n &= A_g - A_h \\ &= 2.63 \text{ in}^2 - 0.66 \text{ in}^2 \\ &= 1.97 \text{ in}^2 \end{aligned}$$

Because the two plates are in full contact with each other, the reduction factor, U, is 1.0, and the effective area and the net area are identical. Calculate the nominal strength based on the gross section yielding.

$$T_n = A_g F_y = (2.63 \text{ in}^2)\left(36 \ \frac{\text{kips}}{\text{in}^2}\right) = 94.68 \text{ kips}$$

Calculate the design strength (LRFD) and the allowable strength (ASD) based on the gross section yielding.

LRFD	ASD
$T_u = \phi_t T_n = (0.90)(94.68 \text{ kips})$ $= 85.21 \text{ kips}$	$T_a = \dfrac{T_n}{\Omega_t} = \dfrac{94.68 \text{ kips}}{1.67}$ $= 56.69 \text{ kips}$

Calculate the nominal strength based on the net section rupture.

$$T_n = A_e F_u = (1.97 \text{ in}^2)\left(58 \ \frac{\text{kips}}{\text{in}^2}\right) = 114.26 \text{ kips}$$

Because all elements in each member are in contact, $A_e = A_n$. Calculate the design strength (LRFD) and the allowable strength (ASD) based on the net section rupture.

LRFD	ASD
$T_u = \phi_t T_n = (0.75)(114.26 \text{ kips})$ $= 85.70 \text{ kips}$	$T_a = \dfrac{T_n}{\Omega_t} = \dfrac{114.26 \text{ kips}}{2}$ $= 57.13 \text{ kips}$

Calculate the nominal shear capacity of the six bolts. Obtain the available shear strength per bolt from *AISC Manual* Table 7-1 (the bolts are in single shear).

LRFD	ASD
$\phi_v r_n = 17.9 \text{ kips/bolt}$	$\dfrac{r_n}{\Omega_v} = 11.9 \text{ kips/bolt}$
$P_u = n(\phi_v r_n) = (6 \text{ bolts})\left(17.9 \ \dfrac{\text{kips}}{\text{bolt}}\right)$ $= 107.40 \text{ kips}$	$P_a = n\left(\dfrac{r_n}{\Omega_v}\right) = (6 \text{ bolts})\left(11.9 \ \dfrac{\text{kips}}{\text{bolt}}\right)$ $= 71.40 \text{ kips}$

Use *AISC Manual* Table 7-5 to find the nominal capacity based on the bolts bearing on the steel plates. Use F_u = 58 ksi and $L_e \geq L_{e,\text{full}}$ (2.0 in ≥ $1^{15}/_{16}$ in). *AISC Manual* Table 7-5 gives the capacity in units of kips per inch of thickness, so multiply this by the plate thickness to get the capacity per bolt. Then, multiply by the number of bolts to get the total capacity.

LRFD	ASD
$\phi_v r_n = \left(78.3 \ \frac{\text{kips}}{\text{in thickness}}\right) \times (0.375 \text{ in}) = 29.36 \text{ kips} \quad [\text{per bolt in bearing}]$	$\frac{r_n}{\Omega_t} = \left(52.2 \ \frac{\text{kips}}{\text{in thickness}}\right) \times (0.375 \text{ in}) = 19.58 \text{ kips} \quad [\text{per bolt in bearing}]$
$P_u = n(\phi_v r_n) = (6 \text{ bolts})\left(29.36 \ \frac{\text{kips}}{\text{bolt}}\right) = 176.16 \text{ kips}$	$P_a = n\left(\frac{r_n}{\Omega_v}\right) = (6 \text{ bolts})\left(19.58 \ \frac{\text{kips}}{\text{bolt}}\right) = 117.45 \text{ kips}$

Calculate the nominal resistance to block shear rupture. As shown in the problem illustration, shear rupture may occur along lines A-B-C-D and E-F-G-H as tension failure occurs along line A-E. Use Eq. 9.1 to calculate the nominal resistance to block shear rupture, using a U_{bs} of 1.0. (See also *AISC Specification* Fig. C-J4.2.)

$$R_n = 0.60F_u A_{nv} + U_{bs} F_u A_{nt} \leq 0.60F_y A_{gv} + U_{bs} F_u A_{nt}$$

Calculate the gross and net areas required for the block shear equation. The gross shear area is the total length of the shear rupture paths multiplied by the plate thickness.

$$A_{gv} = (2)(3 \text{ in} + 3 \text{ in} + 2 \text{ in})(0.375 \text{ in}) = 6.00 \text{ in}^2$$

The net shear area is the gross shear area minus the area of the holes in the path. The shear rupture paths go through all of holes B, C, F, and G, but only half of holes A and E, so five holes are counted.

$$\begin{aligned} A_{hv} &= n_{\text{holes}} t (d_{\text{bolt}} + 0.125 \text{ in}) \\ &= (5)(0.375 \text{ in})(0.75 \text{ in} + 0.125 \text{ in}) \\ &= 1.64 \text{ in}^2 \\ A_{nv} &= A_{gv} - A_{hv} = 6.00 \text{ in}^2 - 1.64 \text{ in}^2 \\ &= 4.36 \text{ in}^2 \end{aligned}$$

The tension failure path is from the center of hole A to the center of hole E, so it is 3 in long. The gross tension area is this length multiplied by the plate thickness.

$$A_{gt} = (3 \text{ in})(0.375 \text{ in}) = 1.125 \text{ in}^2$$

The net tension area is the gross tension area minus the area of the holes in the path. The path includes half of hole A and half of hole E, so one hole is counted.

$$\begin{aligned} A_{ht} &= n_{\text{holes}} t \left(d_{\text{bolt}} + 0.125 \text{ in} \right) \\ &= (1)(0.375 \text{ in})(0.75 \text{ in} + 0.125 \text{ in}) \\ &= 0.33 \text{ in}^2 \\ A_{nt} &= A_{gt} - A_{ht} \\ &= 1.125 \text{ in}^2 - 0.33 \text{ in}^2 \\ &= 0.80 \text{ in}^2 \end{aligned}$$

From Eq. 9.1,

$$R_n \le \begin{cases} 0.60 F_u A_{nv} + U_{\text{bs}} F_u A_{nt} \\ = (0.60)\left(58 \ \dfrac{\text{kips}}{\text{in}^2}\right)(4.36 \text{ in}^2) + (1)\left(58 \ \dfrac{\text{kips}}{\text{in}^2}\right)(0.80 \text{ in}^2) \\ = 198.13 \text{ kips} \\ 0.60 F_y A_{gv} + U_{\text{bs}} F_u A_{nt} \\ = (0.60)\left(36 \ \dfrac{\text{kips}}{\text{in}^2}\right)(6.00 \text{ in}^2) + (1)\left(58 \ \dfrac{\text{kips}}{\text{in}^2}\right)(0.80 \text{ in}^2) \\ = 176.00 \text{ kips} \quad [\text{controls}] \end{cases}$$

Calculate the design block shear and the allowable block shear strengths.

LRFD	ASD
$R_u = \phi_t R_n = (0.75)(176.00 \text{ kips})$ $= 132.00 \text{ kips}$	$R_a = \dfrac{R_n}{\Omega_t} = \dfrac{176.00 \text{ kips}}{2}$ $= 88.00 \text{ kips}$

Examine the calculated design strengths.

failure mode	LRFD (kips)	ASD (kips)
gross section yielding	85.21	56.69
net section fracture	85.70	57.13
single shear on bolts	107.40	71.40
bolt bearing on plates	176.16	117.45
block shear strength	132.00	88.00

For both LRFD and ASD, gross section yielding is the lowest value. Therefore, gross section yielding is the limiting value for the available strength of the assembly.

Example 9.2

Lap Splice Connection of Plate and Angle

A steel angle is fastened to a steel gusset plate as shown, with $^3/_4$ in diameter Group A bolts with threads excluded from the shear plane inserted in standard-size holes.

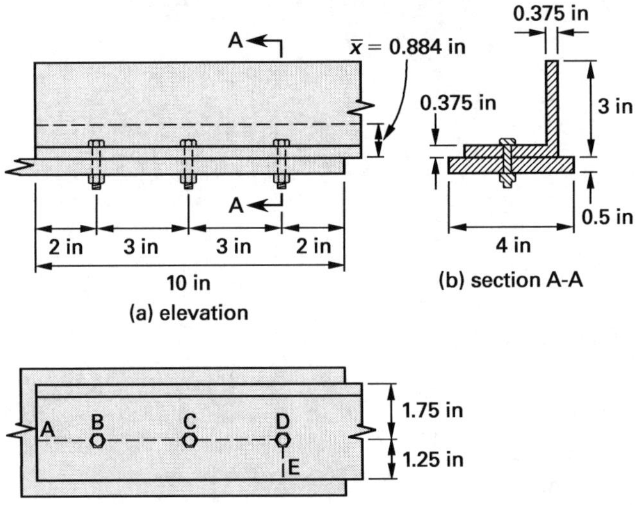

(a) elevation

(b) section A-A

(c) top elevation

(not to scale)

Section properties

L3 × 3 × $^3/_8$ in

$A = 2.11\ \text{in}^2$

$I_x = I_y = 1.75\ \text{in}^4$

$S_x = S_y = 0.825\ \text{in}^3$

$r_x = r_y = 0.910\ \text{in}$

$Z_x = Z_y = 1.48\ \text{in}^3$

$r_z = 0.581\ \text{in}$

$\bar{x} = 0.884\ \text{in}$

plate width = 4 in

plate thickness = $^1/_2$ in

Material properties

ASTM A36 steel angle and plate

$F_y = 36$ ksi

$F_u = 58$ ksi

Determine the design strength (LRFD) and allowable strength (ASD) of the assembly.

Solution

The following possible modes of failure must be evaluated to determine the least value tensile load at which the assembly will fail.

- gross section yielding on plate and angle
- net section fracture on plate and angle
- single shear on bolts
- bolt bearing on plate
- bolt bearing on angle
- shear rupture on plate
- block shear rupture of angle on line A-B-C-D-E

Calculate the nominal strength of the plate based on the gross section yielding.

$$\begin{aligned} A_{g,\text{plate}} &= bt = (4\ \text{in})(0.5\ \text{in}) \\ &= 2.00\ \text{in}^2 \end{aligned}$$

$$\begin{aligned} T_n = F_y A_{g,\text{plate}} &= \left(36\ \frac{\text{kips}}{\text{in}^2}\right)(2.00\ \text{in}^2) \\ &= 72.00\ \text{kips} \end{aligned}$$

Calculate the design strength (LRFD) and the allowable strength (ASD) of the plate based on the gross section yielding.

LRFD	ASD
$T_u = \phi_t T_n = (0.90)(72.00\ \text{kips})$ $= 64.80\ \text{kips}$	$T_a = \frac{T_n}{\Omega_t} = \frac{72.00\ \text{kips}}{1.67}$ $= 43.11\ \text{kips}$

Calculate the nominal strength of the plate based on the net section rupture. Because the plate is in full contact with the angle, $A_{n,\text{plate}} = A_{e,\text{plate}}$.

$$\begin{aligned} A_{n,\text{plate}} = A_{e,\text{plate}} &= A_{g,\text{plate}} - A_h \\ &= 2.00\ \text{in}^2 - (0.875\ \text{in})(0.50\ \text{in}) \\ &= 1.56\ \text{in}^2 \end{aligned}$$

The nominal strength is

$$\begin{aligned} T_n = A_{e,\text{plate}} F_u &= (1.56\ \text{in}^2)\left(58\ \frac{\text{kips}}{\text{in}^2}\right) \\ &= 90.48\ \text{kips} \end{aligned}$$

Calculate the design strength (LRFD) and the allowable strength (ASD) of the plate based on the net section rupture.

LRFD	ASD
$T_u = \phi_t T_n = (0.75)(90.48\ \text{kips})$ $= 67.86\ \text{kips}$	$T_a = \frac{T_n}{\Omega_t} = \frac{90.48\ \text{kips}}{2.00}$ $= 45.24\ \text{kips}$

Calculate the nominal strength of the angle based on the gross section yielding.

$$\begin{aligned} T_n = F_y A_{g,\text{angle}} &= \left(36\ \frac{\text{kips}}{\text{in}^2}\right)(2.11\ \text{in}^2) \\ &= 75.96\ \text{kips} \end{aligned}$$

Calculate the design strength (LRFD) and the allowable strength (ASD) of the angle based on the gross section yielding.

LRFD	ASD
$T_u = \phi_t T_n = (0.90)(75.96\ \text{kips})$ $= 68.36\ \text{kips}$	$T_a = \dfrac{T_n}{\Omega_t} = \dfrac{75.96\ \text{kips}}{1.67}$ $= 45.49\ \text{kips}$

Calculate the effective net area. Not all the elements of the angle are in contact with the steel plate; therefore shear lag occurs and the net area, A_n, must be multiplied by the appropriate reduction factor, U. *AISC Specification* Table D3.1, case 8, gives $U = 0.60$.

From case 2, where L is the length of the connection measured between the centers of the first and last holes,

$$\begin{aligned} U &= 1 - \frac{\overline{x}}{L} \\ &= 1 - \frac{0.884\ \text{in}}{6\ \text{in}} \\ &= 0.85 \end{aligned}$$

It is permissible to use the larger of the two U-values, so $U = 0.85$ controls.

$$\begin{aligned} A_{n,\text{angle}} &= A_{g,\text{angle}} - A_h \\ &= 2.11\ \text{in}^2 - (0.875\ \text{in})(0.375\ \text{in}) \\ &= 1.78\ \text{in}^2 \end{aligned}$$

$$A_{e,\text{angle}} = UA_{n,\text{angle}} = (0.85)(1.78\ \text{in}^2) = 1.51\ \text{in}^2$$

Calculate the nominal strength of the angle based on the net section rupture.

$$\begin{aligned} T_n = A_{e,\text{angle}} F_u &= (1.51\ \text{in}^2)\left(58\ \frac{\text{kips}}{\text{in}^2}\right) \\ &= 87.58\ \text{kips} \end{aligned}$$

Calculate the design strength (LRFD) and the allowable strength (ASD) based on the net section rupture.

LRFD	ASD
$T_u = \phi_t T_n = (0.75)(87.58\ \text{kips})$ $= 65.69\ \text{kips}$	$T_a = \dfrac{T_n}{\Omega_t} = \dfrac{87.58\ \text{kips}}{2.00}$ $= 43.79\ \text{kips}$

Calculate the design strength (LRFD) and the allowable strength (ASD) of the bolts in single shear. (Refer to *AISC Manual* Table 7-1.)

LRFD	ASD
$\phi_v r_n = 22.5 \text{ kips/bolt}$	$\dfrac{r_n}{\Omega_v} = 15.0 \text{ kips/bolt}$
$\phi_v R_n = n(\phi_v r_n)$ $= (3 \text{ bolts})\left(22.5 \dfrac{\text{kips}}{\text{bolt}}\right)$ $= 67.50 \text{ kips}$	$\dfrac{R_n}{\Omega_v} = n\left(\dfrac{r_n}{\Omega_v}\right)$ $= (3 \text{ bolts})\left(15.0 \dfrac{\text{kips}}{\text{bolt}}\right)$ $= 45.00 \text{ kips}$

Calculate the design strength (LRFD) and the allowable strength (ASD) of the bolts bearing on the plate and on the angle. (Refer to *AISC Manual* Table 7-5. The bearing capacity is in kips per inch of thickness.) Because the angle is thinner than the plate, it is the governing criterion. It is therefore not necessary to calculate the bolt bearing capacity on the plate, which is $^1/_8$ in thicker than the angle.

LRFD	ASD
$\phi_v r_n = \left(78.3 \dfrac{\text{kips}}{\text{in thickness}}\right)(0.375 \text{ in})$ $= 29.36 \text{ kips}$ [per bolt in bearing]	$\dfrac{r_n}{\Omega_t} = \left(52.2 \dfrac{\text{kips}}{\text{in thickness}}\right)(0.375 \text{ in})$ $= 19.58 \text{ kips}$ [per bolt in bearing]
$\phi_v R_n = n(\phi_v r_n)$ $= (3 \text{ bolts})\left(29.36 \dfrac{\text{kips}}{\text{bolt}}\right)$ $= 88.08 \text{ kips}$	$\dfrac{R_n}{\Omega_v} = n\left(\dfrac{r_n}{\Omega_v}\right)$ $= (3 \text{ bolts})\left(19.58 \dfrac{\text{kips}}{\text{bolt}}\right)$ $= 58.74 \text{ kips}$

From Eq. 9.1, the nominal block shear resistance of the angle along the line A-B-C-D is

$$R_n = 0.60F_u A_{nv} + U_{\text{bs}} F_u A_{nt} \le 0.60F_y A_{gv} + U_{\text{bs}} F_u A_{nt}$$

Calculate the gross and net areas required for the block shear equation. The gross shear area is the total length of the shear rupture paths multiplied by the plate thickness.

$$\begin{aligned} A_{gv} &= (2 \text{ in} + 3 \text{ in} + 3 \text{ in})(0.375 \text{ in}) \\ &= 3.00 \text{ in}^2 \end{aligned}$$

The net shear area is the gross shear area minus the area of the holes in the path. The shear rupture path goes through holes B and C but only half of hole D, so 2.5 holes are counted.

$$\begin{aligned} A_{hv} &= n_{\text{holes}} t \left(d_{\text{bolt}} + 0.125 \text{ in}\right) \\ &= (2.5)(0.375 \text{ in})(0.75 \text{ in} + 0.125 \text{ in}) \\ &= 0.82 \text{ in}^2 \\ A_{nv} &= A_{gv} - A_{hv} = 3.00 \text{ in}^2 - 0.82 \text{ in}^2 \\ &= 2.18 \text{ in}^2 \end{aligned}$$

The tension failure path is from the center of hole D to the point E, so it is 1.25 in long. The gross tension area is this length multiplied by the plate thickness.

$$A_{gt} = (1.25 \text{ in})(0.375 \text{ in}) = 0.469 \text{ in}^2$$

The net tension area is the gross tension area minus the area of the holes in the path. The path includes half of hole D.

$$\begin{aligned} A_{ht} &= n_{\text{holes}} t \left(d_{\text{bolt}} + 0.125 \text{ in}\right) \\ &= (0.5)(0.375 \text{ in})(0.75 \text{ in} + 0.125 \text{ in}) \\ &= 0.16 \text{ in}^2 \\ A_{nt} &= A_{gt} - A_{ht} \\ &= 0.469 \text{ in}^2 - 0.16 \text{ in}^2 \\ &= 0.31 \text{ in}^2 \end{aligned}$$

Using Eq. 9.1, calculate the nominal block shear resistance of the angle along the line A-B-C-D.

$$R_n \le \begin{cases} 0.60F_uA_{nv} + U_{\text{bs}}F_uA_{nt} \\ \quad = (0.60)\left(58 \ \dfrac{\text{kips}}{\text{in}^2}\right)(2.18 \text{ in}^2) + (1)\left(58 \ \dfrac{\text{kips}}{\text{in}^2}\right)(0.31 \text{ in}^2) \\ \quad = 93.84 \text{ kips} \\ 0.60F_yA_{gv} + U_{\text{bs}}F_uA_{nt} \\ \quad = (0.60)\left(36 \ \dfrac{\text{kips}}{\text{in}^2}\right)(3.00 \text{ in}^2) + (1)\left(58 \ \dfrac{\text{kips}}{\text{in}^2}\right)(0.31 \text{ in}^2) \\ \quad = 82.78 \text{ kips} \quad [\text{controls}] \end{cases}$$

Calculate the design strength (LRFD) and the allowable strength (ASD) resistance to the block shear rupture.

LRFD	ASD
$\phi R_n = (0.75)(82.78 \text{ kips})$ $= 62.09 \text{ kips}$	$\frac{R_n}{\Omega} = \frac{82.78 \text{ kips}}{2.00}$ $= 41.39 \text{ kips}$

Examine the calculated design strengths.

failure mode	LRFD (kips)	ASD (kips)
gross section yielding, plate	64.80	43.11
net section rupture, plate	67.86	45.24
gross section yielding, angle	68.36	45.49
net section rupture, angle	65.69	43.79
bolts in single shear	67.50	45.00
bolts in bearing, angle controls	88.08	58.74
block shear rupture	62.09	41.39

For both LRFD and ASD, block shear rupture gives the lowest value and therefore governs the design.

8. BRACKET CONNECTION WITH ECCENTRIC SHEAR

Steel plates of varying sizes and thicknesses are often bolted to the face of a column flange to support a load beyond the toe of the column flange. When a connection is loaded in this manner, the bolts are subjected to shear forces resulting from the axial load as well as shear forces resulting from the rotational moment caused by the eccentricity of the load. The maximum shear force on a connector is the resultant of the shear forces on the x- and y-axes.

There are two common methods used to analyze an eccentric load placed on a group of fasteners. The instantaneous center of rotation method is more accurate but more difficult. The elastic method is simpler, but its results can be excessively conservative.

The Instantaneous Center of Rotation Method

The *instantaneous center of rotation method* makes use of the fact that the combined shear forces from the axial load and the rotational moment are equivalent to the force that would be produced by rotation alone about some point, which must be found. This point is called the *instantaneous center of rotation,* and its location depends on where and in which direction the axial load is applied and on how the bolts are arranged.

In this method, the resistance force of each fastener is assumed to act in a direction that is perpendicular to a line from the center of the fastener to the instantaneous center. The coefficients in Table 7-7 through Table 7-14 in the *AISC Manual* are for use with the instantaneous center of rotation method. Without these tabulated coefficients, this method of analysis is an iterative process best performed by a computer.

The Elastic Method

The *elastic method* of analysis, sometimes called the *vector analysis method*, is easier to perform and produces conservative results. However, it does not produce a consistent safety factor, and results may be excessively conservative.

In this method, it is assumed that each fastener supports

- an equal share of the vertical component of the load
- an equal share of the horizontal component (if any) of the load
- a proportional share (depending on the fastener's distance from the centroid of the group) of the eccentric moment portion of the load

Example 9.3

Bracket Connection with Eccentric Load

The plate bracket shown supports a dead load of 5 kips and a live load of 15 kips. The bracket is secured to a flange of a wide-flange column as shown, with six $^3/_4$ in diameter Group A bolts with threads in the shear plane. Assume that the bracket plate and column flange are satisfactory.

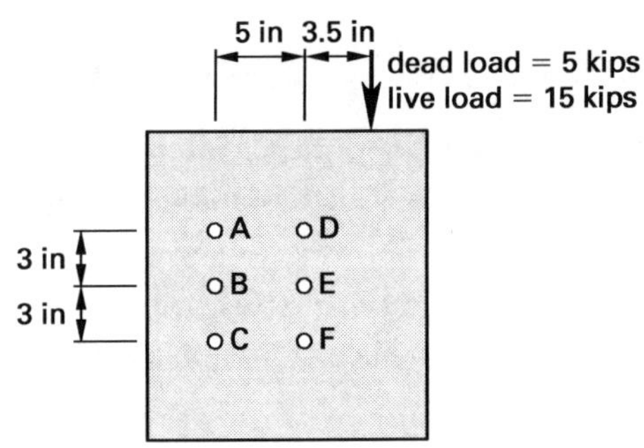

Determine whether the connection is satisfactory to support the given loads.

Solution

The required force is

LRFD	ASD
$P_u = 1.2D + 1.6L$ $= (1.2)(5 \text{ kips}) + (1.6)(15 \text{ kips})$ $= 30 \text{ kips}$	$P_a = D + L$ $= 5 \text{ kips} + 15 \text{ kips}$ $= 20 \text{ kips}$

Calculate the eccentricity from the centroid of the bolt group to the load center.

$$e_x = e = \frac{5 \text{ in}}{2} + 3.5 \text{ in} = 6 \text{ in}$$

From *AISC Manual* Table 7-8, determine the coefficient, C, for the bolt group. The angle is 0°, the bolt spacing is $s = 3$ in, the horizontal component of eccentricity is $e_x = 6$ in, and the number of bolts per vertical row is $n = 3$. From the table, $C = 2.62$. Determine the available shear strength of the bolts using *AISC Manual* Table 7-1.

$$\phi_v r_n = 17.9 \text{ kips}$$

$$\frac{r_n}{\Omega} = 11.9 \text{ kips}$$

Determine the minimum required coefficient, $C_{\min}$, using the formulas from *AISC Manual* Table 7-8.

LRFD	ASD
$C_{\min} = \frac{P_u}{\phi_v r_n}$ $= \frac{30 \text{ kips}}{17.9 \text{ kips}}$ $= 1.67 \quad [\le 2.62, \text{so OK}]$	$C_{\min} = \frac{\Omega P_a}{r_n} = \frac{P_a}{\frac{r_n}{\Omega}}$ $= \frac{20 \text{ kips}}{11.9 \text{ kips}}$ $= 1.68 \quad [\le 2.62, \text{so OK}]$

Use the elastic analysis method (vector analysis) to determine whether $^3/_4$ in diameter Group A bolts are satisfactory for the bracket connection. Analyze the bracket using the allowable stress design (ASD) and load and resistance factor design (LRFD) methods. The eccentricity of the load is

$$e = \frac{5 \text{ in}}{2} + 3.5 \text{ in} = 6.0 \text{ in}$$

Calculate the moment created by the eccentricity.

LRFD	ASD
$M = P_u e$ $= (30 \text{ kips})(6 \text{ in})$ $= 180 \text{ in-kips}$	$M = P_a e$ $= (20 \text{ kips})(6 \text{ in})$ $= 120 \text{ in-kips}$

Calculate the sum of the squares of the distances of the bolts in the group from the center of gravity of the bolt group (this is similar to the polar moment of inertia). x is the horizontal component of the distance to the center of gravity, which is 2.5 in for each bolt. y is the vertical component, which is zero for two of the bolts and 3 in for the others.

Because $d^2 = x^2 + y^2$,

$$\begin{aligned}\sum d^2 &= \sum x^2 + \sum y^2 \\ &= (6)(2.5 \text{ in})^2 + (4)(3 \text{ in})^2 + (2)(0 \text{ in})^2 \\ &= 73.5 \text{ in}^2\end{aligned}$$

Calculate the vertical shear (downward force) on each bolt based on the axial load only.

LRFD	ASD
$R_v = \dfrac{P_u}{n} = \dfrac{30 \text{ kips}}{6 \text{ bolts}} = 5 \text{ kips/bolt}$	$R_v = \dfrac{P_a}{n} = \dfrac{20 \text{ kips}}{6 \text{ bolts}} = 3.33 \text{ kips/bolt}$

Calculate the horizontal shear component on each bolt due to the moment. The force will be to the right for fasteners A and D and to the left for fasteners C and F. Because fasteners B and E are at the neutral axis, there will be no force on them due to this component.

LRFD	ASD
$R_x = \dfrac{My}{\sum d^2} = \dfrac{(180 \text{ in-kips})(3 \text{ in})}{73.5 \text{ in}^2} = 7.35 \text{ kips}$	$R_x = \dfrac{My}{\sum d^2} = \dfrac{(120 \text{ in-kips})(3 \text{ in})}{73.5 \text{ in}^2} = 4.90 \text{ kips}$

Calculate the vertical shear component on each bolt due to the moment. The force will be upward on fasteners A, B, and C and downward on fasteners D, E, and F.

LRFD	ASD
$R_y = \dfrac{Mx}{\sum d^2} = \dfrac{(180 \text{ in-kips})(2.5 \text{ in})}{73.5 \text{ in}^2} = 6.12 \text{ kips}$	$R_y = \dfrac{Mx}{\sum d^2} = \dfrac{(120 \text{ in-kips})(2.5 \text{ in})}{73.5 \text{ in}^2} = 4.08 \text{ kips}$

The resultant shear force on each bolt can be calculated with

$$R = \sqrt{R_x^2 + (R_y + R_v)^2}$$

Calculating the component vector forces shows that the maximum shear force occurs in bolt D. Calculate the required resistance force at that location.

LRFD	ASD
$R_4 = \sqrt{R_x^2 + (R_y + R_v)^2}$ $= \sqrt{(7.35 \text{ kips})^2 + (6.12 \text{ kips} + 5 \text{ kips})^2}$ $= 13.33 \text{ kips}$	$R_4 = \sqrt{R_x^2 + (R_y + R_v)^2}$ $= \sqrt{(4.90 \text{ kips})^2 + (4.08 \text{ kips} + 3.33 \text{ kips})^2}$ $= 8.88 \text{ kips}$

From *AISC Manual* Table 7-1, an ASTM A325-N bolt in single shear has the following resistance capacities. For LRFD,

$$\phi_v r_n = 17.9 \text{ kips} \quad [> 13.33 \text{ kips, so OK}]$$

For ASD,

$$r_n/\Omega_v = 11.9\text{kips} \quad [> 8.88 \text{ kips, so OK}]$$

The connection is satisfactory to support the loads.

9. COMBINED SHEAR AND TENSION IN BEARING TYPE CONNECTIONS

Combined shear and tension in bearing type connections is covered in Sec. J3.7 of the *AISC Specification.* Section J3.9 covers combined shear and tension for slip-critical connections and contains different equations from those in Sec. J3.7.

When the required stress in either shear or tension is less than or equal to 20% of the corresponding available stress, the effects of the combined stress need not be investigated.

When it is necessary to investigate the effects of combined shear and tensile forces, use Eq. 9.2.

$$R_n = F'_{nt} A_b \quad [\text{AISC Eq. J3-2}] \qquad 9.2$$

For LRFD, use Eq. 9.3 with $\phi = 0.75$.

$$F'_{nt} = 1.3F_{nt} - \left(\frac{F_{nt}}{\phi F_{nv}}\right) f_v \le F_{nt} \quad [\text{AISC Eq. J3-3a}] \qquad 9.3$$

For ASD, use Eq. 9.4 with $\Omega = 2.00$.

$$F'_{nt} = 1.3F_{nt} - \left(\frac{\Omega F_{nt}}{F_{nv}}\right) f_v \le F_{nt} \quad [\text{AISC Eq. J3-3b}] \qquad 9.4$$

10. BRACKET CONNECTION WITH SHEAR AND TENSION

Another common type of bracket connection is similar to a seated connection. An angle is typically bolted to the flange of a column and then a load is applied to the outstanding leg of the angle. A variation on this uses a piece of a WT member bolted to the column flange. A load is then applied to the outstanding web of the WT.

The fasteners of these connections are subjected to shear and tension. The design assumption is that each fastener supports an equal percentage of the direct shear load. These connections are usually constructed with pretensioned fasteners (Group A or Group B bolts) and the neutral axis is assumed to be at the centroid of the group of fasteners. Therefore, the tensile force a fastener receives is proportional to its distance from the neutral axis.

Example 9.4

Bracket Subjected to Shear and Tension

A piece of WT section is bolted to a W column section with two rows of five bolts as shown. The bolts are $^7/_8$ in diameter Group A bolts with the threads in the shear plane. The bracket supports the dead and live loads shown. Assume the neutral axis is located at the center of gravity of the bolt group (case II in *AISC Manual* Part 7).

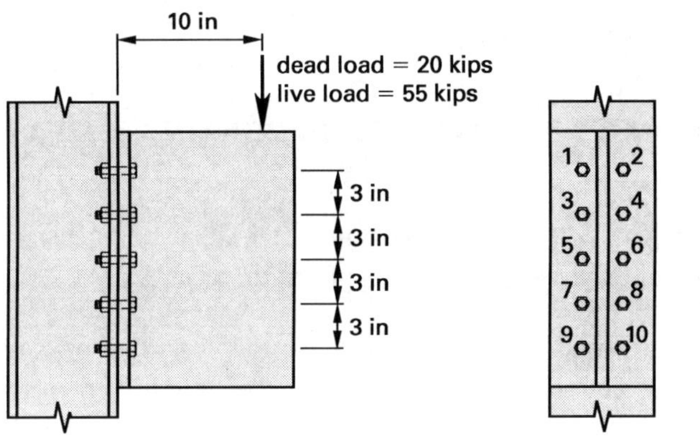

(a) side elevation (b) front elevation

Determine whether the bolts are satisfactory for resisting the combined effects of shear and tension.

Solution

The total shear load is

LRFD	ASD
$P_u = 1.2D + 1.6L$ $= (1.2)(20 \text{ kips}) + (1.6)(55 \text{ kips})$ $= 112 \text{ kips}$	$P_a = D + L$ $= 20 \text{ kips} + 55 \text{ kips}$ $= 75 \text{ kips}$

Calculate the shear load per bolt, on the design basis that each bolt receives equal shear.

LRFD	ASD
$P_v = \frac{P_u}{n} = \frac{112 \text{ kips}}{10 \text{ bolts}} = 11.20 \text{ kips/bolt}$	$P_v = \frac{P_a}{n} = \frac{75 \text{ kips}}{10 \text{ bolts}} = 7.5 \text{ kips/bolt}$

Calculate the moment created by the eccentricity.

LRFD	ASD
$M = P_u e = (112 \text{ kips})(10 \text{ in}) = 1120 \text{ in-kips}$	$M = P_a e = (75 \text{ kips})(10 \text{ in}) = 750 \text{ in-kips}$

Calculate the shear stress in the bolts. From *AISC Manual* Table 7-1, the nominal area of $^7/_8$ in diameter bolts is 0.601 in^2.

LRFD	ASD
$f_v = \frac{P_v}{A_b} = \frac{11.20 \text{ kips}}{0.601 \text{ in}^2} = 18.64 \text{ ksi}$	$f_v = \frac{P_v}{A_b} = \frac{7.5 \text{ kips}}{0.601 \text{ in}^2} = 12.48 \text{ ksi}$

From *AISC Specification* Table J3.2, the nominal shear stress per bolt is 54 ksi. The available shear strength per bolt is

LRFD	ASD
$\phi F_{nv} = (0.75)\left(54 \ \frac{\text{kips}}{\text{in}^2}\right) = 40.5 \text{ ksi}$	$\frac{F_{nv}}{\Omega} = \frac{54 \ \frac{\text{kips}}{\text{in}^2}}{2.00} = 27 \text{ ksi}$

Calculate the moment of inertia of the group of fasteners.

$$
\begin{aligned}
I &= \sum A_b y^2 \\
&= (4 \text{ bolts})\left(0.6013 \ \frac{\text{in}^2}{\text{bolt}}\right)(6 \text{ in})^2 + (4 \text{ bolts})\left(0.6013 \ \frac{\text{in}^2}{\text{bolt}}\right)(3 \text{ in})^2 \\
&\quad + (2 \text{ bolts})\left(0.6013 \ \frac{\text{in}^2}{\text{bolt}}\right)(0 \text{ in})^2 \\
&= 108.23 \text{ in}^4
\end{aligned}
$$

Calculate the tensile stress on bolts 1 and 2.

LRFD	ASD
$f_{t,12} = \frac{My}{I} = \frac{(1120 \text{ in-kips})(6 \text{ in})}{108.23 \text{ in}^4}$ $= 62.09 \text{ ksi}$	$f_{t,12} = \frac{My}{I} = \frac{(750 \text{ in-kips})(6 \text{ in})}{108.23 \text{ in}^4}$ $= 41.58 \text{ ksi}$

From *AISC Specification* Table J3.2, the nominal tensile stress per bolt is 90 ksi. The available tensile strength per bolt is

LRFD	ASD
$\phi F_{nt} = (0.75)\left(90 \frac{\text{kips}}{\text{in}^2}\right) = 67.50 \text{ ksi}$	$\frac{F_{nt}}{\Omega} = \frac{90 \frac{\text{kips}}{\text{in}^2}}{2.00} = 45.00 \text{ ksi}$

Calculate the tensile stress on bolts 3 and 4.

LRFD	ASD
$f_{t,34} = \frac{My}{I} = \frac{(1120 \text{ in-kips})(3 \text{ in})}{108.23 \text{ in}^4}$ $= 31.04 \text{ ksi}$	$f_{t,34} = \frac{My}{I} = \frac{(750 \text{ in-kips})(3 \text{ in})}{108.23 \text{ in}^4}$ $= 20.78 \text{ ksi}$

Determine whether the combined effects of shear and tension must be investigated. The shear ratio is

LRFD	ASD
$\frac{f_v}{\phi F_{nv}} = \frac{18.64 \frac{\text{kips}}{\text{in}^2}}{40.5 \frac{\text{kips}}{\text{in}^2}}$ $= 0.46 \quad [> 0.20]$	$\frac{f_v}{\frac{F_{nv}}{\Omega}} = \frac{12.48 \frac{\text{kips}}{\text{in}^2}}{27 \frac{\text{kips}}{\text{in}^2}}$ $= 0.46 \quad [> 0.20]$

The tension ratio is

LRFD	ASD
$\frac{f_{t,12}}{\phi F_{nt}} = \frac{62.09 \frac{\text{kips}}{\text{in}^2}}{67.5 \frac{\text{kips}}{\text{in}^2}}$ $= 0.92 \quad [> 0.20]$	$\frac{f_{t,12}}{\frac{F_{nt}}{\Omega}} = \frac{41.58 \frac{\text{kips}}{\text{in}^2}}{45.00 \frac{\text{kips}}{\text{in}^2}}$ $= 0.92 \quad [> 0.20]$

Both the shear and tension ratios exceed 20%. If either exceeds 20%, the effect of the combined stresses cannot be neglected. Check the stresses for bearing type connection with combined shear and tension with the threads included in the shear plane.

LRFD	ASD
From Eq. 9.3, $$F'_{nt} = 1.3F_{nt} - \left(\frac{F_{nt}}{\phi F_{nv}}\right) f_v \le F_{nt}$$ $$= (1.3)\left(90\ \frac{\text{kips}}{\text{in}^2}\right) - \left(\frac{90\ \frac{\text{kips}}{\text{in}^2}}{40.5\ \frac{\text{kips}}{\text{in}^2}}\right)\left(18.64\ \frac{\text{kips}}{\text{in}^2}\right)$$ $$= 75.57\ \text{ksi} \quad [\le 90\ \text{ksi}]$$	From Eq. 9.4, $$F'_{nt} = 1.3F_{nt} - \left(\frac{\Omega F_{nt}}{F_{nv}}\right) f_v \le F_{nt}$$ $$= (1.3)\left(90\ \frac{\text{kips}}{\text{in}^2}\right) - \left(\frac{(2.0)\left(90\ \frac{\text{kips}}{\text{in}^2}\right)}{54\ \frac{\text{kips}}{\text{in}^2}}\right) \times \left(12.48\ \frac{\text{kips}}{\text{in}^2}\right)$$ $$= 75.40\ \text{ksi} \quad [\le 90\ \text{ksi}]$$

From Eq. 9.2, the nominal tension resistance capacity is

$$R_n = F'_{nt}A_b = \left(75.57\ \frac{\text{kips}}{\text{in}^2}\right)(0.601\ \text{in}^2) = 45.42\ \text{kips}$$

The design tension strength is

$$\phi R_n = (0.75)(45.42\ \text{kips}) = 34.06\ \text{kips}$$

The tensile load on bolts 1 and 2 is

$$f_{t,12}A_b = \left(62.09\ \frac{\text{kips}}{\text{in}^2}\right)(0.601\ \text{in}^2) = 37.32\ \text{kips}$$

The tension load on a bolt, 37.32 kips, exceeds the design tension strength of a bolt, 31.73 kips. The calculated tensile stress is approximately 9.6% greater than the design strength. Assuming that the threads of the connectors are excluded from the shear plane, the calculated stress will be approximately 3.8% greater than that permitted. Two possible solutions would be to use 1 in diameter bolts or to increase the vertical spacing between the bolts.

10 Welded Connections

Nomenclature

a	ratio of horizontal eccentricity to characteristic length of weld group, e_x/l	–
A	cross-sectional area	in^2
b	width	in
B	width of HSS member measured 90° to plane of connection	in
B	width of member	in
B_{ep}	effective width of plate as defined in *AISC Specification* Sec. K1.3b	in
B_p	plate width taken perpendicular to connection	in
C	coefficient from *AISC Manual* Table 8-8	–
C_1	electrode strength coefficient from *AISC Manual* Table 8-3 (1.0 for E70XX)	–
D	outside diameter of round HSS	in
D	number of sixteenths of an inch in fillet weld size	–
e	eccentricity	in
E	modulus of elasticity	lbf/in^2
F	strength or stress	lbf/in^2
F_{EXX}	tensile strength of weld metal	lbf/in^2
F_n	nominal strength or stress	lbf/in^2
F_u	specified minimum tensile strength	lbf/in^2
F_y	specified minimum yield stress	lbf/in^2
h	for a rectangular HSS member, the clear distance between flanges less inside corner radii	in
H	overall height of rectangular HSS member measured in place of connection	in
I	moment of inertia	in^4
k	outside corner radius of HSS member	in
k	ratio of leg lengths in weld group as defined in *AISC Manual* Table 8-8	–
l	characteristic length of weld group	in

L	length	in
N	bearing length of load measured parallel to axis of HSS member	in
P	force or tensile strength	lbf
Q_f	chord-stress interaction parameter as defined in *AISC Specification* Sec. K2.2	–
r	radius of gyration	in
R	strength or resistance	lbf
S	elastic section modulus	in^3
t	thickness	in
T	tensile strength	lbf/in^2
U	shear lag factor	–
V	shear strength	lbf
w	weld size	in
w	width of welded member	in
$\overline{x}$, $\overline{y}$	connection eccentricity	in
Z	plastic section modulus	in^3

Symbols

β	width ratio as defined in *AISC Specification* Sec. K2.1	–
ϕ	resistance factor (LRFD)	–
Ω	safety factor (ASD)	–

Subscripts

a	required (ASD)
b or BM	base metal
calc	calculated
e	effective
g	gross
h	holes
max	maximum
min	minimum
n	nominal
p	plate
req	required
t	tensile
u	required (LRFD)
v	shear

w	weld or weld metal
x	x-axis, strong axis, or horizontal component
y	y-axis, weak axis, or vertical component

1. GENERAL

Welded connections are used frequently because of their simplicity. They have fewer parts and weigh less than other connections, particularly when the welding is performed in the shop. Combining shop-welded and field-bolted elements is usually the most economical method. The shop-welded parts of a connection frequently reduce required bolting clearances for field erection.

A properly designed and executed weld can be stronger than the base metal. The weld's design is as important as its execution in achieving a good connection. Improperly made welds, though they may appear to be good, can be worthless.

When designing a weld, it is important to specify the type, number, and size of only the welds needed to obtain the necessary strength. Welding in excess of what is needed increases assembly costs and may reduce the ductility of the connection.

AISC Specification Sec. J1 and Sec. J2 provide the requirements for welded connections. *AISC Manual* Parts 8, 10, 11, and 12 contain many tables that can help in designing and analyzing connections.

2. TYPES OF WELDS

Figure 10.1 shows some of the most common weld types and the standard symbols used to indicate them on drawings.

Figure 10.1 *Weld Types*

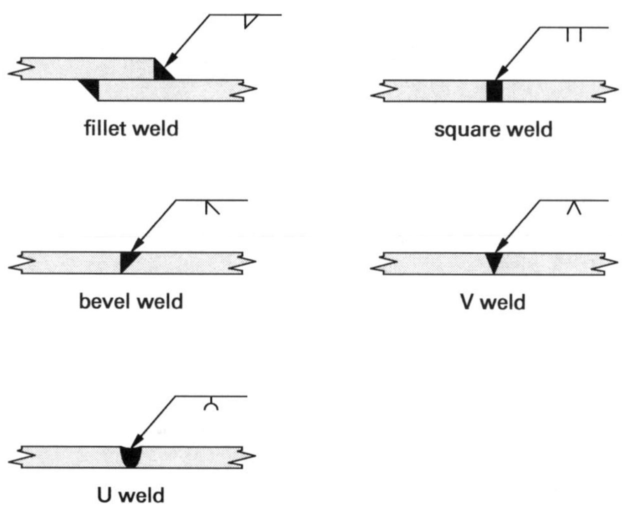

The *fillet weld* is the most common and economical weld used in structural steel. Fillet welds are often used for lightly loaded connections. This type of weld needs little or no preparation of the material to be joined.

Groove welds are often used for heavier loads because they can be designed to develop the full strength of the elements being joined. Groove welds are further classified by the type of joint preparation used to receive the weld. Types of groove welds include square, bevel, V, J, U, flare bevel, and flare V.

Plug welds and *slot welds* are less commonly used than fillet and groove welds. They are used primarily to transmit shear in lapped joints and to prevent buckling of elements in built-up members.

3. WELD ECONOMY

Economy in welding is achieved by using properly designed welds. In general, smaller, longer welds are more economical than heavier, shorter welds.

The strength of a fillet weld, for example, varies with its size, and yet the volume of weld metal varies with the *square* of the weld's size. This means that a $^1/_2$ in fillet weld contains four times the volume of weld metal as a $^1/_4$ in fillet weld, but is only twice as strong. What is more, the $^1/_2$ in fillet weld needs four passes of the rod while the $^1/_4$ in fillet weld needs only one, so there is also a significant difference in labor costs. Table 10.1 gives the number of welding rod passes needed to deposit fillet welds of some common sizes.

Table 10.1 *Passes Needed to Form Fillet Welds*

fillet weld size (in)	number of rod passes
$^3/_{16}$	1
$^1/_4$	1
$^5/_{16}$	1
$^3/_8$	3
$^7/_{16}$	4
$^1/_2$	4
$^5/_8$	6
$^3/_4$	8

4. MAXIMUM AND MINIMUM SIZE FILLET WELDS

AISC Specification Sec. J2 gives the requirements for various types of welds. In addition, almost all the provisions set forth in the American Welding Society's *Structural Welding Code—Steel* (AWS D1.1) apply to buildings and other structures that are constructed with structural steel.

The maximum permitted size of a fillet weld along the edge of connected parts is

- for material less than $^1/_4$ in thick, not greater than the thickness of the material
- for material $^1/_4$ in thick or more, not greater than the thickness of the material minus $^1/_{16}$ in

The effective area of a fillet weld is taken as the throat thickness multiplied by its effective length. The effective throat thickness is the shortest distance from the root of the weld to the face surface of the weld. For an equal leg fillet weld, the throat thickness is the size of the leg multiplied by $\sqrt{2}$.

The effective length of a fillet weld must be at least four times its nominal size. If a fillet weld is shorter than this, then the weld's strength capacity must be reduced proportionately. The minimum permitted sizes of fillet welds are given in Table 10.2.

Table 10.2 *Minimum Sizes of Fillet Welds*

material thickness of thinner part joined	minimum size* (in)
up to $^1/_4$ in (inclusive)	$^1/_8$
more than $^1/_4$ in to $^1/_2$ in	$^3/_{16}$
more than $^1/_2$ in to $^3/_4$ in	$^1/_4$
more than $^3/_4$ in	$^5/_{16}$

*leg dimension of fillet weld, single-pass weld used

Source: *AISC Manual* Table J2.4

To prevent overstressing the base material at a fillet weld, the *AISC Specification* puts a maximum limit on the size of a fillet weld. The capacity of a linear inch of weld cannot exceed the allowable tensile strength or shear strength of a linear inch of the connected part. The following formulas are used to determine the minimum thickness of the connected element.

When the base member is in tension, use

$$t = \frac{0.707wF_{vw}}{F_{tb}} \qquad 10.1$$

When the base member is in shear, use

$$t = \frac{0.707wF_{vw}}{F_{vb}} \qquad 10.2$$

5. INTERMITTENT FILLET WELDS

An intermittent fillet weld can be used to transfer stress across a joint or faying surface when a continuous fillet weld of the smallest permitted size would provide more strength than is required. The minimum weld length is four times the weld size but no less than $1^1/_2$ in.

For built-up tension members, the maximum spacing of intermittent fillet welds is 300 times the radius of gyration of the smaller member being welded (from *AISC Specification* Sec. D4).[1]

For built-up compression members, the maximum spacing of intermittent fillet welds (from *AISC Specification* Sec. E6.2) is

- to connect two rolled shapes, 24 in
- when fasteners are not staggered along adjacent gage lines, $0.75/\sqrt{E/F_y}$ times the thickness of the outside plate or 12 in, whichever is greater
- when fasteners are staggered along adjacent gage lines, $1.12/\sqrt{E/F_y}$ times the thickness of the outside plate or 18 in, whichever is greater

6. WELD STRENGTH

Weld strength is a function of the strength of the base material, the strength of the weld metal, the welding process used, and the weld penetration. Different types of welding electrodes, or rods, exist to meet the different requirements for strength and the welding process being used.

Shielded metal arc welding (SMAW) is the oldest and most common form of welding used. SMAW is frequently referred to as *manual stick welding.* Other welding processes include submerged arc welding (SAW), gas metal arc welding (GMAW), and flux core arc welding (FCAW).

The nominal resistance of a weld, R_n, is the lowest value of the base metal strength according to the limit states of tensile rupture, shear rupture, and yielding, determined as follows. For the base metal,

$$R_n = F_{n,\mathrm{BM}} A_{\mathrm{BM}} \quad [\text{AISC Eq. J2-2}] \tag{10.3}$$

For the weld metal,

$$R_n = F_{nw} A_{we} \quad [\text{AISC Eq. J2-3}] \tag{10.4}$$

F_{nw} and A_{we} are the nominal strength and effective cross-sectional area of the weld, respectively.

[1]In the *AISC Manual: LRFD*, third edition, the maximum spacing of intermittent fillet welds was limited to

- for painted or unpainted members not subject to corrosion, 24 times the thickness of the thinner element or plate or 12 in, whichever is greater
- for unpainted members of weathering steel subject to atmospheric corrosion, 14 times the thickness of the thinner element or plate or 7 in, whichever is greater

These criteria are still in use in some places.

For LRFD, the required strength, R_u, must be less than or equal to the design strength, ϕR_n. For ASD, the required strength, R_a, must be less than or equal to the allowable strength, R_n/Ω.

The properties that affect the strength of a weld include

- strength of the weld metal
- type of weld
- welding position
- effective weld size
- effective throat thickness
- relationship of weld metal strength to base metal strength

Use the following tables in the *AISC Manual* to determine weld strengths.

Table J2.1	Effective Throat of Partial-Joint-Penetration Groove Welds
Table J2.2	Effective Weld Sizes of Flare Groove Welds
Table J2.3	Minimum Effective Throat Thickness of Partial-Joint Penetration Groove Welds
Table J2.5	Available Strength of Welded Joints

AISC Manual Table J2.5 provides the applicable resistance factors, ϕ, and safety factors, Ω, for the various types of welds, load type, and direction.

Tension members connected by welds are subject to shear lag effects similar to bolted connections. Cases 3 and 4 in Table 4.1 specify the shear lag factor, U, that should be used in Eq. 10.5 to calculate the effective weld strength, R_{ne}.

$$R_{ne} = UF_{nw}A_{we} \tag{10.5}$$

7. FILLET WELD STRENGTH

The cross-section of a standard fillet weld is a right triangle with equal legs. The effective throat thickness of the weld is the distance from the heel of the weld (at the right angle) to the face of the weld, measured in the direction normal to the face. Occasionally a fillet weld will have unequal legs. Figure 10.2 shows fillet welds with equal and unequal legs.

The strength of a fillet weld is usually given in pounds per $^1/_{16}$ in of nominal weld size per inch of length. The E70 electrode is commonly used for laying fillet welds on ASTM A36 and ASTM A992 steels and has an available weld stress of 70 ksi. The limit state is shear rupture through the weld throat. The strength of a fillet weld is

$$F_w = 0.60F_{\text{EXX}} \tag{10.6}$$

The nominal shear strength of a fillet weld is

$$V_n = 0.707wF_{nw}L \tag{10.7}$$

Figure 10.2 *Equal Leg and Unequal Leg Fillet Welds*

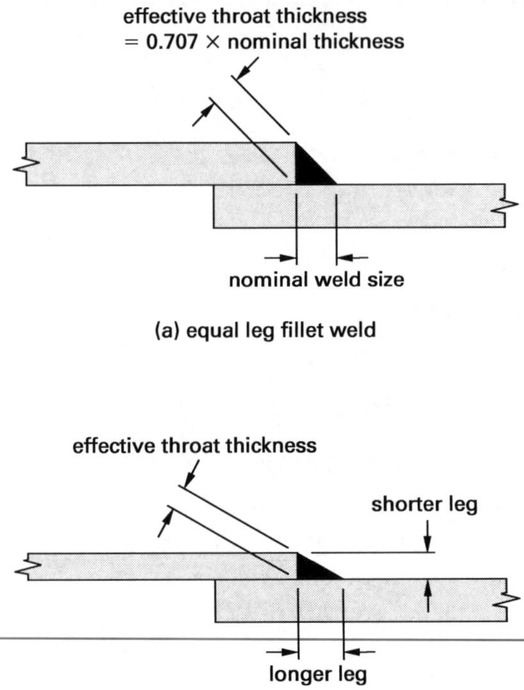

For a weld stress of 70 ksi, then, the nominal shear strength is

$$\begin{aligned} V_n &= 0.707wF_{nw}L \\ &= 0.707w(0.60F_{\text{EXX}})L \\ &= (0.707)\left(\frac{1}{16}\text{ in}\right)(0.60)\left(70\ \frac{\text{kips}}{\text{in}^2}\right)(1\text{ in}) \\ &= 1.86\text{ kips}\quad \begin{bmatrix}\text{per 1/16 in of weld}\\ \text{per inch of length}\end{bmatrix} \end{aligned}$$

For fillet welds, the resistance factor is $\phi = 0.75$ and the safety factor is $\Omega = 2.00$. The fillet weld strength per sixteenth inch for the E70 electrode, then, is

LRFD	ASD
$\phi V_n = (0.75)(1.86\text{ kips}) = 1.40\text{ kips}\ \begin{bmatrix}\text{per 1/16 in of weld}\\ \text{per inch of length}\end{bmatrix}$	$\frac{V_n}{\Omega} = \frac{1.86\text{ kips}}{2} = 0.93\text{ kips}\ \begin{bmatrix}\text{per 1/16 in of weld}\\ \text{per inch of length}\end{bmatrix}$

For LRFD, the *AISC Manual* uses a value of 1.392 kips per $^1/_{16}$ in per inch of length; for ASD, it uses a value of 0.928 kips per $^1/_{16}$ in per inch of length.

Example 10.1
Welded Lap Splice

A welded lap splice connection is shown.

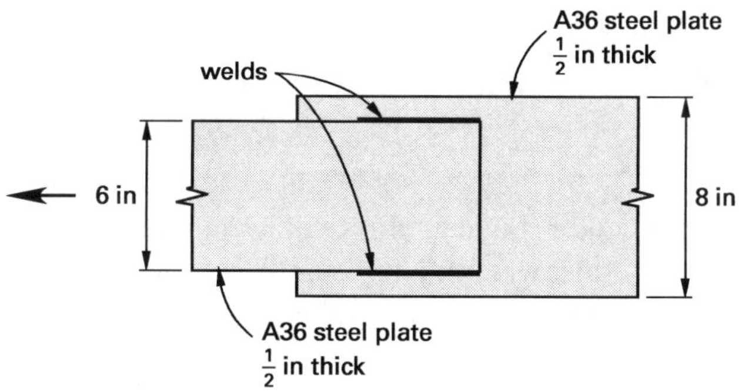

Material properties

ASTM A36 steel

$F_y = 36$ ksi

$F_u = 58$ ksi

weld material = E70XX

Determine the weld length required to develop the maximum tensile force of the assembly for permissible weld sizes between the minimum and maximum size fillet welds.

Solution

The nominal tensile capacity, P_n, is governed by the member with the lesser gross cross-sectional area. As the plates have the same thickness, the narrower one has the smaller area.

$$\begin{aligned} A_g &= bt = (6 \text{ in})\left(\frac{1}{2} \text{ in}\right) \\ &= 3 \text{ in}^2 \end{aligned}$$

For the tension members, the shear lag factor, U, is 1.0 because there are no member elements that are not in contact; therefore, $A_e = A_g$. (The ratio of weld length to weld separation, however, may require $U < 1.0$.) From Eq. 4.2,

$$\begin{aligned} P_n &= F_y A_g = F_y A_e \\ &= \left(36 \ \frac{\text{kips}}{\text{in}^2}\right)(3 \text{ in}^2) \\ &= 108 \text{ kips} \end{aligned}$$

Calculate the design strength (LRFD) and the allowable strength (ASD) required for the welds.

LRFD	ASD
$R_u = \phi P_n = (0.90)(108 \text{ kips})$ $= 97.20 \text{ kips}$	$R_a = \dfrac{P_n}{\Omega} = \dfrac{108 \text{ kips}}{1.67}$ $= 64.67 \text{ kips}$

Determine the minimum and maximum allowable weld sizes for $^1/_2$ in material. From Table 10.2, the minimum fillet weld size for $^1/_2$ in thick material is $^3/_{16}$ in. The material thickness is greater than $^1/_4$ in, so the maximum weld size is $^1/_{16}$ in less than the material thickness.

$$\begin{aligned} w_{max} &= t_p - \frac{1}{16} \text{ in} \\ &= \frac{1}{2} \text{ in} - \frac{1}{16} \text{ in} \\ &= \frac{7}{16} \text{ in} \end{aligned}$$

The permitted weld sizes, then, are $^3/_{16}$ in through $^7/_{16}$ in. Determine the weld resistance capacity, R_w, for each possible weld size by multiplying the number of sixteenths of an inch by the weld strength. For LRFD, the weld strength is 1.392 kips per $^1/_{16}$ in of weld per inch of length. For ASD, the weld strength is 0.928 kips per $^1/_{16}$ in of weld per inch of length.

weld size (in)	LRFD weld resistance capacity (kips/in)	ASD weld resistance capacity (kips/in)
$^3/_{16}$	(3)(1.392) = 4.18	(3)(0.928) = 2.78
$^1/_4$	(4)(1.392) = 5.57	(4)(0.928) = 3.71
$^5/_{16}$	(5)(1.392) = 6.96	(5)(0.928) = 4.64
$^3/_8$	(6)(1.392) = 8.36	(6)(0.928) = 5.57
$^7/_{16}$	(7)(1.392) = 9.74	(7)(0.928) = 6.50

Calculate the required weld lengths assuming a shear lag factor, U, of 1.0.

weld size (in)	LRFD $L=\dfrac{R_u}{R_w}$	ASD $L=\dfrac{R_a}{R_w}$
$^3/_{16}$	$\dfrac{97.20 \text{ kips}}{4.18 \dfrac{\text{kips}}{\text{in}}} = 23.25 \text{ in}$	$\dfrac{64.67 \text{ kips}}{2.78 \dfrac{\text{kips}}{\text{in}}} = 23.26 \text{ in}$
$^1/_4$	$\dfrac{97.20 \text{ kips}}{5.57 \dfrac{\text{kips}}{\text{in}}} = 17.45 \text{ in}$	$\dfrac{64.67 \text{ kips}}{3.71 \dfrac{\text{kips}}{\text{in}}} = 17.43 \text{ in}$
$^5/_{16}$	$\dfrac{97.20 \text{ kips}}{6.96 \dfrac{\text{kips}}{\text{in}}} = 13.97 \text{ in}$	$\dfrac{64.67 \text{ kips}}{4.64 \dfrac{\text{kips}}{\text{in}}} = 13.94 \text{ in}$
$^3/_8$	$\dfrac{97.20 \text{ kips}}{8.36 \dfrac{\text{kips}}{\text{in}}} = 11.63 \text{ in}$	$\dfrac{64.67 \text{ kips}}{5.57 \dfrac{\text{kips}}{\text{in}}} = 11.61 \text{ in}$
$^7/_{16}$	$\dfrac{97.20 \text{ kips}}{9.74 \dfrac{\text{kips}}{\text{in}}} = 9.98 \text{ in}$	$\dfrac{64.67 \text{ kips}}{6.50 \dfrac{\text{kips}}{\text{in}}} = 9.95 \text{ in}$

The length of each longitudinal weld should be 0.50 of the total longitudinal length required. Therefore, the required weld lengths for LRFD or ASD are

weld size (in)	length per side, L (in)
$^3/_{16}$	12
$^1/_4$	9
$^5/_{16}$	7
$^3/_8$	6
$^7/_{16}$	5

As mentioned earlier, the ratio of weld length to weld separation may require a shear lag factor, U, of less than 1.0. From Eq. 4.11, the effective area of a tensile member with a welded connection is

$$A_e = A_g U$$

From Table 4.1, case 4, the shear lag factor, U, is equal to 1.0 only when $L \geq 2w$, where w is the width of the welded member. In this case, $2w = (2)(6 \text{ in}) = 12$ in, so U is 1.0 only when L is at least 12 in. U is less than 1.0 for lower values of L, so the weld lengths must be increased for all weld sizes except the $^3/_{16}$ in weld.

The weld lengths just calculated should be increased by dividing by the appropriate shear lag factor. When $2w > L \geq 1.5w$ (that is, when the length is less than 12 in but at least 9 in), then $U = 0.87$, and when $1.5w > L \geq w$ (that is, when the length is less than 9 in but at least 6 in), then $U = 0.75$.

weld size (in)	length per side, L (in)
$^3/_{16}$	12
$^1/_4$	$\frac{9}{0.87} = 10.34$
$^5/_{16}$	$\frac{7}{0.87} = 8.05$
$^3/_8$	$\frac{6}{0.75} = 8$
$^7/_{16}$	$\frac{5}{0.75} = 6.67$

Example 10.2
Angle-to-Plate Welded Connection

A welded lap splice connection is shown.

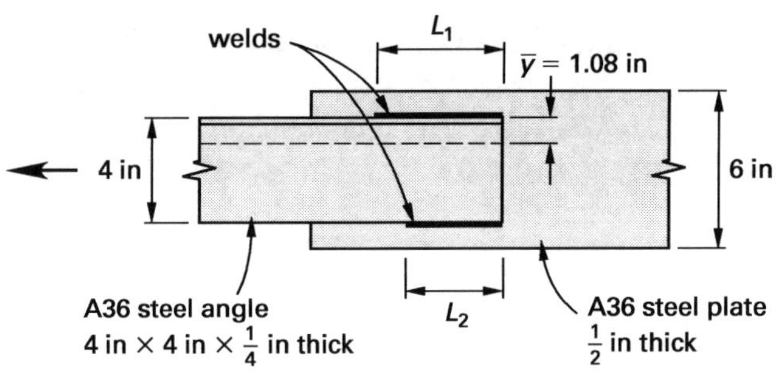

Section properties

$L4 \times 4 \times {}^1/_4$

$A = 1.93$ in^2

$I_x = I_y = 3.0$ in^4

$S_x = S_y = 1.03$ in^3

$r_x = r_y = 1.25$ in

$\overline{y} = \overline{x} = 1.08$ in

$Z_x = Z_y = 1.82$ in^3

Material properties

ASTM A36 steel

$F_y = 36$ ksi

$F_u = 58$ ksi

weld material: E70XX

Determine the weld lengths, L_1 and L_2 in the illustration, that are needed in order to develop the full tensile capacity of the angle.

Solution

The nominal tensile capacity is governed by the member with the lesser gross area.

$$\begin{aligned} P_n &= F_y A \\ &= \left(36\ \frac{\text{kips}}{\text{in}^2}\right)(1.93\ \text{in}^2) \\ &= 69.48\ \text{kips} \end{aligned}$$

Calculate the design strength (LRFD) and the allowable strength (ASD) required for the welds.

LRFD	ASD
$R_u = \phi P_n = (0.90)(69.48\ \text{kips})$ $= 62.53\ \text{kips}$	$R_a = \dfrac{P_n}{\Omega} = \dfrac{69.48\ \text{kips}}{1.67}$ $= 41.60\ \text{kips}$

From Table 10.2, the minimum size fillet weld that can be applied to the $^1/_2$ in thick plate is $^3/_{16}$ in. This is also the maximum size weld that can be applied to the toe because the specification requires that the weld be $^1/_{16}$ in less than the leg thickness to compensate for the radius of the toe of the angle.

Calculate the total required length of a $^3/_{16}$ in weld, assuming that the shear lag factor, U, is 1.0.

LRFD	ASD
$L = \dfrac{\phi P_n}{R_w} = \dfrac{(0.90)(69.48\ \text{kips})}{(3)\left(1.392\ \frac{\text{kips}}{\text{in}}\right)}$ $= 14.97\ \text{in}$	$L = \dfrac{\frac{P_n}{\Omega}}{R_w} = \dfrac{\frac{69.48\ \text{kips}}{1.67}}{(3)\left(0.928\ \frac{\text{kips}}{\text{in}}\right)}$ $= 14.94\ \text{in}$

The minimum weld length for a fillet weld is four times the nominal size of the weld; therefore, the minimum length of a $^3/_{16}$ in weld is $^3/_4$ in. The length of the longitudinal welds in relation to the transverse distance of 4 in between the welds must also be considered in order to determine the shear lag factor, U, in accordance with Table 4.1, case 4.

Calculate the lengths of welds L_1 and L_2 so that the centroid of the weld group coincides with the centroid of the tensile load (welds balanced about the neutral axis). *AISC Specification* Sec. J1.7 does not require balanced welds for single or double angles or similar members with small eccentricities that are statically loaded.

$$\begin{aligned} L_1 &= \left(\frac{4 \text{ in} - \bar{y}}{4 \text{ in}}\right)L \\ &= \left(\frac{4 \text{ in} - 1.08 \text{ in}}{4 \text{ in}}\right)(14.97 \text{ in}) \\ &= 10.93 \text{ in} \quad [\text{use 11 in}] \end{aligned}$$

$$\begin{aligned} L_2 &= \left(\frac{\bar{y}}{4 \text{ in}}\right)L \\ &= \left(\frac{1.08 \text{ in}}{4 \text{ in}}\right)(14.97 \text{ in}) \\ &= 4.04 \text{ in} \quad [\text{use 4 in}] \end{aligned}$$

Use Table 4.1, case 2, to calculate the shear lag factor, U, for the welded connection due to the eccentric load.

$$\begin{aligned} U &= 1 - \frac{\bar{x}}{L_1} \le 0.9 \\ &= 1 - \frac{1.08 \text{ in}}{11 \text{ in}} \\ &= 0.90 \quad [\le 0.90] \end{aligned}$$

Therefore, divide the calculated weld length by $U = 0.90$ to get the required weld length.

$$L_{1,\text{req}} = \frac{L_{1,\text{calc}}}{U} = \frac{11 \text{ in}}{0.90} = 12.22 \text{ in}$$

$$L_{2,\text{req}} = \frac{L_{2,\text{calc}}}{U} = \frac{4 \text{ in}}{0.90} = 4.44 \text{ in}$$

The required weld length for L_1 is 13 in. The required weld length for L_2 is 5 in.

8. WELDED BRACKET WITH ECCENTRIC SHEAR

Two methods are commonly used in designing or analyzing eccentrically loaded connections, the elastic method and the instantaneous center of rotation method. The latter is more accurate, but requires an iterative process or the use of design aids such as the tables in the *AISC Manual*.

The available strength of a weld group, ϕR_n or R_n/Ω, is determined using $\phi = 0.75$ and $\Omega = 2.00$. The nominal strength of the weld group in kips is found with Eq. 10.8.

$$R_{n,\mathrm{kips}} = CC_1Dl_{\mathrm{in}} \quad [\text{AISC Table 8-8}] \tag{10.8}$$

Refer to *AISC Manual* Table 8-8 to determine the appropriate coefficient, C, for the eccentrically loaded weld group, and refer to *AISC Manual* Table 8-3 to determine the electrode strength coefficient, C_1. D is the number of sixteenths of an inch in the fillet weld size, and l is the characteristic length of the weld group in inches. (The formula is not dimensionally consistent.)

The available strength must be no less than the required strength, so

$$\begin{aligned} P_u &\le \phi R_n \\ &\le \phi CC_1Dl \quad [\text{LRFD}] \end{aligned} \tag{10.9}$$

$$\begin{aligned} P_a &\le \frac{R_n}{\Omega} \\ &\le \frac{CC_1Dl}{\Omega} \quad [\text{ASD}] \end{aligned} \tag{10.10}$$

The minimum required value for C, D, or l can be found if the available strength and the values of the other variables are known. For LRFD,

$$C_{\mathrm{min}} = \frac{P_u}{\phi C_1Dl} \tag{10.11}$$

$$D_{\mathrm{min}} = \frac{P_u}{\phi CC_1l} \tag{10.12}$$

$$l_{\mathrm{min}} = \frac{P_u}{\phi CC_1D} \tag{10.13}$$

For ASD,

$$C_{\mathrm{min}} = \frac{\Omega P_a}{C_1Dl} \tag{10.14}$$

$$D_{\mathrm{min}} = \frac{\Omega P_a}{CC_1l} \tag{10.15}$$

$$l_{\mathrm{min}} = \frac{\Omega P_a}{CC_1D} \tag{10.16}$$

Example 10.3

Welded Bracket with Eccentric Shear

The steel bracket shown is welded to the face of a column flange. The bracket supports a 5 kip dead load and a 15 kip live load having an eccentricity of 12 in.

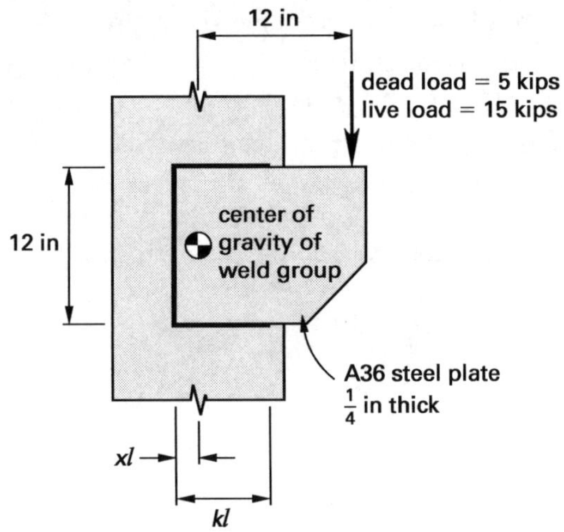

Section properties

plate thickness = $^1/_4$ in

height = 12 in

Material properties

ASTM A36 steel

F_y = 36 ksi

F_u = 58 ksi

weld material: E70XX

Determine the weld size and the horizontal weld length, *kl*, required to support the design loads.

Solution

AISC Manual Tables 8-4 through 8-11 provide assistance in designing the eccentric loads on weld groups. Calculate the required design strengths.

LRFD	ASD
$R_u = 1.2D + 1.6L$ $= (1.2)(5.0 \text{ kips}) + (1.6)(15.0 \text{ kips})$ $= 30.0 \text{ kips}$	$R_a = D + L$ $= 5.0 \text{ kips} + 15.0 \text{ kips}$ $= 20.0 \text{ kips}$

The maximum size fillet weld that can be used is equal to the thickness of the plate bracket, $^1/_4$ in. To meet the code requirements, the minimum size fillet weld that can be used is $^1/_8$ in. With a $^1/_4$ in fillet weld, the minimum flange thickness is $^3/_4$ in.

From the figure, the characteristic length, l, is 12 in. Eccentricity, e_x, is also 12 in, so

$$a = \frac{e_x}{l} = \frac{12 \text{ in}}{12 \text{ in}} = 1.0$$

D, the number of sixteenths of an inch of weld size, is 2, 3, or 4. From *AISC Manual* Table 8-3, the electrode strength coefficient, C_1, is 1.0 for E70XX electrodes. Use Eq. 10.9 through Eq. 10.16 to calculate the minimum coefficient required, C_{min}. Determine k and kl from *AISC Manual* Table 8-8. For a $^1/_8$ in weld, $D = 2$ and the minimum weld length is 0.5 in.

LRFD	ASD
$C_{min} = \frac{P_u}{\phi C_1 D l}$ $= \frac{30.00 \text{ kips}}{(0.75)(1.0)(2)(12 \text{ in})}$ $= 1.67$	$C_{min} = \frac{\Omega P_a}{C_1 D l}$ $= \frac{(2.0)(20.0 \text{ kips})}{(1)(2)(12 \text{ in})}$ $= 1.67$

From *AISC Manual* Table 8-8, with $a = 1.0$ and $C_{min} = 1.67$, $k = 0.5$. Then $kl = (0.5)(12 \text{ in}) = 6.0$ in.

For a $^3/_{16}$ in weld, $D = 3$ and the minimum weld length is 0.75 in.

LRFD	ASD
$C_{min} = \frac{P_u}{\phi C_1 D l}$ $= \frac{30.0 \text{ kips}}{(0.75)(1.0)(3)(12 \text{ in})}$ $= 1.11$	$C_{min} = \frac{\Omega P_a}{C_1 D l}$ $= \frac{(2)(20.0 \text{ kips})}{(1)(3)(12 \text{ in})}$ $= 1.11$

From *AISC Manual* Table 8-8, with $a = 1.0$ and $C_{min} = 1.11$, $k = 0.3$. Then $kl = (0.3)(12 \text{ in}) = 3.6$ in (use 3.75 in).

For a $^1/_4$ in weld, $D = 4$ and the minimum weld length is 1.0 in.

LRFD	ASD
$C_{min} = \frac{P_u}{\phi C_1 D l}$ $= \frac{30.0 \text{ kips}}{(0.75)(1.0)(4)(12 \text{ in})}$ $= 0.83$	$C_{min} = \frac{\Omega P_a}{C_1 D l}$ $= \frac{(2)(20.0 \text{ kips})}{(1)(4)(12 \text{ in})}$ $= 0.83$

From *AISC Manual* Table 8-8, with $a = 1.0$ and $C_{\text{min}} = 0.83$, $k = 0.2$. Then $kl = (0.2)(12 \text{ in}) = 2.4$ in (use 2.5 in).

The weld sizes are $^1/_8$ in, $^3/_{16}$ in, and $^1/_4$ in. The horizontal lengths to be used are: for a $^1/_8$ in weld, 6.0 in; for a $^3/_{16}$ in weld, 3.6 in; for a $^1/_4$ in weld, 2.5 in.

9. DESIGN OF HSS AND BOX MEMBER CONNECTIONS

AISC Specification Chap. K governs the design of HSS and box member connections. The design of these connections is complex, but the many tables in Chap. K organize the needed information and facilitate the process. The chapter is divided into the following sections.

K1	Concentrated Forces on HSS
K2	HSS-to-HSS Truss Connections
K3	HSS-to-HSS Moment Connections
K4	Welds of Plates and Branches to Rectangular HSS

One reason for the complexity is that there are many different criteria that may apply, depending on

- whether the HSS member is round or rectangular (including square)
- the member's wall thickness and slenderness ratio
- whether the load is applied by a plate or another HSS
- the type of connection: T, Y, cap plate, cross, or K with gap
- the angle of the load to the member: axial, transverse, or longitudinal

Moreover, the criteria apply only within certain limits, and the limits of applicability differ depending on the type of connection.

The following tables in *AISC Specification* Chap. K give the formulas for calculating the available strength for connections under various circumstances.

Table K1.1	Available Strengths of Plate-to-Round HSS Connections
Table K1.2	Available Strengths of Plate-to-Rectangular HSS Connections
Table K2.1	Available Strengths of Round HSS-to-HSS Truss Connections
Table K2.2	Available Strengths of Rectangular HSS-to-HSS Truss Connections
Table K3.1	Available Strengths of Round HSS-to-HSS Moment Connections
Table K3.2	Available Strengths of Rectangular HSS-to-HSS Moment Connections
Table K4.1	Effective Weld Properties for Connections to Rectangular HSS

Each of the tables for available strengths is accompanied by an auxiliary table that lists the limits of applicability for the formulas in the table. For example, *AISC Specification* Table K1.2A, Limits of Applicability of Table K1.2, gives ranges for HSS wall slenderness, material strength, and other parameters. Plate-to-rectangular HSS connections must be within the ranges given in this table.

For example, suppose a connection is to be designed for a $^1/_4$ in × 5 in plate to be joined perpendicularly to a rectangular HSS6 × 6 × $^1/_4$ across the width of the HSS. The steel is ASTM A500 Grade B. Begin by checking the limits of applicability in *AISC Specification* Table K1.2A.

- The plate load angle, θ, must be at least 30°. (It is 90°, so OK.)
- For a transverse branch connection, the HSS wall slenderness ratio, B/t or H/t, for the loaded wall must be no more than 35. (From AISC Table 1-12, it is 22.8, so OK.)
- For a transverse branch connection, the width ratio, B_p/B, must be at least 0.25 and no more than 1.0. (B_p/B = 6 in/5 in = 0.83, so OK.)
- The material strength, F_y, must be no greater than 52 ksi. (From Table 2-4, for a rectangular HSS of A500 Grade B steel, F_y = 46 ksi, so OK.)
- Ductility, F_y/F_u, must be no more than 0.8. (From Table 2-4, F_y/F_u = 46 ksi/58 ksi = 0.79, so OK.)

All the limits of applicability are met. The connection may be designed using the appropriate equations in Table K1.2. This is a transverse plate cross-connection; for transverse plate T- and cross-connections, the following limit states and equations apply.

For the limit state of local yielding of the plate,

$$R_n = \left(\frac{10}{\frac{B}{t}} \right) F_y t B_p \leq F_{yp} t_p B_p \quad [\text{AISC Eq. K1-7}] \qquad 10.17$$

ϕ = 0.95 for LRFD, and Ω = 1.58 for ASD.

For the limit state of HSS shear yielding (punching), when $0.85B \leq B_p \leq B - 2t$,

$$R_n = 0.6F_y t\left(2t_p + 2B_{ep}\right) \quad [\text{AISC Eq. K1-8}] \qquad 10.18$$

ϕ = 1.00 for LRFD, and Ω = 1.50 for ASD.

For the limit state of local yielding of HSS sidewalls, when B_p/B = 1.0,

$$R_n = 2F_y t\left(5k + l_b\right) \quad [\text{AISC Eq. K1-9}] \qquad 10.19$$

ϕ = 0.75 for LRFD, and Ω = 1.50 for ASD.

For the limit state of local crippling of HSS sidewalls, when $B_p/B = 1.0$ and the plate is in compression, there are two formulas. For a T-connection,

$$R_n = 1.6t^2\left(1+\frac{3l_b}{H-3t}\right)\sqrt{EF_y}Q_f \quad \text{[AISC Eq. K1-10]} \tag{10.20}$$

$\phi = 0.75$ for LRFD, and $\Omega = 2.00$ for ASD.

For a cross-connection,

$$R_n = \left(\frac{48t^3}{H-3t}\right)\sqrt{EF_y}Q_f \quad \text{[AISC Eq. K1-11]} \tag{10.21}$$

$\phi = 0.90$ for LRFD, and $\Omega = 1.67$ for ASD.

Example 10.4

Beam Moment Connection to HSS Column

A beam-column moment connection is shown.

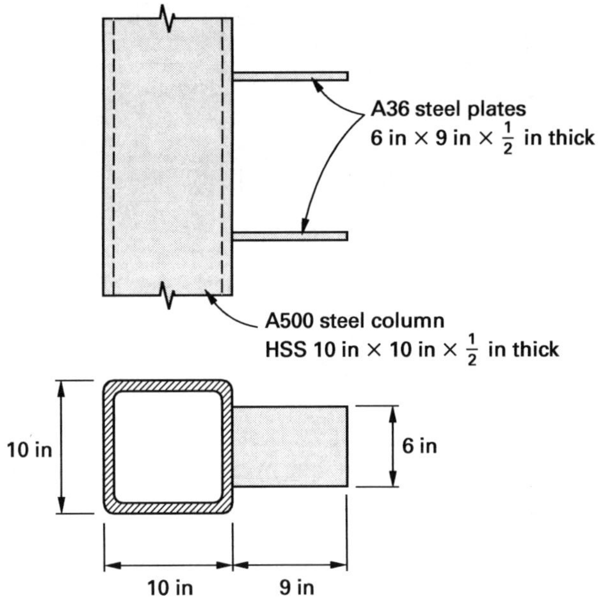

Section properties

$HSS10 \times 10 \times {}^1/_2$

$t = 0.465$ in

$A = 17.2$ in^2

$b/t = 18.5$

$h/t = 18.5$

$I = 256$ in^4

$S = 51.2$ in^3

$r = 3.86$ in

$Z = 60.7$ in^3

flat width = $7^3/_4$ in

plate thickness = $^1/_2$ in

plate width = 6 in

plate length = 9 in

Material properties, HSS

ASTM A500, grade B steel

$F_y = 46$ ksi

$F_u = 58$ ksi

Material properties, plate

ASTM A36 steel

$F_y = 36$ ksi

$F_u = 58$ ksi

Beam-plate connection

$4\,^3/_4$ in diameter ASTM A325X bolts for each flange

Determine the weld size to develop plate capacity, and determine whether the column must be reinforced for the punching shear or web.

Solution

Check the limits of applicability in Table K1.2A.

- plate load angle: $\theta \geq 30°$. (It is 90°, so OK.)
- HSS wall slenderness for a transverse branch connection: $B/t \leq 35$. (It is 18.5, so OK.)
- width ratio for a transverse branch connection: $0.25 \leq B_p/B \leq 1.0$. (B_p/B = 6 in/10 in = 0.60, so OK.)
- material strength: $F_y \leq 52$ ksi. (It is 46 ksi, so OK.)
- ductility: $F_y/F_u \leq 0.8$. (F_y/F_u = 46 ksi/58 ksi = 0.79, so OK.)

All limits of applicability are met. The gross area of the plate is

$$A_g = t_p B_p = \left(\frac{1}{2} \text{ in}\right)(6 \text{ in}) = 3.00 \text{ in}^2$$

From Eq. 4.6, the net area is

$$\begin{aligned} A_n &= A_g - A_h \\ &= 3.00 \text{ in}^2 - (2 \text{ holes})\left(\frac{1}{2} \text{ in}\right)\left(\frac{13}{16} \text{ in} + \frac{1}{16} \text{ in}\right) \\ &= 2.13 \text{ in}^2 \end{aligned}$$

Use Eq. 4.8 to calculate the effective area. Because all elements are in contact, U is 1.0 (from Table 4.1, case 1).

$$A_e = UA_n = (1.0)(2.13 \text{ in}^2) = 2.13 \text{ in}^2$$

Calculate the tensile capacity of the plate for the limit state of yielding on the gross area. The nominal tensile capacity is

$$T_n = F_y A_g = \left(36 \ \frac{\text{kips}}{\text{in}^2}\right)(3 \text{ in}^2) = 108.00 \text{ kips}$$

The required tensile capacity is

LRFD	ASD
$T_u \le \phi_t T_n = (0.90)(108.00 \text{ kips})$ $\le 97.20 \text{ kips}$	$T_a \le \dfrac{T_n}{\Omega_t} = \dfrac{108.00 \text{ kips}}{1.67}$ $\le 64.67 \text{ kips}$

Calculate the tensile capacity of the plate for the limit state of rupture on the net effective area. The nominal tensile capacity is

$$T_n = F_u A_e = \left(58 \ \frac{\text{kips}}{\text{in}^2}\right)(2.13 \text{ in}^2) = 123.54 \text{ kips}$$

The required tensile capacity is

LRFD	ASD
$T_u \le \phi_t T_n$ $\le (0.75)(123.54 \text{ kips})$ $\le 92.66 \text{ kips} \quad [\text{controls}]$	$T_a \le \dfrac{T_n}{\Omega_t} = \dfrac{123.54 \text{ kips}}{2.00}$ $\le 61.77 \text{ kips} \quad [\text{controls}]$

Rupture on the net effective area controls. Check the limits of applicability. From *AISC Manual* Sec. K1.2, the limit for strength is $F_y \le 52$ ksi. For the HSS, $F_y = 46$ ksi, so the strength limit is OK. The limit for ductility is $F_y/F_u \le 0.8$. For the HSS,

$$\frac{F_y}{F_u} = \frac{46 \ \dfrac{\text{kips}}{\text{in}^2}}{58 \ \dfrac{\text{kips}}{\text{in}^2}} = 0.79 \quad [\le 0.8, \text{ so OK}]$$

Therefore, the ductility limit is OK. The limit for the plate width to HSS width ratio, from *AISC Specification* Sec. K1.3b, is $0.25 < B_p/B \le 1.0$. For the HSS,

$$\frac{B_p}{B} = \frac{6 \text{ in}}{10 \text{ in}} = 0.60$$

Therefore, the limit for ratio of plate width to HSS width is OK. The limit for the width-to-thickness ratio of a loaded HSS wall, from *AISC Specification* Sec. K1.3b, is $B/t \leq 35$. From the section properties, $B/t = 18.5$, so this is OK.

Compute the nominal strength using Eq. 10.17.

$$R_n \leq \begin{cases} \left(\dfrac{10F_y t}{\dfrac{B}{t}}\right)B_p = \left(\dfrac{(10)\left(46\ \dfrac{\text{kips}}{\text{in}^2}\right)(0.465\ \text{in})}{18.5}\right)(6\ \text{in}) \\ \quad = 69.37\ \text{kips}\quad [\text{controls}] \\ F_{yp}t_pB_p = \left(36\ \dfrac{\text{kips}}{\text{in}^2}\right)(0.5\ \text{in})(6\ \text{in}) \\ \quad = 108\ \text{kips} \end{cases}$$

Therefore, the nominal strength is OK. Calculate design strength (LRFD) and allowable strength (ASD).

LRFD	ASD
$\phi R_n = (0.95)(69.37\ \text{kips})$ $= 65.90\ \text{kips}$	$\dfrac{R_n}{\Omega} = \dfrac{69.37\ \text{kips}}{1.58}$ $= 43.91\ \text{kips}$

The limit state for shear yielding (punching) need not be checked if

$$\begin{aligned} B_p &< 0.85B \\ 6\ \text{in} &< (0.85)(10\ \text{in}) \\ &< 8.5\ \text{in} \end{aligned}$$

So, shear yielding does not need to be checked. However, use Eq. 10.17 to prove that shear yielding does not govern. First, use Eq. 10.19 to calculate the effective limiting width for shear yielding.

$$B_{ep} \leq \begin{cases} \dfrac{10B_p}{\dfrac{B}{t}} = \dfrac{(10)(6\ \text{in})}{18.5} = 3.24\ \text{in}\quad [\text{controls}] \\ B_p = 8.5\ \text{in} \end{cases}$$

Second, use Eq. 10.18 to calculate the resistance to shear yielding.

$$\begin{aligned} R_n &= 0.6F_y t\left(2t_p + 2B_{ep}\right) \\ &= (0.6)\left(46\ \frac{\text{kips}}{\text{in}^2}\right)(0.456\ \text{in})\big((2)(0.5\ \text{in})+(2)(3.24\ \text{in})\big) \\ &= 94.14\ \text{kips} \end{aligned}$$

Third, calculate the design strength (LRFD) and the allowable strength (ASD).

LRFD	ASD
$\phi R_n = (0.95)(94.14\ \text{kips}) = 89.43\ \text{kips}$	$\frac{R_n}{\Omega} = \frac{94.14\ \text{kips}}{1.58} = 59.58\ \text{kips}$

Summarize the limit states.

limit state	LRFD (kips)	ASD (kips)
tension on gross area of plate	97.20	64.67
tension on net effective area of plate	92.66	61.77
local yielding on HSS wall	65.90	43.91
shear yielding (punching) on HSS wall	89.43	59.58

The local yielding on the HSS wall is the governing limit state. Therefore, as long as the design tensile load on the plate does not exceed 65.90 kips or the allowable load does not exceed 43.91 kips, the column does not have to be reinforced.

The plate width is 6 in. Therefore, the weld needed to develop the design strength would need to resist 65.90 kips/6 in = 10.98 kips/in. To develop the allowable strength, the weld would need to resist 59.58 kips/6 in = 9.93 kips/in.

To obtain the needed weld capacity, a complete penetration weld must be used. The $^1/_2$ in plate thickness is insufficient for the application of $^3/_8$ in fillet welds to the top and bottom surfaces.

11 Plate Girders

Nomenclature

a	clear distance between transverse stiffeners	in
a_w	ratio defined in *AISC Specification* Eq. F4-11, equal to $h_c t_w / b_{fc} t_{fc}$ but no greater than 10	–
A	cross-sectional area	in^2
A_{fg}	gross tension flange area defined in *AISC Specification* Sec. D3.1	in^2
A_{fn}	net tension flange area defined in *AISC Specification* Sec. D3.2	in^2
b	width	in
c	distance to extreme fiber	in
C	compressive force	lbf
C_b	lateral-torsional buckling modification factor	–
C_v	web shear coefficient	–
d	depth or distance	in
D_s	factor defined in *AISC Specification* Sec. G3.3 (1.0 for stiffeners in pairs, 1.8 for single angle stiffeners, 2.4 for single plate stiffeners)	–
E	modulus of elasticity	lbf/in^2
F	strength or stress	lbf/in^2
F_u	specified minimum tensile strength	lbf/in^2
F_y	specified minimum yield stress	lbf/in^2
h	height of web between flanges	in
h_c	distance defined in *AISC Specification* Sec. B4.2	in
h_o	distance between flange centroids	in
I	moment of inertia	in^4
j	factor defined in *AISC Specification* Eq. G2-6	–
k	distance from outer face of flange to web toe of fillet	in
k_c	coefficient for slender unstiffened elements	–
k_v	web plate buckling coefficient	–
K	effective length factor	–
L	length	in
L_b	length between braces or braced points	in

L_p	limiting unbraced length for full plastic moment	in
L_r	limiting unbraced length for inelastic lateral-torsional buckling	in
M	flexural strength or moment	in-lbf
n	number of items	–
N	length of bearing	in
P	concentrated load	lbf
Q	statical moment	in^3
r	radius of gyration	in
r_t	effective radius of gyration for lateral bucking	–
R	reaction	lbf
R_{pg}	bending strength reduction factor	–
S	elastic section modulus	in^3
t	thickness	in
T	tensile force	lbf
V	shear strength or shear stress	lbf
w	load per unit length	lbf/in
Y_t	hole reduction coefficient (1.0 if $F_y/F_u \leq 0.8$, otherwise 1.1)	–

Symbols

λ	limiting width-to-thickness ratio	–
λ_p	limiting width-to-thickness ratio for compactness	–
λ_r	limiting width-to-thickness ratio for noncompactness	–
τ	shear stress	lbf/in^2
ϕ	resistance factor (LRFD)	–
Ω	safety factor (ASD)	–

Subscripts

a	required (ASD)
c	compressive
cr	critical
cross	cross-shaped column
D	dead load
e	elastic critical buckling (Euler)
f	flange
fc	compression flange
ft	tension flange

g	gross
gir	girder
h	horizontal
L	live load
max	maximum
min	minimum
n	net or nominal
o	with respect to the origin
r or req	required
s	steel
st	stiffener
u	required (LRFD)
v	shear
w	web
x	x-axis, strong axis, or horizontal component
xc	about x-axis referred to compression flange
xt	about x-axis referred to tension flange
y	y-axis, weak axis, or vertical component
yc	about y-axis referred to compression flange

1. GENERAL

Plate girders are built-up I-shaped sections that consist of a web member and flanges at each end of the web. When rolled wide-flange sections will not meet the requirements of a project, using plate girders may be necessary or more economical.

While they usually are not thought of as plate girders, the columns and frames of pre-engineered metal buildings are fabricated from plate steel to meet the requirements of the particular project. These elements are designed in a manner similar to plate girders, using the appropriate sections of *AISC Specification* Chap. F. Plate girders are either regular or hybrid. In a *regular plate girder*, all the steel used has the same yield strength. The flanges of a *hybrid girder* have a higher yield strength than the web, putting higher-strength steel at the point of maximum stress. Both types of girders can have uniform or tapered web depths.

The *AISC Manual* no longer places the design specifications for plate girders in a separate chapter. The flexural requirements for plate girders are specified in *AISC Specification* Chap. F. The requirements for shear are covered in Chap. G.

Plate girders may be stiffened (with transverse stiffeners) or unstiffened. (See Fig. 11.1.)

Figure 11.1 *Unstiffened and Stiffened Plate Girders*

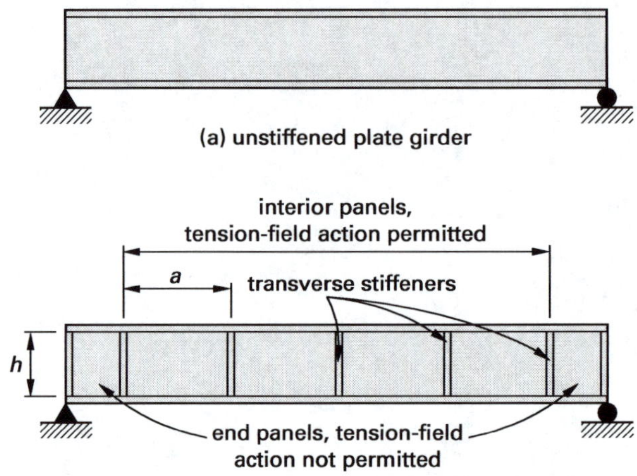

(b) stiffened plate girder

Stiffened plate girders may or may not be designed for *tension-field action* to resist the shear. A stiffened plate girder with tension-field action acts like a Pratt truss, with the web plate resisting diagonal tension; the vertical stiffeners are in compression and also add stability to the web plate. (See Fig. 11.2.) Tension-field action is this truss-like behavior, and designing for it is an economical way to increase the strength of the girder because the stability added by the stiffeners allows the web plate to be thinner and lighter than would otherwise be necessary. When tension-field action is used, however, the stiffeners must then be designed to have a larger moment of inertia.

Figure 11.2 *Tension-Field Action in a Stiffened Plate Girder*

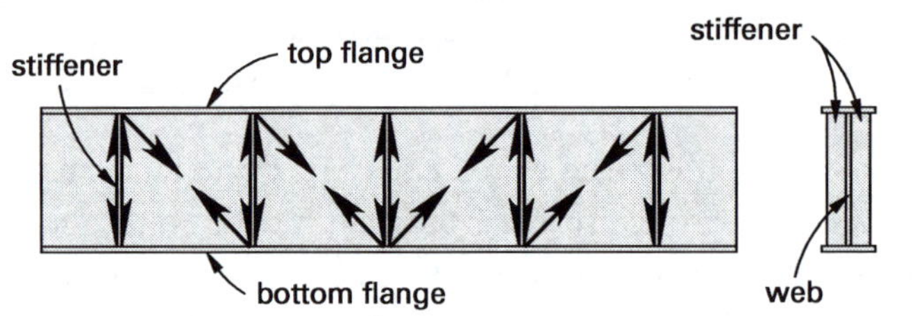

(a) elevation (b) side view

Tension-field action is not permitted in the design of end panels. When designing or analyzing rolled sections, the overall depth of the member (measured between the outside faces of the flanges) is used to resist the shear force. For plate girders, only the girder web (measured between the inside faces of the flanges) is used to resist shear.

2. PLATE GIRDER PROPORTIONING LIMITS

Flange Proportions

The proportioning limits for plate girders are given in *AISC Specification* Sec. F13.

If there are holes in the tension flange for a bolted splice connection or other attachments, Sec. F13.1 requires that the limit state for tensile rupture be checked as follows.

- If $F_u A_{fn} \geq Y_t F_y A_{fg}$, the limit state of tensile rupture doesn't apply.
- If $F_u A_{fn} < Y_t F_y A_{fg}$, the nominal flexural strength at the holes in the tension flange is no more than

$$M_n = \left(\frac{F_u A_{fn}}{A_{fg}}\right) S_x \quad \text{[AISC Eq. F13-1]} \qquad 11.1$$

Singly symmetrical I-shaped members must satisfy Eq. 11.2.

$$0.1 \leq \frac{I_{yc}}{I_y} \leq 0.9 \quad \text{[AISC Eq. F13-2]} \qquad 11.2$$

In Eq. 11.2, I_y is the moment of inertia about the y-axis, and I_{yc} is the moment of inertia about the y-axis referred to the compression flange.

Web Proportions

An I-shaped member with a web height-to-thickness ratio $h/t_w > 5.70\sqrt{E/F_y}$ is considered to have a slender web (*AISC Specification* Table B4.1b, case 15) and must be designed in accordance with *AISC Specification* Sec. F5.

The web thickness for a plate girder designed under *AISC Specification* Sec. F5 and Table B4.1b, case 15, has an upper limit of

$$t_w < \frac{h}{5.70\sqrt{\dfrac{E}{F_y}}} \qquad 11.3$$

For unstiffened girders,

$$\frac{h}{t_w} \leq 260 \quad \text{[AISC Sec. F13.2]} \qquad 11.4$$

$$\frac{A_w}{A_{fc}} \leq 10 \quad \text{[AISC Sec. F13.2]} \qquad 11.5$$

I-shaped members with slender webs must also satisfy the following limits, where a is the clear distance between transverse stiffeners, and h is the height of the web between flanges.

For $a/h \leq 1.5$,

$$\left(\frac{h}{t_w}\right)_{\max} = 12.0\sqrt{\frac{E}{F_y}} \quad \text{[AISC Eq. F13-3]} \tag{11.6}$$

For $a/h > 1.5$,

$$\left(\frac{h}{t_w}\right)_{\max} = \frac{0.40E}{F_y} \quad \text{[AISC Eq. F13-4]} \tag{11.7}$$

The available shear strength in the girder web is a function of the a/h ratio. The following tables in the *AISC Manual* may be used to determine the available shear strength in the web, $\phi V_n/A_w$ (LRFD) or $V_n/\Omega A_w$ (ASD).

- Table 3-16a (F_y = 36 ksi, tension-field action not included)
- Table 3-16b (F_y = 36 ksi, tension-field action included)
- Table 3-17a (F_y = 50 ksi, tension-field action not included)
- Table 3-17b (F_y = 50 ksi, tension-field action included)

3. FLEXURAL STRENGTH

Plate girders are doubly or singly symmetric I-shaped members with slender webs, so the flexural design of plate girders is governed by *AISC Specification* Sec. F5 and the following equations. The nominal flexural strength, M_n, must be calculated separately for up to four different limit states: compression flange yielding, lateral-torsional buckling, compression flange local buckling, and tension flange yielding. (Not every limit state is applicable in every case.) The lowest of the resulting values governs.

For the limit state of *compression flange yielding*, use Eq. 11.8.

$$M_n = R_{pg}F_yS_{xc} \quad \text{[AISC Eq. F5-1]} \tag{11.8}$$

For the limit state of *lateral-torsional buckling*, use Eq. 11.9.

$$M_n = R_{pg}F_{\text{cr}}S_{xc} \quad \text{[AISC Eq. F5-2]} \tag{11.9}$$

In Eq. 11.8 and Eq. 11.9, R_{pg} is the bending strength reduction factor and is equal to

$$R_{pg} = 1 - \left(\frac{a_w}{1200 + 300a_w}\right)\left(\frac{h_c}{t_w} - 5.7\sqrt{\frac{E}{F_y}}\right) \leq 1.0 \quad \text{[AISC Eq. F5-6]} \tag{11.10}$$

In Eq. 11.10,

$$a_w = \frac{h_c t_w}{b_{fc} t_{fc}} \le 10 \quad \text{[AISC Eq. F4-12 and Sec. F5.2]} \qquad 11.11$$

Two formulas are used in different circumstances to determine the value of the critical stress, F_{cr}, in Eq. 11.9. To determine which of these formulas should be used, compare the length between braces or braced points, L_b, with the values of L_p and L_r as defined in Eq. 11.12 and 11.13.

$$L_p = 1.1 r_t \sqrt{\frac{E}{F_y}} \quad \text{[AISC Eq. F4-7]} \qquad 11.12$$

$$L_r = \pi r_t \sqrt{\frac{E}{0.7F_y}} \quad \text{[AISC Eq. F5-5]} \qquad 11.13$$

In Eq. 11.12 and Eq. 11.13, r_t is the effective radius of gyration for lateral buckling and is equal to

$$r_t = \frac{b_{fc}}{\sqrt{12\left(\frac{h_o}{d} + \left(\frac{a_w}{6}\right)\left(\frac{h^2}{h_o d}\right)\right)}} \quad \text{[AISC Eq. F4-11]} \qquad 11.14$$

a_w is defined as in Eq. 11.11.

Depending on how L_b compares with L_p and L_r, use either Eq. 11.15 or Eq. 11.16 to determine F_{cr}.

- When $L_b \le L_p$, the limit state of lateral-torsional buckling does not apply.
- When $L_p < L_b \le L_r$, the critical stress is

$$F_{cr} = C_b\left(F_y - 0.3F_y\left(\frac{L_b - L_p}{L_r - L_p}\right)\right) \le F_y \quad \text{[AISC Eq. F5-3]} \qquad 11.15$$

- When $L_b > L_r$,

$$F_{cr} = \frac{C_b \pi^2 E}{\left(\frac{L_b}{r_t}\right)^2} \le F_y \quad \text{[AISC Eq. F5-4]} \qquad 11.16$$

In Eq. 11.16, r_t is defined as in Eq. 11.14.

For the limit state of *compression flange local buckling*, use Eq. 11.17.

$$M_n = R_{pg} F_{cr} S_{xc} \quad \text{[AISC Eq. F5-7]} \tag{11.17}$$

Either Eq. 11.18 or Eq. 11.20 is used to determine the value of F_{cr} in Eq. 11.17, depending on whether the flanges are noncompact or slender.

- If the section has compact flanges, the limit state of compression flange local buckling doesn't apply.
- If the flange section is noncompact, then

$$F_{cr} = \left(F_y - 0.3F_y \left(\frac{\lambda - \lambda_{pf}}{\lambda_{rf} - \lambda_{pf}} \right) \right) \quad \text{[AISC Eq. F5-8]} \tag{11.18}$$

In Eq. 11.18, λ_{pf} and λ_{rf} are equal to λ_p and λ_r, respectively, from *AISC Specification* Table B4.1 as applied to the flange. λ is

$$\lambda = \frac{b_{fc}}{2t_{fc}} \quad \text{[AISC Sec. F5.3]} \tag{11.19}$$

- If the flange section is slender, then

$$F_{cr} = \frac{0.9Ek_c}{\left(\frac{b_f}{2t_f} \right)^2} \quad \text{[AISC Eq. F5-9]} \tag{11.20}$$

In Eq. 11.20, k_c is the coefficient for slender unstiffened elements and is defined as

$$k_c = \frac{4}{\sqrt{\frac{h}{t_w}}} \quad [0.35 \le k_c \le 0.76] \quad \text{[AISC Sec. F5.3]} \tag{11.21}$$

For *tension flange yielding*, use Eq. 11.22.

$$M_n = F_y S_{xt} \quad \text{[AISC Eq. F5-10]} \tag{11.22}$$

Equation 11.22 should be used only when $S_{xt} < S_{xc}$. When $S_{xt} \ge S_{xc}$, the limit state of tension flange yielding doesn't apply.

4. SHEAR STRENGTH

Whether a plate girder will be unstiffened or stiffened must be decided in the early design stages. Using stiffeners reduces the total steel weight but increases fabrication costs. Visual aspects and long-term maintenance costs must also be considered.

If the web height-to-thickness ratio $h/t_w > 260$, transverse stiffeners are required.

Once the decision has been made to use a stiffened girder, then, in cases where the web plate will be supported on all four sides by the flanges and stiffeners, further economies can be gained by using tension-field action. However, per *AISC Specification* Sec. G3.1, the use of tension-field action is not permitted in some conditions.

- for end panels in all members with transverse stiffeners
- for a member for which $a/h > 3.0$ or $a/h > (260/(h/t))^2$
- where $2A_w/(A_{fc} + A_{ft}) > 2.5$
- where $h/b_{fc} > 6.0$ or $h/b_{ft} > 6.0$

Tension-Field Action Prohibited

When tension-field action cannot be used, the nominal shear strength, V_n, is determined by Eq. 11.23 with $\phi = 0.90$ (LRFD) or $\Omega = 1.67$ (ASD).

$$V_n = 0.6F_yA_wC_v \quad \text{[AISC Eq. G2-1]} \qquad 11.23$$

The value of the web shear coefficient, C_v, depends on the height-to-thickness ratio.

- If $h/t_w \le 1.10\sqrt{k_vE/F_y}$, then

$$C_v = 1.0 \quad \text{[AISC Eq. G2-3]} \qquad 11.24$$

- If $1.10\sqrt{k_vE/F_y} < h/t_w \le 1.37\sqrt{k_vE/F_y}$, then

$$C_v = \frac{1.10\sqrt{\frac{k_vE}{F_y}}}{\frac{h}{t_w}} \quad \text{[AISC Eq. G2-4]} \qquad 11.25$$

- If $1.37\sqrt{k_vE/F_y} < h/t_w$, then

$$C_v = \frac{1.51k_vE}{\left(\frac{h}{t_w}\right)^2 F_y} \quad \text{[AISC Eq. G2-5]} \qquad 11.26$$

In Eq. 11.24 through Eq. 11.26, k_v is the web plate buckling coefficient. $k = 5.0$ for

- unstiffened webs with $h/t_w < 260$
- stiffened webs when $a/h > 3.0$ or $a/h > (260/(h/t))^2$

For other stiffened webs,

$$k_v = 5 + \frac{5}{\left(\frac{a}{h}\right)^2} \quad \text{[AISC Sec. G2.1]} \tag{11.27}$$

The stems of tees are sometimes used as stiffeners. In these cases, $k_v = 1.2$.

Tension-Field Action Permitted

When tension-field action is permitted, use Eq. 11.28 or Eq. 11.29 to determine the nominal shear strength, with $\phi = 0.90$ (LRFD) and $\Omega = 1.67$ (ASD). k_v and C_v are determined as before (Eq. 11.24 through Eq. 11.27).

- If $h/t_w \leq 1.10\sqrt{k_v E/F_y}$, then

$$V_n = 0.6F_y A_w \quad \text{[AISC Eq. G3-1]} \tag{11.28}$$

- If $h/t_w > 1.10\sqrt{k_v E/F_y}$, then

$$V_n = \left(0.6F_y A_w\right)\left(C_v + \frac{1 - C_v}{1.15\sqrt{1 + \left(\frac{a}{h}\right)^2}}\right) \quad \text{[AISC Eq. G3-2]} \tag{11.29}$$

Transverse Stiffeners Without Tension-Field Action

Transverse stiffeners are not required when $h/t_w \leq 2.46\sqrt{E/F_y}$ or when the required shear strength (see Eq. 11.23) is less than or equal to the available shear strength.

When transverse stiffeners are used, they must have a minimum moment of inertia as determined by Eq. 11.30. When stiffeners are used in pairs, this moment of inertia is taken about an axis in the web center; when single stiffeners are used, the moment of inertia is taken about the face of the web plate.

$$I_{st} \geq bt_w^3 j \quad \text{[AISC Eq. G2-7]} \tag{11.30}$$

In Eq. 11.30, b is the smaller of the dimensions a and h, and j is

$$j = \frac{2.5}{\left(\frac{a}{h}\right)^2} - 2 \geq 0.5 \quad \text{[AISC Eq. G2-8]} \tag{11.31}$$

Transverse stiffeners can be terminated short of the tension flange as long as bearing is not needed to transmit a concentrated load. Stiffener-to-web welds should be terminated at a distance between $4t_w$ and $6t_w$ from the near toe of the web-to-flange weld.

When single stiffeners are used, they should be attached to the compression flange. If intermittent fillet welds are used, the clear distance between the welds should not be more than $16t_w$ and not more than 10 in.

Transverse Stiffeners with Tension-Field Action

Transverse stiffeners with tension-field action must meet the preceding requirements for stiffeners without tension-field action, and they must meet the requirements in Eq. 11.32 and Eq. 11.33 as well.

$$\left(\frac{b}{t}\right)_{\text{st}} \le 0.56\sqrt{\frac{E}{F_{y,\text{st}}}} \quad \text{[AISC Eq. G3-3]} \qquad 11.32$$

In Eq. 11.32, $(b/t)_{\text{st}}$ is the width-to-thickness ratio of the stiffener, and $F_{y,\text{st}}$ is the specified minimum yield stress of the stiffener material.

$$I_{\text{st}} \ge I_{\text{st1}} + (I_{\text{st2}} - I_{\text{st1}})\left(\frac{V_r - V_{c1}}{V_{c2} - V_{c1}}\right) \quad \text{[AISC Eq. G3-4]} \qquad 11.33$$

In Eq. 11.33, V_r is the larger of the required shear strengths in the adjacent web panels, using either LRFD or ASD load combinations.

V_{c1} and V_{c2} both stand for the smaller of the available shear strengths in the adjacent web panels; for V_{c1}, nominal shear strength is defined as in Eq. 11.23, and for V_{c2}, nominal shear strength is defined as in Eq. 11.28 or Eq. 11.29, whichever applies.

For pairs of stiffeners, I_{st} is the moment of inertia taken about the center of the web; for a single stiffener, I_{st} is the moment of inertia taken about the face in contact with the web plate. I_{st1} is the minimum moment of inertia required for the stiffener for resistance of web shear buckling, as determined by Eq. 11.30. I_{st2} is the minimum moment of inertia required for the stiffener for resisting the full web shear buckling and web tension field, and is equal to

$$I_{\text{st2}} = \left(\frac{h^4\rho_{\text{st}}^{1.3}}{40}\right)\left(\frac{F_{yw}}{E}\right)^{1.5} \quad \text{[AISC Eq. G3-5]} \qquad 11.34$$

In Eq. 11.34, ρ_{st} is either 1.0 or the ratio of the specified minimum yield stresses of the web and stiffener, $F_{yw}/F_{y,\text{st}}$, whichever is greater.

Example 11.1
Plate Girder Design

Design a plate steel girder to support both concentrated and uniform loads as shown. Each concentrated load (points B, C, and D) consists of a dead load of 60 kips and a live load of 120 kips. The uniform load consists of a dead load of 0.6 kip/ft and a live load of 1.2 kips/ft.

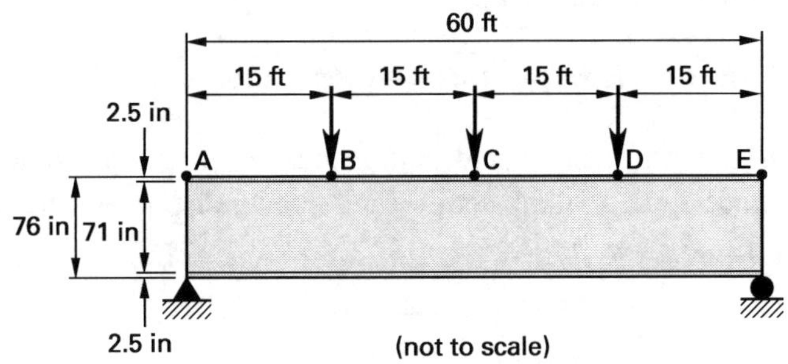

Section properties	Material properties
total depth = 76 in	all plate steel, $F_y = 50$ ksi
web depth = 71 in	
girder braced at both ends and at concentrated loads	

Solution

Combine the dead and live loads to simplify the calculations.

LRFD	ASD
$w_u = 1.2w_D + 1.6w_L = (1.2)\left(0.60\ \frac{\text{kip}}{\text{ft}}\right) + (1.6)\left(1.20\ \frac{\text{kips}}{\text{ft}}\right) = 2.64\ \text{kips/ft}$	$w_a = w_D + w_L = 0.60\ \frac{\text{kip}}{\text{ft}} + 1.20\ \frac{\text{kips}}{\text{ft}} = 1.80\ \text{kips/ft}$
$P_u = 1.2P_D + 1.6P_L = (1.2)(60\ \text{kips}) + (1.6)(120\ \text{kips}) = 264\ \text{kips}$	$P_a = P_D + P_L = 60\ \text{kips} + 120\ \text{kips} = 180\ \text{kips}$

Calculate the end reactions at points A and E. By symmetry, $R_A = R_E$, so each is equal to the sum of the loads divided by two.

LRFD	ASD
$R_A = R_E = \frac{3P_u + w_u L}{2}$ $= \frac{(3)(264\text{ kips}) + \left(2.64\ \frac{\text{kips}}{\text{ft}}\right)(60\text{ ft})}{2}$ $= 475.20\text{ kips}$	$R_A = R_E = \frac{3P_a + w_a L}{2}$ $= \frac{(3)(180\text{ kips}) + \left(1.80\ \frac{\text{kips}}{\text{ft}}\right)(60\text{ ft})}{2}$ $= 324.00\text{ kips}$

Find the bending moments at concentrated loads B and D. By symmetry, $M_B = M_D$.

LRFD	ASD
$M_B = M_D = R_A L_{AB} - \frac{w_u L_{AB}^2}{2}$ $= (475.20\text{ kips})(15\text{ ft}) - \frac{\left(2.64\ \frac{\text{kips}}{\text{ft}}\right)(15\text{ ft})^2}{2}$ $= 6831\text{ ft-kips}$	$M_B = M_D = R_A L_{AB} - \frac{w_u L_{AB}^2}{2}$ $= (324.00\text{ kips})(15\text{ ft}) - \frac{\left(1.80\ \frac{\text{kips}}{\text{ft}}\right)(15\text{ ft})^2}{2}$ $= 4657.50\text{ ft-kips}$

Calculate the bending moment at the concentrated load C.

LRFD	ASD
$M_C = (475.20\text{ kips})(30\text{ ft}) - (264\text{ kips})(15\text{ ft}) - \frac{\left(2.64\ \frac{\text{kips}}{\text{ft}}\right)(30\text{ ft})^2}{2}$ $= 9108\text{ ft-kips}$	$M_C = (324.00\text{ kips})(30\text{ ft}) - (180\text{ kips})(15\text{ ft}) - \frac{\left(1.80\ \frac{\text{kips}}{\text{ft}}\right)(30\text{ ft})^2}{2}$ $= 6210\text{ ft-kips}$

Calculate the upper and lower shear values at the concentrated loads B and C.

LRFD	ASD
$V_{\text{B,upper}} = R_{\text{A}} - w_u L_{\text{AB}}$ $= 475.20 \text{ kips} - \left(2.64 \dfrac{\text{kips}}{\text{ft}}\right)(15 \text{ ft})$ $= 435.60 \text{ kips}$	$V_{\text{B,upper}} = R_{\text{A}} - w_a L_{\text{AB}}$ $= 324.00 \text{ kips} - \left(1.80 \dfrac{\text{kips}}{\text{ft}}\right)(15 \text{ ft})$ $= 297.00 \text{ kips}$
$V_{\text{B,lower}} = V_{\text{B,upper}} - P_{u,\text{B}}$ $= 435.60 \text{ kips} - 264 \text{ kips}$ $= 171.60 \text{ kips}$	$V_{\text{B,lower}} = V_{\text{B,upper}} - P_{a,\text{B}}$ $= 297.00 \text{ kips} - 180 \text{ kips}$ $= 117 \text{ kips}$
$V_{\text{C,upper}} = V_{\text{B,lower}} - w_u L_{\text{BC}}$ $= 171.60 \text{ kips} - \left(2.64 \dfrac{\text{kips}}{\text{ft}}\right)(15 \text{ ft})$ $= 132 \text{ kips}$	$V_{\text{C,upper}} = V_{\text{B,lower}} - w_a L_{\text{BC}}$ $= 117 \text{ kips} - \left(1.8 \dfrac{\text{kips}}{\text{ft}}\right)(15 \text{ ft})$ $= 90 \text{ kips}$
$V_{\text{C,lower}} = V_{\text{C,upper}} - P_{u,\text{C}}$ $= 132 \text{ kips} - 264 \text{ kips}$ $= -132 \text{ kips}$	$V_{\text{C,lower}} = V_{\text{C,upper}} - P_{a,\text{C}}$ $= 90 \text{ kips} - 180 \text{ kips}$ $= -90 \text{ kips}$

Calculate the governing h/t_w ratios. For unstiffened girders, use Eq. 11.4.

$$\frac{h}{t_w} \leq 260$$

$$t_{w,\text{min}} = \frac{h}{260} = \frac{71 \text{ in}}{260} = 0.27 \text{ in}$$

For stiffened girders with $a/h \leq 1.5$, use Eq. 11.6.

$$\left(\frac{h}{t_w}\right)_{\text{max}} = 12.0\sqrt{\frac{E}{F_y}}$$

$$t_{w,\text{min}} = \frac{h}{12.0\sqrt{\dfrac{E}{F_y}}} = \frac{71 \text{ in}}{12.0\sqrt{\dfrac{29{,}000 \dfrac{\text{kips}}{\text{in}^2}}{50 \dfrac{\text{kips}}{\text{in}^2}}}}$$

$$= 0.2456 \text{ in}$$

For stiffened girders with $a/h > 1.5$, use Eq. 11.7.

$$\left(\frac{h}{t_w}\right)_{\max} = \frac{0.40E}{F_y}$$

$$t_{w,\min} = \frac{hF_y}{0.40E} = \frac{(71\text{ in})\left(50\ \frac{\text{kips}}{\text{in}^2}\right)}{(0.40)\left(29{,}000\ \frac{\text{kips}}{\text{in}^2}\right)} = 0.31\text{ in}$$

Use Eq. 11.3 to determine the maximum web thickness allowable for the slender web member.

$$t_w < \frac{h}{5.70\sqrt{\frac{E}{F_y}}} = \frac{71\text{ in}}{5.70\sqrt{\frac{29{,}000\ \frac{\text{kips}}{\text{in}^2}}{50\ \frac{\text{kips}}{\text{in}^2}}}} = 0.52\text{ in}$$

The web thickness must be

$$0.25\text{ in} \le t_w < 0.52\text{ in}$$

Try a $^3/_8 \times 71$ in the web.

$$A_w = t_w h = (0.375\text{ in})(71\text{ in}) = 26.625\text{ in}^2$$

Calculate the moment of inertia of the web.

$$I_w = \frac{t_w h^3}{12} = \frac{(0.375\text{ in})(71\text{ in})^3}{12} = 11{,}184.72\text{ in}^4$$

Determine the approximate flange area required based on the tension flange yielding. This does not consider the flexural contribution of the web and is therefore a conservative approach. Assume a flange thickness of 2.5 in.

LRFD	ASD
$T_u = C_u = \frac{M_C}{d}$ $= \frac{(9108\text{ ft-kips})\left(12\ \frac{\text{in}}{\text{ft}}\right)}{71\text{ in} + 2.5\text{ in}}$ $= 1487\text{ kips}$	$T_a = C_a = \frac{M_C}{d}$ $= \frac{(6210\text{ ft-kips})\left(12\ \frac{\text{in}}{\text{ft}}\right)}{71\text{ in} + 2.5\text{ in}}$ $= 1014\text{ kips}$

LRFD	ASD
$A_f = \dfrac{T_u}{\phi_t F_y} = \dfrac{1487 \text{ kips}}{(0.90)\left(50 \ \dfrac{\text{kips}}{\text{in}^2}\right)}$ $= 33.04 \text{ in}^2$	$A_f = \dfrac{T_a \Omega_t}{F_y} = \dfrac{(1014 \text{ kips})(1.67)}{50 \ \dfrac{\text{kips}}{\text{in}^2}}$ $= 33.87 \text{ in}^2$

Determine the approximate width of the flange.

LRFD	ASD
$b_f = \dfrac{A_f}{t_f} = \dfrac{33.04 \text{ in}^2}{2.5 \text{ in}}$ $= 13.22 \text{ in} \quad [\text{use 15 in}]$	$b_f = \dfrac{A_f}{t_f} = \dfrac{33.87 \text{ in}^2}{2.5 \text{ in}}$ $= 13.55 \text{ in} \quad [\text{use 15 in}]$

A 14 in wide flange could be used, but then the stiffeners would have to be significantly thicker to meet the required value for moment of inertia. A slightly wider flange is a more economical choice overall. Try a 15 in wide flange.

Use *AISC Specification* Table B4.1b, case 10, to check the flange for compactness.

$$\lambda_p = 0.38\sqrt{\frac{E}{F_y}} = 0.38\sqrt{\frac{29{,}000 \ \frac{\text{kips}}{\text{in}^2}}{50 \ \frac{\text{kips}}{\text{in}^2}}} = 9.15$$

$$\frac{b_f}{2t_f} = \frac{15 \text{ in}}{(2)(2.5 \text{ in})} = 3 \quad [< \lambda_p, \text{ so compact}]$$

The flanges are compact. Calculate the moment of inertia of the two flanges.

$$\begin{aligned} I_f &= 2\left(I_o + A_f d^2\right) \\ &= 2\left(\frac{bh^3}{12} + bhd^2\right) \\ &= (2)\left(\frac{(15 \text{ in})(2.5 \text{ in})^3}{12} + (15 \text{ in})(2.5 \text{ in})\left(\frac{76 \text{ in}}{2} - \frac{2.5 \text{ in}}{2}\right)^2\right) \\ &= 101{,}331 \text{ in}^4 \end{aligned}$$

Add the moments of inertia for the web and the flanges to get the moment of inertia for the entire girder.

$$\begin{aligned} I_x &= I_w + I_f = 11{,}185 \text{ in}^4 + 101{,}331 \text{ in}^4 \\ &= 112{,}516 \text{ in}^4 \end{aligned}$$

Calculate the section moduli for the tension and compression flanges. When a plate girder is not doubly symmetrical, these must be calculated separately. In this case, the section is doubly symmetrical, so the section moduli for the two flanges are equal.

$$S_{xt} = S_{xc} = \frac{I_x}{c} = \frac{112{,}516 \text{ in}^4}{\frac{76 \text{ in}}{2}} = 2961 \text{ in}^3$$

Calculate the bending strength reduction factor, R_{pg}. First, use Eq. 11.11 to find a_w.

$$a_w \le \begin{cases} \dfrac{h_c t_w}{b_{\text{fc}} t_{\text{fc}}} = \dfrac{(71 \text{ in})(0.375 \text{ in})}{(15 \text{ in})(2.5 \text{ in})} = 0.71 \quad [\text{controls}] \\ 10 \end{cases}$$

Use this value in Eq. 11.10 to find R_{pg}.

$$R_{pg} \le \begin{cases} 1-\left(\dfrac{a_w}{1200+300a_w}\right)\left(\dfrac{h_c}{t_w}-5.7\sqrt{\dfrac{E}{F_y}}\right) \\ \quad = 1-\left(\dfrac{0.71}{1200+(300)(0.71)}\right)\left(\dfrac{71 \text{ in}}{0.375 \text{ in}}-5.7\sqrt{\dfrac{29{,}000 \ \frac{\text{kips}}{\text{in}^2}}{50 \ \frac{\text{kips}}{\text{in}^2}}}\right) \\ \quad = 0.97 \quad [\text{controls}] \\ 1.0 \end{cases}$$

Use Eq. 11.8 to calculate the compression flange yielding strength.

$$M_n = R_{pg} F_y S_{xc} = \frac{(0.97)\left(50 \ \frac{\text{kips}}{\text{in}^2}\right)(2961 \text{ in}^3)}{12 \ \frac{\text{in}}{\text{ft}}} = 11{,}967 \text{ ft-kips}$$

Use Eq. 11.14 to calculate the effective radius of gyration for lateral-torsional buckling.

$$r_t = \frac{b_{\text{fc}}}{\sqrt{12\left(\dfrac{h_o}{d}+\left(\dfrac{a_w}{6}\right)\left(\dfrac{h^2}{h_o d}\right)\right)}}$$

$$= \frac{15 \text{ in}}{\sqrt{(12)\left(\dfrac{73.5 \text{ in}}{76 \text{ in}}+\left(\dfrac{0.71}{6}\right)\left(\dfrac{(71 \text{ in})^2}{(73.5 \text{ in})(76 \text{ in})}\right)\right)}}$$

$$= 4.18 \text{ in}$$

Use Eq. 11.12 and Eq. 11.13 to calculate L_p and L_r.

$$L_p = 1.1 r_t \sqrt{\frac{E}{F_y}} = \frac{(1.1)(4.18 \text{ in})\sqrt{\frac{29{,}000 \frac{\text{kips}}{\text{in}^2}}{50 \frac{\text{kips}}{\text{in}^2}}}}{12 \frac{\text{in}}{\text{ft}}}$$

$$= 9.23 \text{ ft}$$

$$L_r = \pi r_t \sqrt{\frac{E}{0.7F_y}} = \frac{\pi(4.18 \text{ in})\sqrt{\frac{29{,}000 \frac{\text{kips}}{\text{in}^2}}{(0.7)\left(50 \frac{\text{kips}}{\text{in}^2}\right)}}}{12 \frac{\text{in}}{\text{ft}}}$$

$$= 31.50 \text{ ft}$$

$L_b - 15$ ft, so $L_p < L_b < L_r$ and latcral-torsional buckling applies.

Calculate the lateral-torsional buckling strength. First, use Eq. 11.15 to compute the critical stress. Take the value of C_b conservatively as 1.0.

$$F_{\text{cr}} \le \begin{cases} C_b\left(F_y - 0.3F_y\left(\frac{L_b - L_p}{L_r - L_p}\right)\right) \\ \quad = (1.0)\left(50 \frac{\text{kips}}{\text{in}^2} - (0.3)\left(50 \frac{\text{kips}}{\text{in}^2}\right)\left(\frac{15 \text{ ft} - 9.23 \text{ ft}}{31.50 \text{ ft} - 9.23 \text{ ft}}\right)\right) \\ \quad = 46.11 \text{ ksi} \quad [\text{controls}] \\ F_y = 50 \text{ ksi} \end{cases}$$

From Eq. 11.9, the strength in resistance to lateral-torsional buckling is

$$M_n = R_{pg} F_{\text{cr}} S_{xc}$$

$$= \frac{(0.97)\left(46.11 \frac{\text{kips}}{\text{in}^2}\right)(2961 \text{ in}^3)}{12 \frac{\text{in}}{\text{ft}}}$$

$$= 11{,}036 \text{ ft-kips}$$

Because the flanges are compact, the limit state of flange local buckling does not apply. Because $S_{xt} \ge S_{xc}$, the limit state of tension yielding does not apply.

The following summarizes the flexural strengths. The required flexural strength is

LRFD	ASD
$M_u = 9108$ ft-kips	$M_a = 6210$ ft-kips

For compression flange yielding, M_n = 11,967 ft-kips and the available flexural strength is

LRFD	ASD
$\phi M_n = (0.90)(11{,}967 \text{ ft-kips})$ $= 10{,}770 \text{ ft-kips} \quad [\geq M_u, \text{so OK}]$	$\frac{M_n}{\Omega} = \frac{11{,}967 \text{ ft-kips}}{1.67}$ $= 7166 \text{ ft-kips} \quad [\geq M_a, \text{so OK}]$

For lateral-torsional buckling, M_n = 11,036 ft-kips, and the available flexural strength is

LRFD	ASD
$\phi M_n = (0.90)(11{,}036 \text{ ft-kips})$ $= 9932.4 \text{ ft-kips} \quad [\geq M_u, \text{so OK}]$	$\frac{M_n}{\Omega} = \frac{11{,}036 \text{ ft-kips}}{1.67}$ $= 6608.4 \text{ ft-kips} \quad [\geq M_a, \text{so OK}]$

Compression flange buckling and tension flange yielding are not applicable limit states.

Bearing Stiffeners

Determine whether bearing stiffeners are required beneath the concentrated loads (R_u = P_u = 264 kips and R_a = P_a = 180 kips). First, use Eq. 6.4 to determine the web local yielding strength. Assume l_b is zero.

$$
\begin{aligned}
R_n &= (5k + l_b) F_{yw} t_w \\
&= \big((5)(2.5 \text{ in}) + 0 \text{ in}\big)\left(50 \ \frac{\text{kips}}{\text{in}^2}\right)(0.375 \text{ in}) \\
&= 234.38 \text{ kips}
\end{aligned}
$$

Check whether this is adequate.

LRFD	ASD
$R_u \leq \phi R_n$ $264 \text{ kips} \leq (1.00)(234.38 \text{ kips})$ $\leq 234.38 \text{ kips} \quad [\text{not OK}]$	$R_a \leq \frac{R_n}{\Omega}$ $180 \text{ kips} \leq \frac{234.38 \text{ kips}}{1.50}$ $\leq 156.25 \text{ kips} \quad [\text{not OK}]$

A stiffener is required.

Determine the web crippling strength using Eq. 6.6.

$$R_n = \left(0.80t_w^2\right)\left(1+3\left(\frac{l_b}{d}\right)\left(\frac{t_w}{t_f}\right)^{1.5}\right)\sqrt{\frac{EF_{yw}t_f}{t_w}}$$

$$= (0.80)(0.375\text{ in})^2\left(1+(3)\left(\frac{0\text{ in}}{76\text{ in}}\right)\left(\frac{0.375\text{ in}}{2.5\text{ in}}\right)^{1.5}\right)$$

$$\times\sqrt{\frac{\left(29{,}000\ \frac{\text{kips}}{\text{in}^2}\right)\left(50\ \frac{\text{kips}}{\text{in}^2}\right)(2.5\text{ in})}{0.375\text{ in}}}$$

$$= 350\text{ kips}$$

Check whether this is adequate.

LRFD	ASD
$R_u \le \phi R_n$ $264\text{ kips} \le (0.75)(350\text{ kips})$ $\le 263\text{ kips}$ [not OK]	$R_a \le \dfrac{R_n}{\Omega}$ $180\text{ kips} \le \dfrac{350\text{ kips}}{2.00}$ $\le 175\text{ kips}$ [not OK]

Bearing stiffeners are required at the concentrated loads (R_u = 264 kips and R_a = 180 kips) to prevent web local yielding. Because the reactions at the ends of the girder are of a greater magnitude (R_A = 475.20 kips for LRFD, and R_A = 324.00 kips for ASD), bearing stiffeners will also be required at those locations. (These bearing stiffeners will be designed in Ex. 11.2.)

Transverse Shear Stiffeners

Determine whether additional stiffeners located between the bearing stiffeners at the ends of the girders and those at the concentrated loads will be required for shear resistance. First, check the web height-to-thickness ratio (see Sec. 11.4).

$$\frac{h}{t_w} = \frac{71\text{ in}}{0.375\text{ in}} = 189.33 \quad [\le 260]$$

h/t_w < 260, so the web plate buckling coefficient is k_v = 5. Determine whether Eq. 11.24, Eq. 11.25, or Eq. 11.26 is the applicable formula for the web shear coefficient, C_v.

$$1.37\sqrt{\frac{k_vE}{F_y}} = (1.37)\sqrt{\frac{(5)\left(29{,}000\ \frac{\text{kips}}{\text{in}^2}\right)}{50\ \frac{\text{kips}}{\text{in}^2}}} = 73.78 \quad [< h/t_w,\ \text{so use Eq. 11.26}]$$

From Eq. 11.26, the web shear coefficient is

$$C_v = \frac{1.51Ek_v}{\left(\frac{h}{t_w}\right)^2 F_y}$$

For stiffened webs, the web plate buckling coefficient is $k_v = 5.0$ when $a/h > 3.0$ and in other cases is determined by Eq. 11.27.

$$k_v = 5 + \frac{5}{\left(\frac{a}{h}\right)^2}$$

As the spacing between transverse stiffeners, a, is not yet established, k_v cannot yet be determined, but a tentative value must be used. An easy place to start is by assuming that $a/h > 3.0$ and $k_v = 5.0$.

$$C_v = \frac{1.51Ek_v}{\left(\frac{h}{t_w}\right)^2 F_y} = \frac{(1.51)\left(29{,}000\ \frac{\text{kips}}{\text{in}^2}\right)(5.0)}{(189.33)^2\left(50\ \frac{\text{kips}}{\text{in}^2}\right)} = 0.12$$

Determine the nominal shear strength using Eq.11.23.

$$A_w = ht_w = (71\ \text{in})(0.375\ \text{in}) = 26.625\ \text{in}^2$$

$$\begin{aligned} V_n &= 0.6F_yA_wC_v = (0.6)\left(50\ \frac{\text{kips}}{\text{in}^2}\right)(26.625\ \text{in}^2)(0.12) \\ &= 95.85\ \text{kips} \end{aligned}$$

Check whether the nominal shear strength is adequate ($V_{u,\text{A}} = R_\text{A}$ for LRFD and $V_{a,\text{A}} = R_\text{A}$ for ASD).

LRFD	ASD
$V_{u,\text{A}} \le \phi_v V_n$ $475.20\ \text{kips} \le (0.90)(95.85\ \text{kips})$ $\le 86.27\ \text{kips}\quad [\text{not OK}]$	$V_{a,\text{A}} \le \frac{V_n}{\Omega_v}$ $324.00\ \text{kips} \le \frac{95.85\ \text{kips}}{1.67}$ $\le 57.40\ \text{kips}\quad [\text{not OK}]$

Stiffeners are required.

To determine the location of the first transverse stiffener (the end panel stiffener), use *AISC Manual* Table 3-17a. (Tension-field action is not permitted in end panels.

However, if tension-field action is used at all, then all transverse stiffeners must be designed using Eq. 11.33.)

$$\frac{h}{t_w} = 189.33 \quad [\text{use } 190]$$

LRFD	ASD
$\frac{\phi_v V_n}{A_w} = \frac{V_{u,\text{A}}}{A_w} = \frac{475.20 \text{ kips}}{26.625 \text{ in}^2} = 17.85 \text{ ksi}$	$\frac{V_n}{\Omega_v A_w} = \frac{V_{a,\text{A}}}{A_w} = \frac{324.00 \text{ kips}}{26.625 \text{ in}^2} = 12.17 \text{ ksi}$

From *AISC Manual* Table 3-17a, for both LRFD and ASD, $a/h = 0.40$. Therefore, a stiffener is required at a maximum of $a = 0.4h = (0.4)(71 \text{ in}) = 28.4$ in; use $a = 28$ in. The end panel, then, will extend from the end of the girder to 28 in from the end. Use *AISC Manual* Table 3-17b to determine the location of successive interior stiffeners. First, determine the shear at 28 in from the end.

LRFD	ASD
$V_{u,28 \text{ in}} = V_{u,\text{A}} - w_u a = 475.20 \text{ kips} - \left(2.64 \frac{\text{kips}}{\text{ft}}\right)\left(\frac{28 \text{ in}}{12 \frac{\text{in}}{\text{ft}}}\right) = 469.05 \text{ kips}$	$V_{a,28 \text{ in}} = V_{a,\text{A}} - w_a a = 324.00 \text{ kips} - \left(1.80 \frac{\text{kips}}{\text{ft}}\right)\left(\frac{28 \text{ in}}{12 \frac{\text{in}}{\text{ft}}}\right) = 319.81 \text{ kips}$

Determine the available shear strength of the end panel. First, find the a/h ratio of the panel.

$$\frac{a}{h} = \frac{28 \text{ in}}{71 \text{ in}} = 0.39$$

$a/h < 3.0$, so use Eq. 11.27 to find the web plate buckling coefficient.

$$k_v = 5 + \frac{5}{\left(\frac{a}{h}\right)^2} = 5 + \frac{5}{(0.39)^2} = 37.87$$

Use Eq. 11.26 to find the web shear coefficient.

$$C_v = \frac{1.51 E k_v}{\left(\frac{h}{t_w}\right)^2 F_y} = \frac{(1.51)\left(29{,}000 \frac{\text{kips}}{\text{in}^2}\right)(37.87)}{(189.33)^2 \left(50 \frac{\text{kips}}{\text{in}^2}\right)} = 0.93$$

Determine the nominal shear strength using Eq. 11.23.

$$V_n = 0.6F_yA_wC_v = (0.60)\left(50\ \frac{\text{kips}}{\text{in}^2}\right)\left(26.625\ \text{in}^2\right)(0.93) = 742.84\ \text{kips}$$

The available shear strength of the end panel is

LRFD	ASD
$V_{u,A} \le \phi_v V_n$ $475.20\ \text{kips} \le (0.90)(742.84\ \text{kips})$ $\le 668.56\ \text{kips}\quad [\text{so OK}]$	$V_{a,A} \le \dfrac{V_n}{\Omega_v}$ $324.00\ \text{kips} \le \dfrac{742.84\ \text{kips}}{1.67}$ $\le 444.81\ \text{kips}\quad [\text{so OK}]$

Continuing to use Table 3-15a would result in four or five more transverse stiffeners between the end panel stiffener and the bearing stiffener at the quarter point of the girder. Taking advantage of tension-field action, however, allows the use of fewer stiffeners spaced further apart.

To take advantage of tension-field action, determine the location of the first interior stiffener using *AISC Manual* Table 3-17b.

$$\frac{h}{t_w} = 189.33\quad [\text{use } 190]$$

LRFD	ASD
$\dfrac{\phi_v V_n}{A_w} = \dfrac{V_{u,28\ \text{in}}}{A_w} = \dfrac{469.05\ \text{kips}}{26.625\ \text{in}^2}$ $= 17.62\ \text{ksi}$	$\dfrac{V_n}{\Omega_v A_w} = \dfrac{V_{a,28\ \text{in}}}{A_w} = \dfrac{319.81\ \text{kips}}{26.625\ \text{in}^2}$ $= 12.01\ \text{ksi}$

From *AISC Manual* Table 3-17a, $a/h = 1.10$. Therefore, the maximum distance between transverse stiffeners is $a = 1.10h = (1.10)(71\ \text{in}) = 78.10\ \text{in}$.

The distance from the end of the girder to the concentrated load is 180 in, and the distance from the end of the girder to the end panel stiffener is 28 in, so there is 152 in between the end panel stiffener and the bearing stiffener beneath the concentrated load. This is less than twice the maximum distance between stiffeners (78.10 in), so only one interior stiffener is needed. If this stiffener were placed at the maximum of 78 in from the end panel stiffener, there would be 74 in between this stiffener and the bearing stiffener. It is better, both for aesthetics and to more evenly distribute the shear load, to make the two interior panels equal by placing the interior stiffener at 76 in.

The stiffeners will be placed as shown.

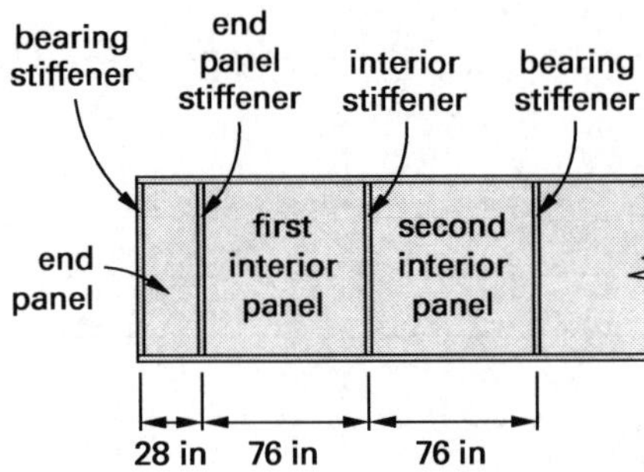

Check that the shear strengths of the panels are adequate. The aspect ratio of the first interior panel is

$$\frac{a}{h} = \frac{76 \text{ in}}{71 \text{ in}} = 1.07$$

$a/h < 3.0$, so use Eq. 11.27 to find the web plate buckling coefficient.

$$k_v = 5 + \frac{5}{\left(\frac{a}{h}\right)^2} = 5 + \frac{5}{(1.07)^2} = 9.37$$

Use Eq. 11.26 to find the web shear coefficient.

$$C_v = \frac{1.51Ek_v}{\left(\frac{h}{t_w}\right)^2 F_y} = \frac{(1.51)\left(29{,}000\ \frac{\text{kips}}{\text{in}^2}\right)(9.37)}{(189.33)^2\left(50\ \frac{\text{kips}}{\text{in}^2}\right)} = 0.23$$

Tension-field action is permitted, so use Eq. 11.29 to determine the available nominal shear strength.

$$V_n = 0.60F_yA_w\left(C_v + \frac{1-C_v}{1.15\sqrt{1+\left(\frac{a}{h}\right)^2}}\right)$$

$$= (0.60)\left(50\ \frac{\text{kips}}{\text{in}^2}\right)(26.625 \text{ in}^2)\left(0.23 + \frac{1-0.23}{1.15\sqrt{1+(1.07)^2}}\right)$$

$$= 548.89 \text{ kips}$$

Check whether this is adequate.

LRFD	ASD
$V_{u,28\text{ in}} \le \phi_v V_n$ $469.05\text{ kips} \le (0.90)(548.89\text{ kips})$ $\le 494.00\text{ kips}$ [so OK]	$V_{a,28\text{ in}} \le \dfrac{V_n}{\Omega_v}$ $319.81\text{ kips} \le \dfrac{548.89\text{ kips}}{1.67}$ $\le 328.68\text{ kips}$ [so OK]

Additional stiffeners are not required for the first interior panel.

The available shear capacities of the second interior panel will be the same as for the first interior panel because the aspect ratios, a/h, and web height-to-thickness ratios, h/t_w, are the same for both panels.

Determine whether stiffeners are required in the panel between the concentrated loads at the quarter point and the midpoint of the girder.

$$\frac{a}{h} = \frac{180\text{ in}}{71\text{ in}} = 2.54$$

$a/h < 3.0$, so use Eq. 11.27 to find the web plate buckling coefficient.

$$k_v = 5 + \frac{5}{\left(\frac{a}{h}\right)^2} = 5 + \frac{5}{(2.54)^2} = 5.78$$

Use Eq. 11.26 to find the web shear coefficient.

$$C_v = \frac{1.51Ek_v}{\left(\frac{h}{t_w}\right)^2 F_y} = \frac{(1.51)\left(29{,}000\ \frac{\text{kips}}{\text{in}^2}\right)(5.78)}{(189.33)^2\left(50\ \frac{\text{kips}}{\text{in}^2}\right)} = 0.14$$

Use Eq. 11.29 to determine the available nominal shear strength.

$$V_n = 0.60F_yA_w\left(C_v + \frac{1-C_v}{1.15\sqrt{1+\left(\frac{a}{h}\right)^2}}\right)$$

$$= (0.60)\left(50\ \frac{\text{kips}}{\text{in}^2}\right)(26.625\text{ in}^2)\left(0.14 + \frac{1-0.14}{1.15\sqrt{1+(2.54)^2}}\right)$$

$$= 330.64\text{ kips}$$

Check whether this is adequate ($V_{u,\text{B}} = V_{B,\text{lower}}$ for LRFD, and $V_{a,\text{B}} = V_{B,\text{lower}}$ for ASD).

LRFD	ASD
$V_{u,\text{B}} \le \phi_v V_n$ $171.6 \text{ kips} \le (0.90)(330.64 \text{ kips})$ $\le 297.58 \text{ kips}$ [so OK]	$V_{a,\text{B}} \le \dfrac{V_n}{\Omega_v}$ $117 \text{ kips} \le \dfrac{330.64 \text{ kips}}{1.67}$ $\le 197.99 \text{ kips}$ [so OK]

Stiffeners are not required for the panels between the concentrated loads.

The following summarizes the shear strengths of the panels between the ends of the girder and the bearing stiffeners at the quarter points.

LRFD	ASD
For each end panel (from end of girder to 28 in from end), $V_u \le \phi_v V_n$ $475.20 \text{ kips} \le 668.55 \text{ kips}$	For each end panel (from end of girder to 28 in from end), $V_a \le \dfrac{V_n}{\Omega_v}$ $324.00 \text{ kips} \le 444.81 \text{ kips}$
For each first interior panel (from 28 in to 104 in from end of girder), $V_u \le \phi_v V_n$ $469.05 \text{ kips} \le 494.00 \text{ kips}$	For each first interior panel (from 28 in to 104 in from end of girder), $V_a \le \dfrac{V_n}{\Omega_v}$ $319.81 \text{ kips} \le 328.68 \text{ kips}$
For each second interior panel (from 104 in to 180 in from end of girder), $V_u \le \phi_v V_n$ $452.31 \text{ kips} \le 494.00 \text{ kips}$	For each second interior panel (from 104 in to 180 in from end of girder), $V_a \le \dfrac{V_n}{\Omega_v}$ $308.39 \text{ kips} \le 328.68 \text{ kips}$

The locations of the stiffeners and the shear diagram for the stiffened girder (with factored shear loads) are shown.

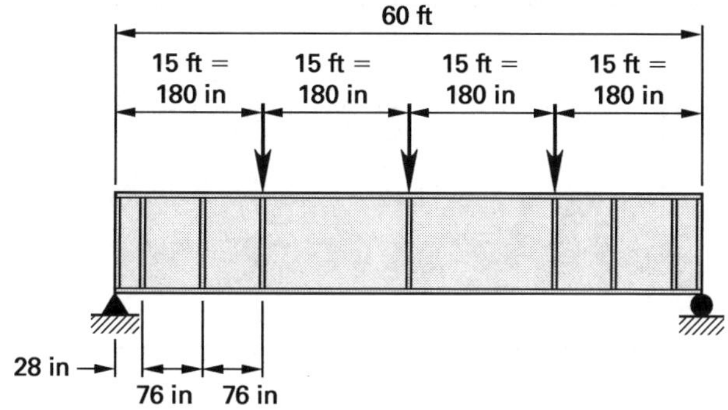

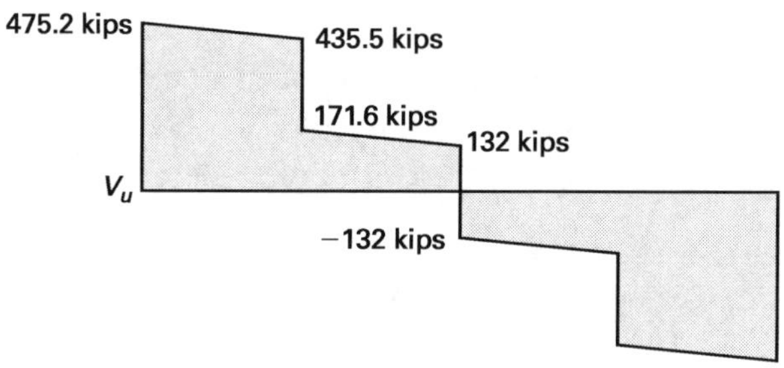

(not to scale)

Transverse Stiffener Design

In placing the stiffeners, advantage was taken of tension-field action between the end panel stiffener and the bearing stiffener. The stiffeners must now be designed to resist the additional load this creates.

Calculate the maximum stiffener width.

$$b_{max} = \frac{b_f - t_w}{2} = \frac{15\text{ in} - 0.375\text{ in}}{2} = 7.31\text{ in} \quad [\text{use } 7.0\text{ in}]$$

Use Eq. 11.32 to calculate the maximum width-to-thickness ratio for each transverse stiffener.

$$\left(\frac{b}{t}\right)_{st} \le 0.56\sqrt{\frac{E}{F_{y,st}}} = 0.56\sqrt{\frac{29{,}000\ \frac{\text{kips}}{\text{in}}}{50\ \frac{\text{kips}}{\text{in}}}} = 13.49$$

Begin by designing the first transverse stiffener (the end panel stiffener). This stiffener must meet the requirement of Eq. 11.33, so in order to find the minimum value of I_{st}, the values of I_{st1}, I_{st2}, V_r, V_{c1}, and V_{c2} must first be found.

Use Eq. 11.30 to calculate I_{st1}, the minimum moment of inertia required for development of web shear buckling for the first transverse stiffener, based on the geometry of the end panel. First, use Eq. 11.31 to determine the factor j.

$$j \geq \begin{cases} \dfrac{2.5}{\left(\dfrac{a}{h}\right)^2} - 2 = \dfrac{2.5}{\left(\dfrac{28\text{ in}}{71\text{ in}}\right)^2} - 2 = 14.07 \quad [\text{controls}] \\ 0.5 \end{cases}$$

From Eq. 11.30, the minimum moment of inertia required is

$$I_{st1} \geq bt_w^3 j = (28\text{ in})(0.375\text{ in})^3(14.07) = 20.78\text{ in}^4$$

Use Eq. 11.34 to calculate I_{st2}, the minimum moment of inertia required for the first transverse stiffener for development of the full web shear buckling plus the web tension field resistance. First, find ρ_{st}, which is the greater of $F_{yw}/F_{y,st}$ and 1.0.

$$\rho_{st} \geq \begin{cases} \dfrac{F_{yw}}{F_{y,st}} = \dfrac{50\ \dfrac{\text{kips}}{\text{in}^2}}{50\ \dfrac{\text{kips}}{\text{in}^2}} = 1.0 \\ 1.0 \end{cases}$$

From Eq. 11.34, the minimum moment of inertia is

$$I_{st2} = \left(\frac{h^4 \rho_{st}^{1.3}}{40}\right)\left(\frac{F_{yw}}{E}\right)^{1.5} = \left(\frac{(71\text{ in})^4(1)^{1.3}}{40}\right)\left(\frac{50\ \dfrac{\text{kips}}{\text{in}^2}}{29{,}000\ \dfrac{\text{kips}}{\text{in}^2}}\right)^{1.5}$$

$$= 45.48\text{ in}^4$$

The required shear strength, V_r, is the greater of the two required shear strengths of the end and first interior panels. From the summary,

LRFD	ASD
For the end panel,	For the end panel,
$V_u = 475.20\text{ kips}$	$V_a = 324.00\text{ kips}$
For the first interior panel,	For the first interior panel,
$V_u = 469.05\text{ kips}$	$V_a = 319.81\text{ kips}$
The end panel controls, so	The end panel controls, so
$V_r = 475.20\text{ kips}$	$V_r = 324.00\text{ kips}$

Calculate V_{c1}, which is the smaller of the available shear strengths in the adjacent web panels, with the nominal shear strength, V_n, defined as in Eq. 11.23. First, calculate the ratio h/t_w, which is the same for both adjacent panels.

$$\frac{h}{t_w} = \frac{71\ \text{in}}{0.375\ \text{in}} = 189.33$$

Use Eq. 11.27 to calculate the web plate buckling coefficient for the end panel.

$$k_v = 5 + \frac{5}{\left(\frac{a}{h}\right)^2} = 5 + \frac{5}{\left(\frac{28\ \text{in}}{71\ \text{in}}\right)^2} = 37.15$$

Determine whether to use Eq. 11.24, Eq. 11.25, or Eq. 11.26 for calculating the web shear coefficient.

$$1.10\sqrt{\frac{k_v E}{F_y}} = 1.10\sqrt{\frac{(37.15)\left(29{,}000\ \frac{\text{kips}}{\text{in}^2}\right)}{50\ \frac{\text{kips}}{\text{in}^2}}} = 161.47 \quad [< h/t_w = 189.33]$$

$$1.37\sqrt{\frac{k_v E}{F_y}} = 1.37\sqrt{\frac{(37.15)\left(29{,}000\ \frac{\text{kips}}{\text{in}^2}\right)}{50\ \frac{\text{kips}}{\text{in}^2}}} = 201.10 \quad [> h/t_w = 189.33]$$

$1.10\sqrt{k_v E/F_y} < h/t_w \le 1.37\sqrt{k_v E/F_y}$, so use Eq. 11.25.

$$C_v = \frac{1.10\sqrt{\frac{k_v E}{F_y}}}{\frac{h}{t_w}} = \frac{161.47}{189.33} = 0.85$$

Use Eq. 11.23 to calculate the nominal shear strength.

$$\begin{aligned} V_n &= 0.6F_y A_w C_v = (0.6)\left(50\ \frac{\text{kips}}{\text{in}^2}\right)\big((0.375\ \text{in})(71\ \text{in})\big)(0.85) \\ &= 678.9\ \text{kips} \quad [\text{end panel}] \end{aligned}$$

Repeat the calculations for the first interior panel. From Eq. 11.27,

$$k_v = 5 + \frac{5}{\left(\frac{a}{h}\right)^2} = 5 + \frac{5}{\left(\frac{76\ \text{in}}{71\ \text{in}}\right)^2} = 9.36$$

Determine whether to use Eq. 11.24, Eq. 11.25, or Eq. 11.26.

$$1.10\sqrt{\frac{k_vE}{F_y}} = 1.10\sqrt{\frac{(9.36)\left(29{,}000\ \frac{\text{kips}}{\text{in}^2}\right)}{50\ \frac{\text{kips}}{\text{in}^2}}} = 81.04\quad [< h/t_w = 189.33]$$

$$1.37\sqrt{\frac{k_vE}{F_y}} = 1.37\sqrt{\frac{(9.36)\left(29{,}000\ \frac{\text{kips}}{\text{in}^2}\right)}{50\ \frac{\text{kips}}{\text{in}^2}}} = 100.94\quad [< h/t_w = 189.33]$$

$1.37\sqrt{k_vE/F_y} < h/t_w$, so use Eq. 11.26 to calculate the web shear coefficient.

$$C_v = \frac{1.51k_vE}{\left(\frac{h}{t_w}\right)^2 F_y} = \frac{(1.51)(9.36)\left(29{,}000\ \frac{\text{kips}}{\text{in}^2}\right)}{(189.33)^2\left(50\ \frac{\text{kips}}{\text{in}^2}\right)} = 0.23$$

Use Eq. 11.23 to calculate the nominal shear strength.

$$\begin{aligned} V_n &= 0.6F_yA_wC_v = (0.6)\left(50\ \frac{\text{kips}}{\text{in}^2}\right)\big((0.375\ \text{in})(71\ \text{in})\big)(0.23) \\ &= 183.7\ \text{kips}\quad [\text{first interior panel}] \end{aligned}$$

The smaller of the two nominal shear strengths is that of the first interior panel. Use this value of V_n to calculate the available shear strength, V_{c1}.

LRFD	ASD
$V_{c1} = \phi V_n = (0.9)(183.7\ \text{kips})$ $= 165.33\ \text{kips}$	$V_{c1} = \frac{V_n}{\Omega} = \frac{183.7\ \text{kips}}{1.67}$ $= 110.00\ \text{kips}$

Calculate V_{c2} in the same way, but with the nominal shear strength defined as in either Eq. 11.28 or Eq. 11.29. Because $h/t_w > 1.10\sqrt{k_vE/F_y}$ for both panels, use Eq. 11.29. k_v and C_v are defined as before.

For the end panel,

$$V_n = (0.6F_yA_w)\left(C_v + \frac{1-C_v}{1.15\sqrt{1+\left(\frac{a}{h}\right)^2}}\right)$$

$$= (0.6)\left(50\ \frac{\text{kips}}{\text{in}^2}\right)\big((0.375\ \text{in})(71\ \text{in})\big)\left(0.85 + \frac{1-0.85}{1.15\sqrt{1+\left(\frac{28\ \text{in}}{71\ \text{in}}\right)^2}}\right)$$

$$= 776.2\ \text{kips} \quad [\text{end panel}]$$

For the first interior panel,

$$V_n = (0.6F_yA_w)\left(C_v + \frac{1-C_v}{1.15\sqrt{1+\left(\frac{a}{h}\right)^2}}\right)$$

$$= (0.6)\left(50\ \frac{\text{kips}}{\text{in}^2}\right)\big((0.375\ \text{in})(71\ \text{in})\big)\left(0.23 + \frac{1-0.23}{1.15\sqrt{1+\left(\frac{76\ \text{in}}{71\ \text{in}}\right)^2}}\right)$$

$$= 548.8\ \text{kips} \quad [\text{first interior panel}]$$

The first interior panel has the smaller nominal shear strength. Use this value to calculate the available shear strength, V_{c2}.

LRFD	ASD
$V_{c2} = \phi V_n = (0.9)(548.8\ \text{kips})$ $= 493.92\ \text{kips}$	$V_{c2} = \frac{V_n}{\Omega} = \frac{548.8\ \text{kips}}{1.67}$ $= 328.62\ \text{kips}$

Use these values in Eq. 11.33 to calculate the minimum moment of inertia required for the first transverse stiffener to resist the full web shear buckling and web tension field.

LRFD	ASD
$I_{st} \geq I_{st1} + (I_{st2} - I_{st1})\left(\dfrac{V_r - V_{c1}}{V_{c2} - V_{c1}}\right)$ $\geq 20.78\text{ in}^4 + (45.48\text{ in}^4 - 20.78\text{ in}^4) \times \left(\dfrac{475.20\text{ kips} - 165.33\text{ kips}}{493.92\text{ kips} - 165.33\text{ kips}}\right)$ $\geq 44.07\text{ in}^4$	$I_{st} \geq I_{st1} + (I_{st2} - I_{st1})\left(\dfrac{V_r - V_{c1}}{V_{c2} - V_{c1}}\right)$ $\geq 20.78\text{ in}^4 + (45.48\text{ in}^4 - 20.78\text{ in}^4) \times \left(\dfrac{324.00\text{ kips} - 110.00\text{ kips}}{328.62\text{ kips} - 110.00\text{ kips}}\right)$ $\geq 44.95\text{ in}^4$

Use the required moment of inertia to determine the stiffener size. Try a 5 in stiffener. The total stiffener width of b = 10.375 in includes the stiffeners on both sides of the web (each 5 in) and the web thickness of 0.375 in.

LRFD	ASD
$I_{st} = \dfrac{tb^3}{12}$ $t = \dfrac{12I_{st}}{b^3} = \dfrac{(12)(44.07)}{(10.375\text{ in})^3}$ $= 0.47\text{ in}$	$I_{st} = \dfrac{tb^3}{12}$ $t = \dfrac{12I_{st}}{b^3} = \dfrac{(12)(44.95)}{(10.375\text{ in})^3}$ $= 0.48\text{ in}$

Use a pair of plates $^1/_2$ in × 5 in. While the moment of inertia required for tension-field action is slightly more than twice that required without tension-field action, the need for fewer stiffeners makes it the more economical design.

The design of the transverse stiffener between panels two and three would follow the same procedure using the appropriate geometry and shear strengths.

Example 11.2
Bearing Stiffener Design

Design bearing stiffeners for the plate girder in Ex. 11.1. (The design of stiffeners is discussed in Chap. 6.)

Solution

End Bearing Stiffeners

Use Eq. 6.50 to calculate the maximum stiffener width.

$$b_{\text{max}} = \frac{b_f - t_w}{2} = \frac{15 \text{ in} - 0.375 \text{ in}}{2} = 7.31 \text{ in} \quad [\text{use 7 in}]$$

Use Eq. 11.32 to calculate the maximum ratio of width to thickness.

$$\left(\frac{b}{t}\right)_{\text{st}} \le 0.56\sqrt{\frac{E}{F_{y,\text{st}}}} = 0.56\sqrt{\frac{29{,}000 \ \frac{\text{kips}}{\text{in}^2}}{50 \ \frac{\text{kips}}{\text{in}}}} = 13.49$$

The minimum thickness for a 7 in wide stiffener is

$$t_{\text{min}} = \frac{b}{\left(\frac{b}{t}\right)_{\text{st}}} = \frac{7 \text{ in}}{13.49} = 0.52 \text{ in} \quad [\text{try 0.625 in}]$$

Try a 7 in by $^5/_8$ in plate. Calculate the gross area of the cross-shaped stiffener column to include a section of the web equal to $12t_w$ (see Eq. 6.49). Using Eq. 6.55,

$$\begin{aligned} A_{g,\text{cross}} &= A_{\text{st}} + 12t_w^2 = n_{\text{st}} b_{\text{st}} t_{\text{st}} + 12t_w^2 \\ &= (2)(7 \text{ in})(0.625 \text{ in}) + (12)(0.375 \text{ in})^2 \\ &= 10.44 \text{ in}^2 \end{aligned}$$

Calculate the moment of inertia for the cross-shaped column section.

$$\begin{aligned} I_{\text{cross}} &= I_{\text{st}} + I_w = \frac{(bd^3)_{\text{st}}}{12} + \frac{(bd^3)_w}{12} \\ &= \frac{t_{\text{st}}(t_w + 2b_{\text{st}})^3}{12} + \frac{(12t_w - t_{\text{st}})t_w^3}{12} \\ &= \frac{(0.625 \text{ in})(0.375 \text{ in} + (2)(7 \text{ in}))^3}{12} \\ &\quad + \frac{((12)(0.375 \text{ in}) - 0.625 \text{ in})(0.375 \text{ in})^3}{12} \\ &= 154.73 \text{ in}^4 \end{aligned}$$

The radius of gyration for the column is

$$r = \sqrt{\frac{I_{\text{cross}}}{A_{g,\text{cross}}}} = \sqrt{\frac{154.73 \text{ in}^4}{10.44 \text{ in}^2}} = 3.85 \text{ in}$$

Calculate the nominal strength of the cross-shaped stiffener column. The effective length factor, K, for the stiffeners is 0.75. The effective slenderness ratio is

$$\frac{KL}{r} = \frac{(0.75)(71 \text{ in})}{3.85 \text{ in}} = 13.83$$

Determine the correct formula for calculating the critical flexural buckling stress.

$$4.71\sqrt{\frac{E}{F_y}} = 4.71\sqrt{\frac{29{,}000 \ \frac{\text{kips}}{\text{in}^2}}{50 \ \frac{\text{kips}}{\text{in}^2}}} = 113.43 \quad [> KL/r, \text{ so use Eq. 6.58}]$$

Use Eq. 6.60 to determine the elastic critical buckling stress.

$$F_e = \frac{\pi^2 E}{\left(\frac{KL}{r}\right)^2} = \frac{\pi^2\left(29{,}000 \ \frac{\text{kips}}{\text{in}^2}\right)}{(13.83)^2} = 1496 \text{ ksi}$$

Use Eq. 6.58 to find the critical flexural buckling stress.

$$F_{\text{cr}} = 0.658^{F_y/F_e} F_y = \left(0.658^{50 \ \frac{\text{kips}}{\text{in}^2} / 1496 \ \frac{\text{kips}}{\text{in}^2}}\right)\left(50 \ \frac{\text{kips}}{\text{in}^2}\right) = 49.31 \text{ ksi}$$

Use Eq. 6.57 to find the nominal compressive strength.

$$P_n = F_{\text{cr}} A_{g,\text{cross}} = \left(49.31 \ \frac{\text{kips}}{\text{in}^2}\right)\left(10.44 \text{ in}^2\right) = 514.80 \text{ kips}$$

Calculate the available strength of the cross-shaped stiffener column.

LRFD	ASD
$P_u \le \phi_c P_n$ $475.20 \text{ kips} \le (0.90)(514.80 \text{ kips})$ $\le 463.32 \text{ kips} \quad [\text{not OK}]$	$P_a \le \dfrac{P_n}{\Omega_c}$ $324.00 \text{ kips} \le \dfrac{514.80 \text{ kips}}{1.67}$ $\le 308.26 \text{ kips} \quad [\text{not OK}]$

The required demand capacities are only slightly less than the calculated available capacities, so it is safe to assume that increasing the stiffener thickness by $^1/_8$ in will be satisfactory. Use 7 in by $^3/_4$ in stiffeners.

The bearing stiffeners beneath the concentrated loads are designed in a similar manner, except that the length of the web to be included in the cross-shaped stiffener column is 25 times the thickness of the girder web.

Intermediate Bearing Stiffeners

Design the intermediate stiffeners. The required area of steel and the moment of inertia were calculated in Ex. 11.1. Size the stiffeners to meet the existing requirements, which are as follows.

$$\begin{aligned} A_{\text{st}} &= 6.00 \text{ in}^2 \\ I_{\text{st}} &= 20.78 \text{ in}^4 \\ b_f &= 15 \text{ in} \\ t_w &= 0.375 \text{ in} \\ h &= 71 \text{ in} \end{aligned}$$

Calculate the stiffener thickness based on the required area and width-to-thickness ratio for the compression elements.

$$t_{\text{st}} = \frac{A_{\text{st}}}{2b_{\text{st}}} = \frac{6.00 \text{ in}^2}{(2)(6 \text{ in})} = 0.50 \text{ in}$$

$$\left(\frac{b}{t}\right)_{\text{st}} \leq 0.56\sqrt{\frac{E}{F_{y,\text{st}}}} = 0.56\sqrt{\frac{29{,}000 \frac{\text{kips}}{\text{in}}}{50 \frac{\text{kips}}{\text{in}}}} = 13.49$$

$$t_{\min} = \frac{b}{\left(\frac{b}{t}\right)_{\text{st}}} = \frac{6 \text{ in}}{13.49} = 0.445 \text{ in} \quad [\text{use } 0.50 \text{ in}]$$

Use 6 in by $^1/_2$ in plate stiffeners on each side of web. Because the intermediate stiffeners are not needed to transmit a concentrated load or reaction, they can be terminated short of the tension flange.

The distance from the inside face of the tension flange to the near toe of the web-to-flange weld must be at least four times but no more than six times the web thickness.

$$\begin{aligned} 4t_w &= (4)(0.375 \text{ in}) = 1.5 \text{ in} \\ 6t_w &= (6)(0.375 \text{ in}) = 2.25 \text{ in} \end{aligned}$$

Use a distance of 2 in, and make the stiffener height 71 in – 2 in = 69 in.

Stiffener-to-Web Weld

Although no longer included in the *AISC Specification*, Eq. G4-3 from the *Manual of Steel Construction: Allowable Stress Design*, ninth edition, is useful for calculating the shear transfer requirements for the intermediate stiffeners.

$$f_{vs} = h\sqrt{\left(\frac{F_y}{340}\right)^3} = (71\text{ in})\sqrt{\left(\frac{50\ \frac{\text{kips}}{\text{in}^2}}{340}\right)^3}$$

$$= 4.00\ \text{kips/in}\quad \left[\begin{array}{l}\text{kips per linear inch of single}\\ \text{stiffener or pair of stiffeners}\end{array}\right]$$

The minimum size fillet weld that can be used for the $^1/_2$ in stiffener and the $^3/_8$ in web is a $^3/_{16}$ in fillet weld that has a capacity of 2.78 kips per inch of weld for E70XX electrodes. One continuous weld for each stiffener, with the two welds placed diagonally opposite one another on the web, will provide 5.56 kips/in.

Girder-Flange-to-Web Weld

The flange thickness is 2.5 in, and the web thickness is $^3/_8$ in. The minimum size fillet weld, based on the thinner element (the web), is $^3/_{16}$ in.

Calculate the horizontal shear stress at the interface between the web and the flange.

$$\tau_h = \frac{VQ}{I_x b} = \frac{(475.2\text{ kips})\left((2.5\text{ in})(15.0\text{ in})\left(38\text{ in} - \frac{2.5\text{ in}}{2}\right)\right)}{(112{,}516\text{ in}^4)(0.375\text{ in})} = 15.52\text{ ksi}$$

Calculate the weld demand capacity per linear inch. (See Sec. 10.7 and Ex. 10.1.)

$$R_{w,\text{req}} = \left(15.52\ \frac{\text{kips}}{\text{in}^2}\right)(0.375\text{ in}) = 5.82\ \text{kips/in}$$

Determine the resistance capacity of the two $^3/_{16}$ in welds on each side of the web.

$$R_w = (2)\left(1.392\ \frac{\text{kips}}{\text{in}}\right)D = (2)\left(1.392\ \frac{\text{kips}}{\text{in}}\right)(3) = 8.352\ \text{kips/in}$$

The two $^3/_{16}$ in welds on each side have more capacity than the demand capacity. Determine the minimum web thickness for a double $^3/_{16}$ in weld.

$$t_{\min} = \frac{6.19D}{F_u} = \frac{\left(6.19\ \frac{\text{kips}}{\text{in}}\right)(3)}{65\ \frac{\text{kips}}{\text{in}^2}} = 0.2857\text{ in}$$

The web has sufficient thickness to accept a double $^3/_{16}$ in on each side.

12 Composite Steel Members

Nomenclature

a	depth of concrete in compression	in
A	area	in^2
A_1	loaded area of concrete	in^2
b	width	in
C	compressive force	lbf
C_1	coefficient defined in *AISC Specification* Eq. I2-7	–
C_2	coefficient defined in *AISC Specification* Sec. I2.2b, equal to 0.85 for rectangular sections and 0.95 for circular sections	–
C_3	coefficient defined in *AISC Specification* Eq. I2-13	–
C_{in}	coefficient defined in *AISC Specification* Sec. I6.3c	–
d	depth of beam	in
D	dead load	lbf/in^2
D	outer diameter	in
E	modulus of elasticity	lbf/in^2
EI	stiffness	$lbf\text{-}in^2$
f_c'	specified compressive strength of concrete	lbf/in^2
F_{cr}	critical stress	lbf/in^2
F_{in}	nominal bond stress, equal to 0.06 ksi	lbf/in^2
F_u	specified minimum tensile strength	lbf/in^2
F_y	specified minimum yield stress	lbf/in^2
I	moment of inertia	in^4
K	effective length factor	–
KL	effective length	in
L	length or span	in
L	live load	lbf/in^2
M	flexural strength or moment	in-lbf
n	number of shear connectors	–
P	strength or load	lbf
P_{no}	nominal axial compressive strength disregarding adjustments due to length	lbf
P_p	nominal bearing strength	lbf

P_r	required force or load	lbf
P_y	axial yield strength	lbf
ΣQ_{cv}	sum of available shear strengths of shear connectors	lbf
Q_n	nominal strength of one stud shear connector	lbf
R	strength	lbf
s	tributary width	in
t	thickness	in
T	tensile force	lbf
V	required shear force introduced to column	lbf
V_r'	required shear force transferred by shear connectors	lbf
w	unit weight	lbf/ft^2 or lbf/ft^3
W	uniformly distributed load	lbf/ft
$Y1$	distance from top of steel beam to plastic neutral axis	in
$Y2$	distance from top of steel beam to concrete flange force	in
Z_x	plastic section modulus about x-axis	in^3

Symbols

Δ	deflection	in
λ	limiting width-to-thickness ratio	–
λ_p	limiting width-to-thickness ratio for compactness	–
λ_r	limiting width-to-thickness ratio for noncompactness	–
ρ_{sr}	reinforcement ratio, A_{sr}/A_g	–
ϕ	resistance factor (LRFD)	–
Ω	safety factor (ASD)	–

Subscripts

a	required (ASD)
b	bending or flexural
B	bearing
c	compression, compressive, or concrete
comp	compression
D	from dead load
des	design
e	effective or elastic buckling
eff	effective
f	flange
flex	flexural

g	gross
LB	lower bound
n	nominal
p	plastic bending
pc	partial composite action
s	steel
sc	stud shear connector
sr	steel reinforcement
t	tensile
u	required (LRFD)

1. GENERAL

A composite steel member consists of a steel member to which concrete is added in such a way that the two materials act together and form a single nonhomogeneous member. The design of composite steel members is governed by *AISC Specification* Chap. I, which is divided into the following sections.

I1	General Provisions
I2	Axial Force
I3	Flexural
I4	Shear
I5	Combined Axial Force and Flexure
I6	Load Transfer
I7	Composite Diaphragms and Collector Beams
I8	Steel Anchors
I9	Special Cases

The use of composite steel beams started in the mid twentieth century and continues to develop. Their design was first covered in the sixth edition of the *AISC Manual* in 1963; the thirteenth (2005) and fourteenth (2010) editions added significant new material.

The *AISC Manual* includes the following types of composite members.

- steel axial compression members
  - — steel members fully encased in concrete
  - — hollow structural sections filled with concrete
- steel flexural members
  - — steel members fully encased in concrete
  - — hollow structural sections filled with concrete
- steel beams anchored to concrete slabs in such a way that they act together to resist bending

The fundamental design concept for a composite steel member is that the concrete resists compression forces and the steel resists tensile forces. The tensile strength of concrete is neglected.

Composite members can have a number of benefits over steel members, including less weight, greater load-bearing capacity, shallower construction depth, and greater system stiffness. Composite construction is more likely to be economical for longer spans and heavier loads, but it can be advantageous for shorter spans as well, depending on the combination of loads and spans.

It's important to consider load effects when designing a composite member, whether axial or flexural. The steel element must be designed to support the load that will be imparted to it before the concrete hardens. The completed member must be designed so that it will support the critical load combination when the concrete reaches its design strength.

2. DESIGN METHODS

The *AISC Manual* permits two types of design and analysis for determining the nominal strength of a composite member: the plastic stress distribution method and the strain-compatibility method.

In the *plastic stress distribution method*, the steel components are assumed to reach a stress of F_y in either tension or compression, while the concrete components are assumed to reach a compressive stress of $0.85f_c'$. (For round hollow structural sections (HSS) members filled with concrete, a stress of $0.95f_c'$ is permitted for the concrete components in uniform compression to account for the confinement of the concrete.)

The *strain-compatibility method* is based on a linear distribution of strains across the section. The maximum concrete compressive strain should be 0.003 in/in. The stress-strain relationships for steel and concrete are obtained from tests or published sources.

3. MATERIAL LIMITATIONS

The following limits generally apply to the steel and concrete in a composite system.

- The compressive strength of regular weight concrete must be at least 3 ksi and no more than 10 ksi.
- The compressive strength of lightweight concrete must be at least 3 ksi and no more than 6 ksi.
- For purposes of calculating column strength, the specified minimum yield stress of steel must be no more than 75 ksi.

Higher strengths may be used in calculations, however, if they are supported by testing or analysis.

Steel headed stud anchors may be headed steel studs or hot-rolled steel channels. Headed steel studs must have a length after installation of at least four stud diameters.

4. AXIAL MEMBERS

The *AISC Manual* recognizes two types of composite axial members.

- encased composite columns (steel columns fully encased in concrete)
- filled composite columns (HSS members filled with concrete)

Fully encased composite steel members are in less common use because of the cost of building concrete formwork to encase the beam. HSS members filled with concrete are a more recent development and were first covered in the *AISC Manual* in the thirteenth edition; they avoid the need for formwork, have better fire resistance than unfilled HSS members, and have aesthetic appeal in exposed structures.

5. ENCASED COMPOSITE COLUMNS

An *encased composite column* consists of concrete encasement around a steel core. Figure 12.1 shows some examples.

Figure 12.1 *Examples of Encased Composite Columns*

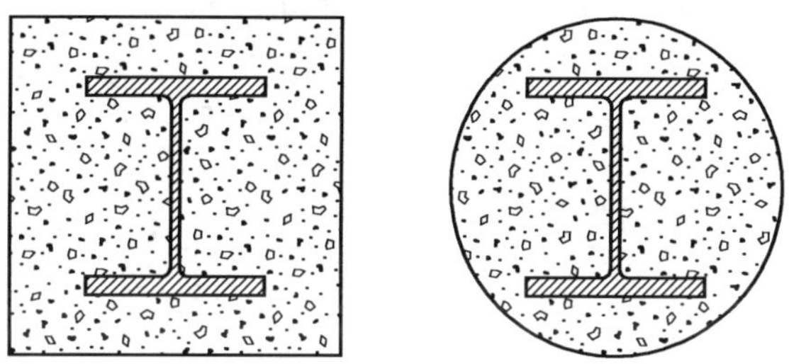

Encased composite columns must meet the following limitations.

- The cross-sectional area of the core must be at least 1% of the total cross-sectional area.
- The concrete encasement must be reinforced with continuous longitudinal bars and lateral ties or spirals.
- The spacing of the transverse reinforcement must be whichever of the following values is smallest: half the smallest dimension of the member, 16 times the diameter of the longitudinal reinforcement, or 48 times the diameter of the lateral reinforcement.
- The continuous longitudinal reinforcement must have a reinforcement ratio of at least 0.004. The reinforcement ratio, ρ_{sr}, is the ratio of the area of continuous steel reinforcement, A_{sr}, to the gross area of the column, A_g.

$$\rho_{sr} = \frac{A_{sr}}{A_g} \quad \text{[AISC Eq. I2-1]} \qquad 12.1$$

Compressive Strength

The nominal compressive strength, P_n, the design compressive strength (LRFD), and the allowable compressive strength (ASD) should be computed in accordance with the following.

For LRFD, with $\phi_c = 0.75$,

$$P_u \le \phi_c P_n \qquad 12.2$$

For ASD, with $\Omega_c = 2.00$,

$$P_a \le \frac{P_n}{\Omega_c} \qquad 12.3$$

Which formula should be used to calculate P_n depends on the relation between the elastic buckling load, P_e, and the nominal compressive strength of the column, disregarding adjustments due to length, P_{no}.

When $P_{no}/P_e \le 2.25$, the nominal compressive strength is

$$P_n = 0.658^{P_{no}/P_e} P_{no} \quad \text{[AISC Eq. I2-2]} \qquad 12.4$$

When $P_{no}/P_e > 2.25$, the nominal compressive strength is

$$P_n = 0.877 P_e \quad \text{[AISC Eq. I2-3]} \qquad 12.5$$

Use Eq. 12.6 and Eq. 12.7 to calculate P_{no} and P_e

$$P_{no} = F_y A_s + F_{y,\text{sr}} A_{\text{sr}} + 0.85 f'_c A_c \quad \text{[AISC Eq. I2-4]} \qquad 12.6$$

$$P_e = \frac{\pi^2 (EI)_{\text{eff}}}{(KL)^2} \quad \text{[AISC Eq. I2-5]} \qquad 12.7$$

In Eq. 12.5, the effective stiffness of the composite section is

$$(EI)_{\text{eff}} = E_s I_s + 0.5 E_s I_{\text{sr}} + C_1 E_c I_c \quad \text{[AISC Eq. I2-6]} \qquad 12.8$$

The coefficient C_1 is

$$C_1 = 0.1 + 2\left(\frac{A_s}{A_c + A_s}\right) \le 0.3 \quad \text{[AISC Eq. I2-7]} \qquad 12.9$$

In any case, the *available compressive strength* of the composite member does not need to be taken as less than the available compressive strength of the steel member alone (see Chap. 7).

In Eq. 12.6, the nominal compressive strength of the composite column is based on the assumption that both the steel section and the reinforced concrete section will reach their ultimate strengths (yield strength for steel, and crushing strength for concrete). The term F_yA_s represents the plastic strength of the steel section; the remaining terms represent the strength of the reinforced concrete.

The modulus of elasticity of concrete is found using Eq. 12.10. This equation is not dimensionally consistent. The weight of the concrete, w_c, must be in pounds-force per cubic foot (pcf), and the compressive strength of the concrete, f_c', must be in pounds-force per square inch (psi). The resulting modulus of elasticity, E_c, is in pounds-force per square inch.

$$E_{c,\text{psi}} = 33w_{c,\text{pcf}}^{1.5}\sqrt{f'_{c,\text{psi}}} \qquad 12.10$$

Tensile Strength

The tensile strength for an encased composite column is based on the tensile strength of its steel only. The relatively small tensile strength of the concrete is neglected. The nominal tensile strength of the composite section is

$$P_n = F_yA_s + F_{y,\text{sr}}A_{\text{sr}} \quad [\text{AISC Eq. I2-8}] \qquad 12.11$$

In calculating the design tensile strength, ϕ_tP_n, use $\phi_t = 0.90$ (LRFD). In calculating the allowable tensile strength, P_n/Ω_t, use $\Omega_t = 1.67$ (ASD).

For LRFD, with $\phi_t = 0.90$, the required tensile strength is

$$P_u \le \phi_tP_n \qquad 12.12$$

For ASD, with $\Omega_t = 1.67$, the required tensile strength is

$$P_a \le \frac{P_n}{\Omega_t} \qquad 12.13$$

6. FILLED COMPOSITE COLUMNS

A *filled composite column* consists of an HSS member filled with concrete. Figure 12.2 shows some examples.

A filled composite column must meet the general requirement that the cross-sectional area of the HSS member is at least 1% of the total cross-sectional area.

In addition, the width-to-thickness ratio (b/t, h/t, or D/t, depending on the shape of the member) must not exceed the limit given by *AISC Specification* Table I1.1A (for members subject to axial compression) or Table I1.1B (for members subject to flexure).

Figure 12.2 *Examples of Filled Composite Columns*

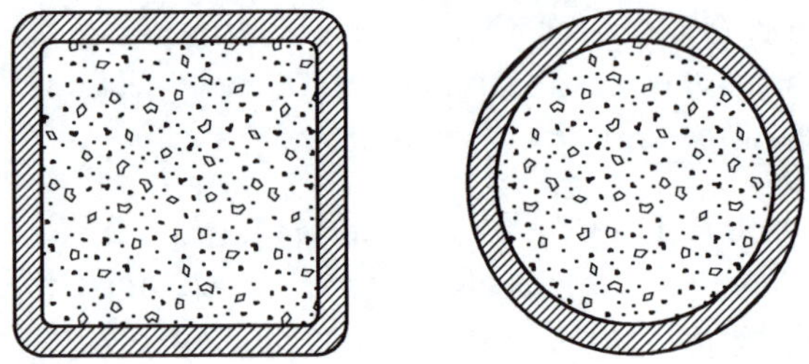

The *AISC Manual* provides the following tables to assist with the design or analysis of filled composite columns.

Table 4-13	Rectangular HSS, $f_c' = 4$ ksi
Table 4-14	Rectangular HSS, $f_c' = 5$ ksi
Table 4-15	Square HSS, $f_c' = 4$ ksi
Table 4-16	Square HSS, $f_c' = 5$ ksi
Table 4-17	Round HSS, $f_c' = 4$ ksi
Table 4-18	Round HSS, $f_c' = 5$ ksi
Table 4-19	Round Pipe, $f_c' = 4$ ksi
Table 4-20	Round Pipe, $f_c' = 5$ ksi

Compressive Strength

The formula for the compressive strength of a filled composite member subject to axial compression depends on whether the section is compact, noncompact, or slender. This classification is made on the basis of the limiting width-to-thickness ratios λ_p and λ_r as given in Eq. 12.14 through Eq. 12.17.

- A section is compact if each of its compression steel elements has a width-to-thickness ratio less than λ_p.
- A section is noncompact if one or more of its compression steel elements has width-to-thickness ratios of λ_p or more, but none has a ratio greater than λ_r.
- A section is slender if one or more of its compression steel elements has a width-to-thickness ratio of λ_r or more.

For the walls of rectangular HSS,

- the width-to-thickness ratio is b/t
- the lower limit for noncompactness is

$$\lambda_p = 2.26\sqrt{E/F_y} \quad \text{[AISC Table I1-1A]} \qquad 12.14$$

- the lower limit for slenderness is

$$\lambda_r = 3.00\sqrt{E/F_y} \quad \text{[AISC Table I1-1A]} \qquad 12.15$$

For round HSS,

- the width-to-thickness ratio is D/t
- the lower limit for noncompactness is

$$\lambda_p = 0.15E/F_y \quad \text{[AISC Table I1-1A]} \qquad 12.16$$

- the lower limit for slenderness is

$$\lambda_r = 0.19E/F_y \quad \text{[AISC Table I1-1A]} \qquad 12.17$$

For *compact sections*, the nominal axial compressive strength is

$$P_{no} = P_p \quad \text{[AISC Eq. I2-9a]} \qquad 12.18$$

In Eq. 12.18, the nominal bearing strength, P_p, is

$$P_p = F_y A_s + C_2 f_c' \left(A_c + A_{sr} \left(\frac{E_s}{E_c} \right) \right) \quad \text{[AISC Eq. I2-9b]} \qquad 12.19$$

In Eq. 12.19, the coefficient C_2 is 0.85 for rectangular sections and 0.95 for round sections.

As with Eq. 12.6, in Eq. 12.19 the nominal compressive strength of the composite column is based on the assumption that both the steel and reinforced concrete sections will reach their ultimate strengths (yield strength for steel, crushing strength for concrete). The term $F_y A_s$ is the plastic strength of the steel section; the rest of the equation represents the strength of the reinforced concrete.

For *noncompact sections*, the nominal axial compressive strength is

$$P_{no} = P_p - \left(\frac{P_p - P_y}{(\lambda_r - \lambda_p)^2} \right) (\lambda - \lambda_p)^2 \quad \text{[AISC Eq. I2-9c]} \qquad 12.20$$

In Eq. 12.20, λ is the member's width-to-thickness ratio, b/t or D/t. λ_p and λ_r are the appropriate limits from Eq. 12.14 through Eq. 12.17. The nominal bearing strength, P_p, is as defined in Eq. 12.19. The axial yield strength, P_y, is

$$P_y = F_y A_s + 0.7 f_c' \left(A_c + A_{sr} \left(\frac{E_s}{E_c} \right) \right) \quad \text{[AISC Eq. I2-9d]} \qquad 12.21$$

For *slender sections*, the nominal axial compressive strength is

$$P_{no} = F_{cr} A_s + 0.7 f_c' \left(A_c + A_{sr} \left(\frac{E_s}{E_c} \right) \right) \quad \text{[AISC Eq. I2-9e]} \qquad 12.22$$

For a rectangular filled section, the critical stress, F_{cr}, in Eq. 12.22 is

$$F_{cr} = \frac{9E_s}{\left(\frac{b}{t}\right)^2} \quad \text{[AISC Eq. I2-10]} \qquad 12.23$$

For a round filled section, F_{cr} in Eq. 12.22 is

$$F_{cr} = \frac{0.72F_y}{\left(\left(\frac{D}{t}\right)\left(\frac{F_y}{E_s}\right)\right)^{0.2}} \quad \text{[AISC Eq. I2-11]} \qquad 12.24$$

In any case, the *available compressive strength* of the composite member does not need to be taken as less than the available compressive strength of the steel member alone (see Chap. 7).

For all sections, the *effective stiffness* of the composite section is

$$(EI)_{eff} = E_s I_s + E_s I_{sr} + C_3 E_c I_c \quad \text{[AISC Eq. I2-12]} \qquad 12.25$$

In Eq. 12.25, the coefficient C_3 is

$$C_3 = 0.6 + 2\left(\frac{A_s}{A_c + A_s}\right) \le 0.9 \quad \text{[AISC Eq. I2-13]} \qquad 12.26$$

Tensile Strength

The design tensile strength, $\phi_t P_n$ (LRFD), and the allowable tensile strength, P_n/Ω_t (ASD), for filled composite columns are determined for the limit state of yielding from the nominal tensile strength as defined in Eq. 12.27.

$$P_n = A_s F_y + A_{sr} F_{y,sr} \quad \text{[AISC Eq. I2-14]} \qquad 12.27$$

For LRFD, with $\phi_t = 0.90$, the required tensile strength is

$$P_u \le \phi_t P_n \qquad 12.28$$

For ASD, with $\Omega_t = 1.67$, the required tensile strength is

$$P_a \le \frac{P_n}{\Omega_t} \qquad 12.29$$

7. LOAD TRANSFER

In order for the steel and concrete in a composite column to work in a unified way to resist an axial load, the longitudinal shear force must be distributed between the two materials so that a state of equilibrium is achieved over the cross section. Some portion of the longitudinal shear force, then, must be transferred through the interface between the two materials.

The *AISC Specification* assumes *plastic stress distribution*, so that the applied external force will be distributed between the steel and reinforced concrete sections in the same proportions as the two materials contribute to the ultimate capacity of the composite column.

An axial load can be applied to a composite column in one of three ways. The entire load can be applied directly to the steel section, the entire load can be applied directly to the concrete fill or concrete encasement, or the load can be applied to both the steel and the concrete.

When the external force is applied directly to the steel section, the force required to be transferred to the concrete, V_r', is calculated as

$$V_r' = P_r\left(1 - \frac{F_y A_s}{P_{no}}\right) \quad \text{[AISC Eq. I6-1]} \qquad 12.30$$

When the external force is applied directly to the concrete, the force required to be transferred to the steel, V_r', is calculated as

$$V_r' = P_r\left(\frac{F_y A_s}{P_{no}}\right) \quad \text{[AISC Eq. I6-2]} \qquad 12.31$$

In both Eq. 12.30 and Eq. 12.31, P_r is the required external force being applied to the composite member. The value for P_{no} is calculated with Eq. 12.6 for encased composite members, and calculated with Eq. 12.18 for filled composite members.

When the external force is applied to both steel and concrete concurrently, V_r' is the force that must be transferred from one material to the other to establish equilibrium across the cross section. In this case, V_r' may be calculated in either of two ways, as the difference between

- the portion of the external force that is applied directly to the concrete and the value given by Eq. 12.30, or
- the portion of the external force that is applied directly to the steel and the value given by Eq. 12.31

Force Transfer Mechanisms

Once it has been determined how much longitudinal shear force must be transferred between the steel and concrete, a means of transferring that force can be selected. There are three mechanisms by which the required transfer of force can be achieved: direct bearing, shear connection, and direct bond interaction. Force transfer mechanisms may not be superimposed; however, it is acceptable to use the mechanism that gives the largest nominal strength.

Direct Bearing

When force is transferred by *direct bearing* from a bearing mechanism within the composite member (for example, internal steel plates within a filled composite member), the nominal bearing strength of the concrete for the limit state of concrete crushing is

$$R_n = 1.7 f'_c A_1 \quad \text{[AISC Eq. I6-3]} \qquad 12.32$$

A_1 is the loaded area of concrete.

For LRFD, with $\phi_B = 0.65$, the required bearing strength is

$$\begin{aligned} R_u &\le \phi_B R_n \\ &\le \phi_B 1.7 f'_c A_1 \end{aligned} \qquad 12.33$$

For ASD, with $\Omega_B = 2.31$, the required bearing strength is

$$\begin{aligned} R_a &\le \frac{R_n}{\Omega_B} \\ &\le \frac{1.7 f'_c A_1}{\Omega_B} \end{aligned} \qquad 12.34$$

Shear Connection

When force is transferred by *shear connection*, the available bearing strength of the shear connectors (steel headed stud anchors or steel channel anchors) is

$$R_n = \sum Q_{cv} \quad \text{[AISC Eq. I6-4]} \qquad 12.35$$

ΣQ_{cv} is the sum of the available shear strengths of the shear connectors.

Direct Bond Interaction

Direct bond interaction may be used only with filled composite members. It may not be used with encased composite members. When force is transferred by direct bond interaction, the available bond strength between the steel and concrete is calculated with Eq. 12.36 or Eq. 12.37.

For a rectangular steel section filled with concrete,

$$R_n = B^2 C_{in} F_{in} \quad \text{[AISC Eq. I6-5]} \qquad 12.36$$

For a round steel section filled with concrete,

$$R_n = 0.25\pi D^2 C_{in} F_{in} \quad \text{[AISC Eq. I6-6]} \qquad 12.37$$

In Eq. 12.36, B is the overall width of the rectangular section along the face that is transferring the load. In Eq. 12.37, D is the outer diameter of the round section. In both equations, the coefficient C_{in} is equal to 2 if the member extends to one side of the load transfer point, and 4 if the member extends to both sides of the load transfer point. F_{in} is nominal bond stress and is taken as 0.06 ksi.

For LRFD, with $\phi = 0.45$, the required bond strength is

$$\begin{aligned} R_u &\le \phi R_n \\ &\le 0.45 R_n \end{aligned} \qquad 12.38$$

For ASD, with $\Omega = 3.33$, the required bearing strength is

$$\begin{aligned} R_a &\le \frac{R_n}{\Omega_B} \\ &\le \frac{R_n}{3.33} \end{aligned} \qquad 12.39$$

Steel Anchors

The following detailing requirements apply to the installation of steel anchors.

- There must be at least 1 in of lateral clear concrete cover.
- The center-to-center spacing of steel headed stud anchors must be at least four diameters in any direction, and may not exceed 32 times the shank diameter.
- The diameter of steel headed stud anchors may not be more than 2.5 times the thickness of the base metal to which they are welded, unless the anchors are welded to a flange directly over a web.
- The center-to-center spacing of steel channel anchors may not be greater than 24 in.

These requirements are absolute limits; *AISC Specification* Sec. I3 contains additional requirements for steel anchors used to transfer loads for encased composite members, filled composite members, composite diaphragms, and collector beams.

Example 12.1

Axially Loaded Concrete-Filled Pipe Composite Section

A 12 in diameter, 30 ft standard steel pipe is filled with concrete. It will be used as a column, laterally braced in both axes top and bottom, and with pinned connections top and bottom.

Section properties

outer diameter = 12.8 in

inner diameter = 12.0 in

$t = 0.375$ in

$t_{\text{des}} = 0.349$

$D/t = 36.5$

$A = 13.7\ \text{in}^2$

$I = 262\ \text{in}^4$

Material properties

pipe is ASTM A53, Grade B

$F_y = 35$ ksi

$F_u = 60$ ksi

concrete is normal weight (150 lbf/ft^3)

$f_c' = 6$ ksi

Determine the nominal strength, P_n, the design strength, $\phi_c P_n$, and the allowable strength, P_n/Ω.

Solution

The gross cross-sectional area of the pipe is

$$A_g = \frac{\pi D^2}{4} = \frac{\pi (12.8\ \text{in})^2}{4} = 128.68\ \text{in}^2$$

The cross-sectional area of the concrete is

$$\begin{aligned} A_c &= A_g - A_s \\ &= 128.68\ \text{in}^2 - 13.7\ \text{in}^2 \\ &= 114.98\ \text{in}^2 \end{aligned}$$

Check the general requirements for filled composite columns. Check that the percentage of steel in the cross-sectional area is at least 1%.

$$\begin{aligned} \%_{\text{steel}} &= \frac{A_s}{A_g} \times 100\% \\ &= \frac{13.7\ \text{in}^2}{128.68\ \text{in}^2} \times 100\% \\ &= 10.65\% \quad [> 1\%, \text{so OK}] \end{aligned}$$

Check that the D/t ratio is no more than $0.15E/F_y$.

$$\frac{D}{t} \leq \frac{0.15E}{F_y}$$

$$36.5 \leq \frac{(0.15)\left(29{,}000\ \frac{\text{kips}}{\text{in}^2}\right)}{35\ \frac{\text{kips}}{\text{in}^2}}$$

$$\leq 124.29\quad [\text{so OK}]$$

Use Eq. 12.10 to determine the modulus of elasticity for the concrete.

$$E_c = 33w_c^{1.5}\sqrt{f_c'}$$

$$= (33)\left(150\ \frac{\text{lbf}}{\text{ft}^3}\right)^{1.5}\sqrt{6000\ \frac{\text{lbf}}{\text{in}^2}}$$

$$= 4{,}695{,}982\ \text{psi}\quad (4696\ \text{ksi})$$

Use Eq. 12.18 and Eq. 12.19 to determine the base strength. A_{sr} is zero because there is no reinforcing steel within the composite column.

$$P_{no} = P_p$$

$$= F_yA_s + C_2f_c'\left(A_c + A_{sr}\left(\frac{E_s}{E_c}\right)\right)$$

$$= \left(35\ \frac{\text{kips}}{\text{in}^2}\right)(13.7\ \text{in}^2) + (0.95)\left(6.0\ \frac{\text{kips}}{\text{in}^2}\right) \times\left(114.98\ \text{in}^2 + (0\ \text{in}^2)\left(\frac{29{,}000\ \frac{\text{kips}}{\text{in}^2}}{4696\ \frac{\text{kips}}{\text{in}^2}}\right)\right)$$

$$= 1135\ \text{kips}$$

The moment of inertia of the concrete is

$$I_c = \frac{\pi D^4}{64} = \frac{\pi(12.0\ \text{in})^4}{64}$$

$$= 1018\ \text{in}^4$$

Find the effective stiffness from Eq. 12.25 and Eq. 12.26.

$$\begin{aligned}C_3 &= 0.6 + 2\left(\frac{A_s}{A_c + A_s}\right) \leq 0.9\\ &= 0.6 + (2)\left(\frac{13.7\ \text{in}^2}{114.98\ \text{in}^2 + 13.7\ \text{in}^2}\right)\\ &= 0.81 \quad [\leq 0.9]\end{aligned}$$

$$\begin{aligned}(EI)_{\text{eff}} &= E_s I_s + E_s I_{\text{sr}} + C_3 E_c I_c\\ &= \left(29{,}000\ \frac{\text{kips}}{\text{in}^2}\right)(262\ \text{in}^4) + \left(29{,}000\ \frac{\text{kips}}{\text{in}^2}\right)(0\ \text{in}^4)\\ &\quad + (0.81)\left(4696\ \frac{\text{kips}}{\text{in}^2}\right)(1018\ \text{in}^4)\\ &= 11{,}470{,}228\ \text{kips-in}^2\end{aligned}$$

Use Eq. 12.7 to calculate the elastic buckling load.

$$\begin{aligned}P_e &= \frac{\pi^2 (EI)_{\text{eff}}}{(KL)^2} = \frac{\pi^2 (11{,}470{,}228\ \text{in}^2\text{-kips})}{\left((1)(30\ \text{ft})\left(12\ \frac{\text{in}}{\text{ft}}\right)\right)^2}\\ &= 873\ \text{kips}\end{aligned}$$

Determine whether Eq. 12.4 or Eq. 12.5 is appropriate to use to calculate P_n.

$$\begin{aligned}\frac{P_{no}}{P_e} &= \frac{1135\ \text{kips}}{873\ \text{kips}}\\ &= 1.30 \quad [\leq 2.25, \text{ so use Eq. 12.4}]\end{aligned}$$

From Eq. 12.4, the nominal compressive strength is

$$\begin{aligned}P_n &= 0.658^{P_{no}/P_e} P_{no}\\ &= (0.658)^{1135\ \text{kips}/873\ \text{kips}}(1135\ \text{kips})\\ &= 658.73\ \text{kips}\end{aligned}$$

Determine the design strength (LRFD) and the allowable strength (ASD) using Eq. 12.2 and Eq. 12.3.

LRFD	ASD
$P_u = \phi_c P_n = (0.75)(658.73\ \text{kips})$ $= 494\ \text{kips}$	$P_a = \frac{P_n}{\Omega_c} = \frac{658.73\ \text{kips}}{2}$ $= 329\ \text{kips}$

8. FLEXURAL MEMBERS

The *AISC Manual* recognizes three types of composite flexural members.

- encased composite beams (steel beams fully encased in concrete)
- filled composite beams (HSS members filled with concrete)
- steel beams with mechanical anchorage to a concrete slab

The first two types are similar to the two types of composite axial members. In the third type, the steel beams are anchored to the slab with shear studs or other types of connectors so that the steel and concrete act together as a single, nonhomogeneous member to resist bending. (See Fig. 12.3.) This form of construction is in common use, and it is generally most cost effective when used with a formed steel deck.

***Figure 12.3** Composite Steel Beams with Formed Steel Deck*

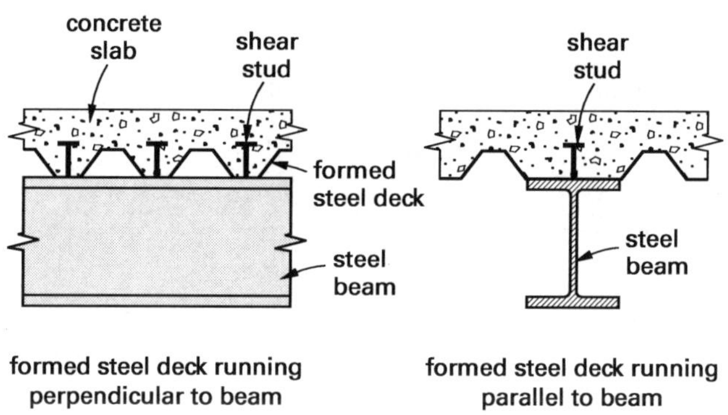

Design Basis

The design of a composite flexural member requires a two-stage design or analysis. In the first stage, the steel member must be designed to support all the loads that will be imparted to it before the concrete has hardened (to 75% of its required strength). The only exceptions are loads supported by temporary shoring, but temporary shoring increases the cost of the installation and consequently is seldom used.

In the second stage, the transformed composite section must be designed to support all the loads, dead and live, that are to be supported after the concrete has hardened. Concrete tensile stresses are ignored.

The following should be considered when a formed steel deck is used in conjunction with composite beams.

- The area taken up by the formed steel deck can carry no compressive force.
- The direction of the deck with respect to the composite beam matters.
- The strength of the shear studs should be adjusted (reduced) to account for the deck.

An effective width, b_e, for the supported portion of the concrete slab is used in designing the composite beam. The effective width of the half-slab on each side of the centerline of the beam is the smallest of

- one-eighth of the beam span (measured from center to center of the supports)
- one-half of the beam spacing (measured from the beam centerline to the centerline of the adjacent beam)
- the distance from the beam centerline to the edge of the slab

The effective width of the entire slab is the sum of the effective widths of its two halves. Figure 12.4 illustrates how the effective width is calculated, both for beams at the edge of the slab and for interior beams.

Figure 12.4 *Effective Concrete Width for Composite Slabs*

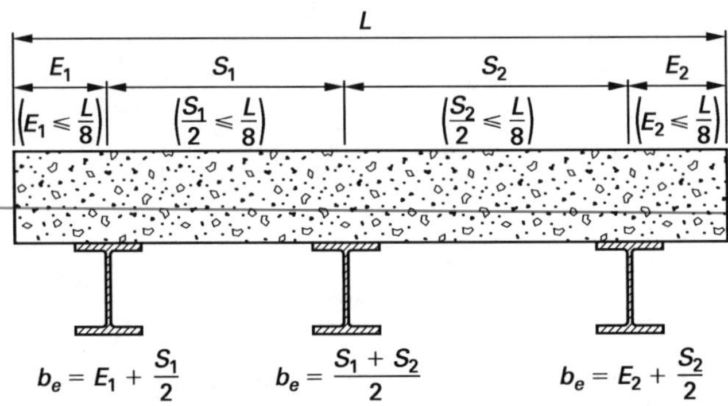

Bottom flange cover plates can be added to the beam to increase its strength or to reduce the depth of the construction. However, this raises labor costs and may reduce the cost effectiveness of the assembly.

The plastic neutral axis (PNA) may be located in the concrete, in the flange of the steel beam, or in the web of the steel beam. The location of the PNA is determined by the compressive force in the concrete, C_c, which is the smallest of the following values.

- A_sF_y (all steel in tension)
- $0.85f_c'A_c$ (all concrete in compression)
- ΣQ_n (maximum force that studs can transfer)

If $A_sF_y < 0.85f_c'A_c$, then steel controls the design and the PNA is in the concrete. If $A_sF_y > 0.85f_c'A_c$, then concrete controls the design and the PNA is in the steel.

Once the location of the PNA has been determined, all element forces can be determined.

- Concrete in compression is stressed to $0.85f_c'$.
- Concrete in tension is ignored.
- Steel in compression is stressed to F_y.
- Steel in tension is stressed to F_y.

When the location of the PNA is known, *AISC Manual* Table 3-19 can be used to find the available strength in flexure. In this table, the available strength in flexure is given for seven possible locations of the PNA; interpolation is used if the PNA is between two of these locations.

Figure 12.5 shows these seven locations. Five are in the beam flange. Locations 1 and 5 are at the top of the steel flange (TFL) and bottom of the steel flange (BFL), respectively. Locations 2, 3, and 4 are equally spaced between the TFL and BFL. The other two locations are in the web. Location 7 is at the point where ΣQ_n is equal to $0.25F_yA_s$, and location 6 is at the point where the value of ΣQ_n is halfway between the values at locations 5 and 7.

$$\sum Q_{n,7} = 0.25F_yA_s \quad 12.40$$

$$\sum Q_{n,6} = \frac{\sum Q_{n,5} + \sum Q_{n,7}}{2} \quad 12.41$$

Figure 12.5 *Plastic Neutral Axis Locations*

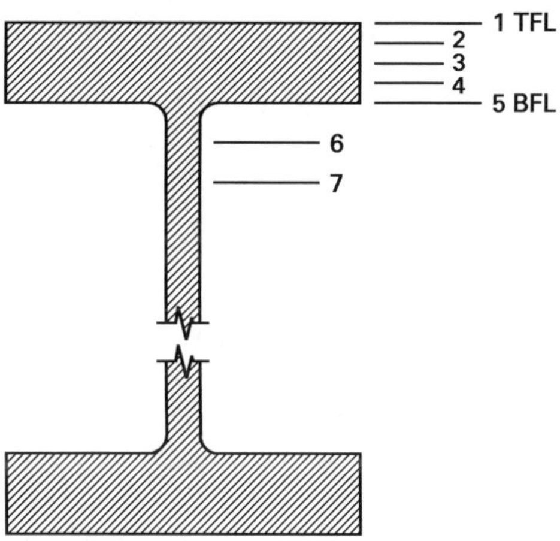

(not to scale)

Shear Studs with Formed Steel Deck

Shear studs cannot be more than $^3/_4$ in in diameter. Also, the diameter of the studs cannot be more than $2^1/_2$ times the thickness of the element to which they are welded. After they are installed, the shear studs must extend at least $1^1/_2$ in above the top of the deck rib and they must be covered by at least $^1/_2$ in of the concrete slab. The slab thickness above the top of the formed steel deck must be at least 2 in. The rib height cannot be more than 3 in, and the average width must be at least 2 in.

AISC Manual Table 3-21 gives the shear capacity for one stud, depending on its diameter, the strength of the concrete used, whether the ribs of the formed steel deck run perpendicular to or parallel to the beam web, and other factors.

Design Procedures

Use the following steps to design a composite steel beam and concrete slab flexural member.

step 1. Determine the required flexural strength.

step 2. Use Eq. 12.42 to calculate a trial moment arm for the distance from the top of the steel beam to the concrete force, $Y2$. (Making the assumption that the depth of the concrete in compression, a, is 1.0 in has proven to be a good starting point for many problems.)

$$Y2 = t - \frac{a}{2} \tag{12.42}$$

step 3. Enter *AISC Manual* Table 3-19 with the required strength and the trial value for $Y2$. Select a beam and a location for the PNA that will provide sufficient available strength. Note the values for the distance from the top of the steel flange to the PNA, $Y1$, and the total horizontal shear capacity, ΣQ_n.

step 4. Determine the effective slab width, b_e, as described earlier in this section.

step 5. Use Eq. 12.43 to determine the depth of the concrete in compression, a.

$$a = \frac{\sum Q_n}{0.85 f_c' b} \tag{12.43}$$

step 6. Use Eq. 12.42 to determine the actual value of $Y2$.

step 7. Use *AISC Manual* Table 3-19 and the value of $Y2$ to find the actual available strength. Interpolate between tabulated values if needed.

step 8. Check that the steel section alone can support the construction loads (loads applied before the concrete hardens) by calculating the beam deflection. For construction economy, assume the use of unshored construction.

step 9. Check the live load deflection, using the lower bound moment of inertia from *AISC Manual* Table 3-20.

step 10. Determine the number and type of shear connectors required. The number of connectors given by Eq. 12.44 is for each side of the point of maximum moment. *AISC Manual* Table 3-21 gives values for Q_n.

$$n_{\text{half}} = \frac{\sum Q_n}{Q_n} \tag{12.44}$$

Example 12.2

Design of Composite Steel Beam with Formed Steel Deck

A simple span composite W shape steel beam spans 40 ft and has a 4 in concrete slab using a 1.5 in formed steel deck as shown. The transverse spacing between beams is 6.0 ft. Dead load (including steel beam weight) is 80 lbf/ft^2. Live load is 150 lbf/ft^2.

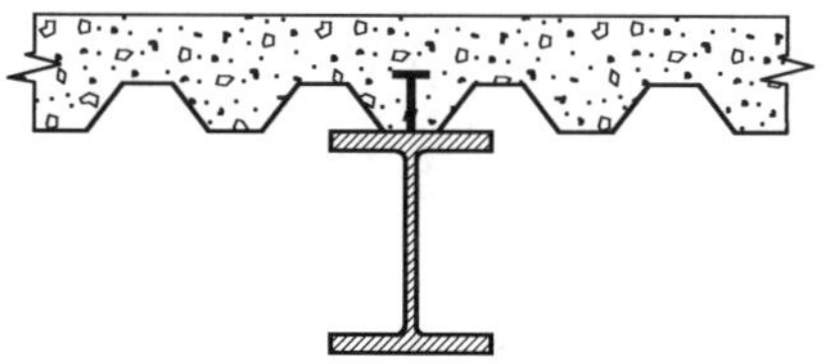

Material properties

ASTM A992 steel	normal weight concrete
$F_y = 50$ ksi	$f_c' = 4000$ ksi
$F_u = 65$ ksi	

Select the beam size required to limit live load deflection to $L/360$ and determine the number of $^3/_4$ in shear studs required.

Solution

Calculate the total load. The tributary width, s, is equal to the spacing between beams, 6.0 ft.

LRFD	ASD
$w_u = 1.2D + 1.6L = (1.2)\left(80\ \frac{\text{lbf}}{\text{ft}^2}\right) + (1.6)\left(150\ \frac{\text{lbf}}{\text{ft}^2}\right) = 336\ \text{lbf/ft}^2$	$w_a = D + L = 80\ \frac{\text{lbf}}{\text{ft}^2} + 150\ \frac{\text{lbf}}{\text{ft}^2} = 230\ \text{lbf/ft}^2$
$W_u = sw_u = \frac{(6\ \text{ft})\left(336\ \frac{\text{lbf}}{\text{ft}^2}\right)}{1000\ \frac{\text{lbf}}{\text{kip}}} = 2.02\ \text{kips/ft}$	$W_a = sw_a = \frac{(6\ \text{ft})\left(230\ \frac{\text{lbf}}{\text{ft}^2}\right)}{1000\ \frac{\text{lbf}}{\text{kip}}} = 1.38\ \text{kips/ft}$

Determine the required flexural strength.

LRFD	ASD
$M_u = \frac{W_u L^2}{8} = \frac{\left(2.02\ \frac{\text{kips}}{\text{ft}}\right)(40\text{ ft})^2}{8} = 404\text{ ft-kips}$	$M_a = \frac{W_a L^2}{8} = \frac{\left(1.38\ \frac{\text{kips}}{\text{ft}}\right)(40\text{ ft})^2}{8} = 276\text{ ft-kips}$

Make a trial selection from *AISC Manual* Table 3-19. Use Eq. 12.42 to calculate a trial moment arm of concrete, $Y2$. Start by assuming that a is 1 in.

$$Y2 = t_{\text{slab}} - \frac{a}{2} = 4\text{ in} - \frac{1\text{ in}}{2} = 3.5\text{ in}$$

From *AISC Manual* Table 3-19, using $Y2$ = 3.5 in and the PNA at location 4, try a W18 × 35.

LRFD	ASD
$\phi_b M_n = 425\text{ ft-kips}$ $[> M_u = 404\text{ ft-kips, so OK}]$	$\frac{M_n}{\Omega_b} = 282\text{ ft-kips}$ $[> M_a = 276\text{ ft-kips, so OK}]$

Determine the effective slab width. As the slab is symmetrical, the effective widths for both halves are the same.

$$b_{e,\text{half}} \le \begin{cases} \frac{L}{8} = \frac{40\text{ ft}}{8} = 5\text{ ft} \\ \frac{s}{2} = \frac{6\text{ ft}}{2} = 3\text{ ft} \quad [\text{controls}] \end{cases}$$

$$b_e = 2b_{e,\text{half}} = (2)(3\text{ ft}) = 6\text{ ft}$$

Use Eq. 12.43 to determine the depth of concrete in compression. For a W18 × 35, ΣQ_n = 323 kips.

$$a = \frac{\sum Q_n}{0.85 f'_c b_e} = \frac{324\text{ kips}}{(0.85)\left(4\ \frac{\text{kips}}{\text{in}^2}\right)(6\text{ ft})\left(12\ \frac{\text{in}}{\text{ft}}\right)} = 1.32\text{ in}$$

The actual depth of a = 1.32 in is greater than the assumed value of 1 in and less than the depth of the concrete above deck, which is 2.5 in.

Use Eq. 12.42 to determine the actual value of $Y2$.

$$\begin{aligned} Y2 &= t - \frac{a}{2} \\ &= 4 \text{ in} - \frac{1.32 \text{ in}}{2} \\ &= 3.34 \text{ in} \end{aligned}$$

Use *AISC Manual* Table 3-19 to determine the actual available strength for a W18 × 35 with PNA at location 4 and $Y2$ = 3.34 in, interpolating between the values for $Y2$ = 3 in and $Y2$ = 3.5 in.

LRFD	ASD
$\phi_b M_n = 420$ ft-kips $[> M_u = 404 \text{ ft-kips, so OK}]$	$\frac{M_n}{\Omega_b} = 279$ ft-kips $[> M_a = 276 \text{ ft-kips, so OK}]$

Therefore, the composite strength of the section is satisfactory to resist the moments created by the full live and dead loads.

Determine whether the W18 × 35 will support the construction loads. The overall slab depth is 4 in and the formed steel deck reduces the amount of concrete in the slab. Conservatively assume 4 in of concrete and a 20 lbf/ft^2 construction load for the workers. Calculate the combined loads.

LRFD	ASD
$\begin{aligned} w_u &= 1.2D + 1.6L \\ &= (1.2)\left(50 \ \frac{\text{lbf}}{\text{ft}^2}\right) + (1.6)\left(20 \ \frac{\text{lbf}}{\text{ft}^2}\right) \\ &= 92 \ \text{lbf/ft}^2 \end{aligned}$	$\begin{aligned} w_a &= D + L \\ &= 50 \ \frac{\text{lbf}}{\text{ft}^2} + 20 \ \frac{\text{lbf}}{\text{ft}^2} \\ &= 70 \ \text{lbf/ft}^2 \end{aligned}$
$\begin{aligned} W_u &= sw_u + 1.2w_{\text{beam}} \\ &= \frac{(6 \text{ ft})\left(92 \ \frac{\text{lbf}}{\text{ft}^2}\right)}{1000 \ \frac{\text{lbf}}{\text{kip}}} \\ &\quad + (1.2)\left(0.04 \ \frac{\text{kips}}{\text{ft}}\right) \\ &= 0.60 \ \text{kips/ft} \end{aligned}$	$\begin{aligned} W_a &= sw_a + w_{\text{beam}} \\ &= \frac{(6 \text{ ft})\left(70 \ \frac{\text{lbf}}{\text{ft}^2}\right)}{1000 \ \frac{\text{lbf}}{\text{kip}}} + 0.04 \ \frac{\text{kips}}{\text{ft}} \\ &= 0.46 \ \text{kips/ft} \end{aligned}$

Determine the required flexural strength.

LRFD	ASD
$M_u = \dfrac{W_u L^2}{8} = \dfrac{\left(0.60\ \dfrac{\text{kips}}{\text{ft}}\right)(40\ \text{ft})^2}{8} = 120\ \text{ft-kips}$	$M_a = \dfrac{W_a L^2}{8} = \dfrac{\left(0.46\ \dfrac{\text{kips}}{\text{ft}}\right)(40\ \text{ft})^2}{8} = 92.0\ \text{ft-kips}$

Assume that the formed steel welded to the top flange of the beam provides adequate lateral bracing to develop the full plastic moment. From *AISC Manual* Table 3-2, for a W18 × 35,

LRFD	ASD
$\phi_b M_{px} = 249\ \text{ft-kips}$ $[> M_u = 120\ \text{ft-kips, so OK}]$	$\dfrac{M_{px}}{\Omega_b} = 166\ \text{ft-kips}$ $[> M_a = 92\ \text{ft-kips, so OK}]$

Check the deflection of the beam prior to the concrete hardening. Assume that 20 lbf/ft² of the dead load is placed on the beam after the concrete has hardened; therefore, the load on the beam prior to the concrete hardening is $w_D = 60\ \text{lbf/ft}^2$. From *AISC Manual* Table 1-1, for a W18 × 35, $I_x = 510\ \text{in}^4$.

$$W_D = s w_D = \frac{(6\ \text{ft})\left(60\ \dfrac{\text{lbf}}{\text{ft}^2}\right)}{1000\ \dfrac{\text{lbf}}{\text{kip}}} = 0.36\ \text{kips/ft}$$

$$\begin{aligned}\Delta &= \frac{5W_D L^4}{384EI} \\ &= \frac{(5)\left(0.36\ \dfrac{\text{kips}}{\text{ft}}\right)(40\ \text{ft})^4\left(12\ \dfrac{\text{in}}{\text{ft}}\right)^3}{(384)\left(29{,}000\ \dfrac{\text{kips}}{\text{in}^2}\right)(510\ \text{in}^4)} \\ &= 1.40\ \text{in}\end{aligned}$$

To minimize the total amount of deflection, the designer could specify that the beam be furnished with a 1.5 in camber.

Calculate the live load deflection using the lower bound moment of inertia for composite beams in *AISC Manual* Table 3-20. The lower bound moment of inertia for a W18 × 35 with $Y2 = 3.5$ and the PNA at location 4 is 1120 in^4. The design live load is $w_L = 150$ lbf/ft^2, so

$$W_L = sw_L = \frac{(6\ \text{ft})\left(150\ \frac{\text{lbf}}{\text{ft}^2}\right)}{1000\ \frac{\text{lbf}}{\text{kip}}} = 0.90\ \text{kips/ft}$$

$$\begin{aligned}\Delta &= \frac{5W_L L^4}{384EI_{\text{LB}}}\\ &= \frac{(5)\left(0.90\ \frac{\text{kips}}{\text{ft}}\right)(40\ \text{ft})^4\left(12\ \frac{\text{in}}{\text{ft}}\right)^3}{(384)\left(29{,}000\ \frac{\text{kips}}{\text{in}^2}\right)(1120\ \text{in}^4)}\\ &= 1.60\ \text{in}\end{aligned}$$

Calculate the allowable live load deflection of span/360.

$$\Delta = \frac{L}{360} = \frac{(40\ \text{ft})\left(12\ \frac{\text{in}}{\text{ft}}\right)}{360} = 1.33\ \text{in}$$

The live load deflection is 20% greater than that permitted by the *International Building Code*. Possible solutions to this problem are to

- select another composite beam section and check the design
- increase the beam camber from 1.5 in to 1.75 in

If another composite beam section is selected, start the process by calculating the lower bound moment of inertia required to limit the live load deflection to span/360.

$$\begin{aligned}I_{\text{LB,needed}} &= \left(\frac{\Delta_{\text{trial}}}{\Delta_{\text{allowable}}}\right) I_{\text{LB,trial}}\\ &= \left(\frac{1.60\ \text{in}}{1.33\ \text{in}}\right)(1120\ \text{in}^4)\\ &= 1347.37\ \text{in}^4\end{aligned}$$

For this solution, however, assume the decision is made to increase the total camber to 1.75 in. Calculate the number of shear studs required. Use $^3/_4$ in diameter shear studs. Assume there is one stud per rib and that the studs can be placed at the weak position of the deck (a conservative approach). From *AISC Manual* Table 3-21, with these assumptions, the deck perpendicular to the beam, the studs located in the weak position, and $f_c' = 4$ ksi, the nominal strength of the shear studs is $Q_n = 17.2$ kips/stud.

The number of studs required per half length of beam is

$$n_{\text{half}} = \frac{\sum Q_n}{Q_n} = \frac{324 \text{ kips}}{17.2 \dfrac{\text{kips}}{\text{stud}}} = 18.83 \text{ studs} \quad [\text{use } 19]$$

38 studs are required for the full length of the W18 × 35.

Partial Composite Action

On a dollar-per-pound basis, the total cost of installing shear studs can be eight to ten times the cost of the beam. Reducing the number of shear studs, then, can increase the economy of construction significantly. When the full strength of the wide flange is not required in the finished structure but may be needed during construction or to meet serviceability requirements, it may be possible to accomplish this by taking advantage of partial composite action.

Example 12.3
Design of Partial Composite Steel Action

For the span in Ex. 12.2, determine the available flexural strength of the W18 × 35 member if the number of shear studs is limited to 24 studs with $^3/_4$ in diameters.

Solution

Determine the shear capacity of the studs for partial composite action. n_{half} is the number of studs per half length of beam.

$$n_{\text{half}} = \frac{\sum Q_{n,\text{pc}}}{Q_n}$$

$$\begin{aligned}\sum Q_{n,\text{pc}} &= n_{\text{half}} Q_n \\ &= \left(\frac{24 \text{ studs}}{2}\right)\left(17.2 \frac{\text{kips}}{\text{stud}}\right) \\ &= 206.4 \text{ kips}\end{aligned}$$

Use Eq. 12.43 to determine the depth of the concrete in compression.

$$\begin{aligned}a &= \frac{\sum Q_{n,\text{pc}}}{0.85 f'_c b_e} \\ &= \frac{206.4 \text{ kips}}{(0.85)\left(4 \dfrac{\text{kips}}{\text{in}^2}\right)(6 \text{ ft})\left(12 \dfrac{\text{in}}{\text{ft}}\right)} \\ &= 0.843 \text{ in}\end{aligned}$$

The compressive force in the concrete is

$$\begin{aligned} C_c &= 0.85 f_c' a b_e \\ &= (0.85)\left(4\ \frac{\text{kips}}{\text{in}^2}\right)(0.843\ \text{in})(6\ \text{ft})\left(12\ \frac{\text{in}}{\text{ft}}\right) \\ &= 206.4\ \text{kips} \end{aligned}$$

Determine the area of steel required for compression. From Ex. 12.2, ΣQ_n is 324 kips.

$$\begin{aligned} A_{s,\text{comp}} &= \frac{\sum Q_n - \sum Q_{n,\text{pc}}}{2F_y} \\ &= \frac{324\ \text{kips} - 206.4\ \text{kips}}{(2)\left(50\ \frac{\text{kips}}{\text{in}^2}\right)} \\ &= 1.17\ \text{in}^2 \end{aligned}$$

The distance from the top of the steel to the plastic neutral axis is

$$\begin{aligned} Y1 &= \frac{A_{s,\text{comp}}}{b_f} \\ &= \frac{1.17\ \text{in}^2}{6\ \text{in}} \\ &- 0.195\ \text{in} \quad \left[< t_f = 0.425\ \text{in}\right] \end{aligned}$$

The PNA is located within the top flange. Determine the compressive force in the steel.

$$\begin{aligned} C_s &= A_{s,\text{comp}} F_y \\ &= (1.17\ \text{in}^2)\left(50\ \frac{\text{kips}}{\text{in}^2}\right) \\ &= 58.5\ \text{kips} \end{aligned}$$

Determine the tensile force in the steel. From *AISC Manual* Table 1-1, the area of a W18 × 35 is 10.3 in^2.

$$\begin{aligned} T_s &= A_s F_y \\ &= (10.3\ \text{in}^2)\left(50\ \frac{\text{kips}}{\text{in}^2}\right) \\ &= 515.0\ \text{kips} \end{aligned}$$

Determine the nominal flexural strength.

$$M_n = M_{pc} = C_c(\text{moment arm}) + T_s(\text{moment arm}) - C_s(\text{moment arm})$$

$$= C_c\left(t_s - \frac{a}{2}\right) + T_s\left(\frac{d}{2}\right) - C_s\left(\frac{Y1}{2}\right)$$

$$= \frac{(206.4\text{ kips})\left(4\text{ in} - \frac{0.85\text{ in}}{2}\right) + (515.0\text{ kips})\left(\frac{17.7\text{ in}}{2}\right) - (58.5\text{ kips})\left(\frac{0.19\text{ in}}{2}\right)}{12\ \frac{\text{in}}{\text{ft}}}$$

$$= 440.8\text{ ft-kips}$$

Determine the design flexural strength (LRFD) and the allowable flexural strength (ASD).

LRFD	ASD
$\phi_b M_n = (0.90)(440.8\text{ ft-kips})$ $= 396.7\text{ ft-kips}$	$\frac{M_n}{\Omega_b} = \frac{440.8\text{ ft-kips}}{1.67}$ $= 264.0\text{ ft-kips}$

The flexural capacities here are less than those required for Ex. 12.2. However, the effectiveness of partial composite action can be seen by comparing the percentage of decrease in flexural capacity with the percentage of decrease in the number of shear connector studs. The decrease in flexural capacity is

LRFD	ASD
$\%_{\text{flex}} = \frac{420\text{ ft-kips} - 397\text{ ft-kips}}{420\text{ ft-kips}} \times 100\%$ $= 5.48\%$	$\%_{\text{flex}} = \frac{279\text{ ft-kips} - 264\text{ ft-kips}}{279\text{ ft-kips}} \times 100\%$ $= 5.38\%$

The decrease in the number of shear studs is

LRFD	ASD
$\%_{\text{stud}} = \frac{38\text{ studs} - 24\text{ studs}}{38\text{ studs}} \times 100\%$ $= 36.84\%$	$\%_{\text{stud}} = \frac{38\text{ studs} - 24\text{ studs}}{38\text{ studs}} \times 100\%$ $= 36.84\%$

Decreasing the number of shear studs by about 37% results in a decrease in flexural capacity of only about 5%.

The following example shows the value of composite beams as compared to noncomposite beams.

Example 12.4

Design of Noncomposite Beam

Select a noncomposite beam for the loading and span in Ex. 12.2. Assume that the compression flange is adequately braced to develop the full plastic moment.

Solution

Calculate the total load moment.

LRFD	ASD
$M_u = \frac{w_u L^2}{8} = \frac{\left(2.02\ \frac{\text{kips}}{\text{ft}}\right)(40\ \text{ft})^2}{8} = 404\ \text{ft-kips}$	$M_a = \frac{w_a L^2}{8} = \frac{\left(1.38\ \frac{\text{kips}}{\text{ft}}\right)(40\ \text{ft})^2}{8} = 276\ \text{ft-kips}$

Calculate the required plastic section modulus, Z.

LRFD	ASD
$M_u \le \phi_b M_n = \phi_b Z_x F_y$ $Z_x = \frac{M_u}{\phi_b F_y} = \frac{(404\ \text{ft-kips})\left(12\ \frac{\text{in}}{\text{ft}}\right)}{(0.90)\left(50\ \frac{\text{kips}}{\text{in}^2}\right)} = 107.73\ \text{in}^3$	$M_a \le \frac{M_n}{\Omega_b} = \frac{Z_x F_y}{\Omega_b}$ $Z_x = \frac{M_a \Omega_b}{F_y} = \frac{(276\ \text{ft-kips})\left(12\ \frac{\text{in}}{\text{ft}}\right)(1.67)}{\left(50\ \frac{\text{kips}}{\text{in}^2}\right)} = 110.62\ \text{in}^3$

The slight difference in the required plastic section modulus can be attributed to the dead load to live load ratio.

A W21 × 50 has a plastic section modulus of 110 in^3. A W18 × 55 has a plastic section modulus of 112 in^3. These selections do not take into consideration any serviceability criteria such as deflections. Compare this with the fully composite W18 × 35 beam that was selected in Ex. 12.2.

9. COMBINED AXIAL FORCE AND FLEXURE

Section I4 of the *AISC Specification* specifies the requirements for members subject to combined axial and flexural forces. The design compressive strength, $\phi_c P_n$, the allowable compressive strength, P_n/Ω_c, the flexural design strength, $\phi_b M_n$, and the allowable flexural strength, M_n/Ω_b, are determined as follows.

- For axial strength, $\phi_c = 0.75$ (LRFD) and $\Omega_c = 2.00$ (ASD).
- For flexural strength, $\phi_b = 0.90$ (LRFD) and $\Omega_b = 1.67$ (ASD).

The nominal strength of the cross section of a composite member should be determined by either the plastic stress distribution method or the strain-compatibility method.

13 Practice Problems

PRACTICE PROBLEM 1

A plant engineer wants to reinforce an existing W10 × 49 steel column by welding a pair of 11 in × $^1/_2$ in plates across the toes of the W10. The column is part of a braced frame system and has an effective length about both axes of 20 ft.

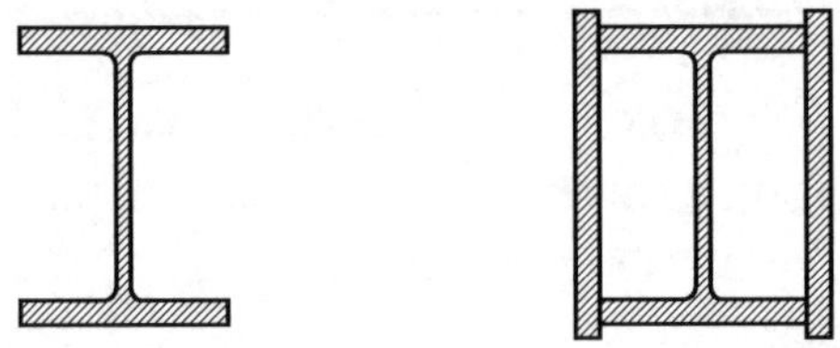

Section properties		Material properties
W10 × 49	$S_x = 54.6$ in^3	column
$A = 14.4$ in^2	$r_x = 4.35$ in	ASTM A992
$d = 10.0$ in	$Z_x = 60.4$ in^3	$F_y = 50$ ksi
$t_w = 0.34$ in	$I_y = 93.4$ in^4	$F_u = 65$ ksi
$b_f = 10.0$ in	$S_y = 18.7$ in^3	plates
$t_f = 0.560$ in	$r_y = 2.54$ in	ASTM A572, grade 50
$I_x = 272$ in^4	$Z_y = 28.3$ in^3	$F_y = 50$ ksi
		$F_u = 65$ ksi

Adding the reinforcement will increase the load-carrying capacity of the column by a factor of most nearly

(A) 1.8

(B) 2.2

(C) 2.6

(D) 2.8

Solution

Determine the available strength of the column without the reinforcing. The effective length, KL, is the same about both axes, so the axis with the smaller radius of gyration will govern the design. $r_x = 4.35$ in and $r_y = 2.54$ in, so the y-axis governs.

From *AISC Manual* Table 4-1, the available strength for the unreinforced column is

$$\phi_c P_n = 337 \text{ kips} \quad [\text{LRFD}]$$

$$\frac{P_n}{\Omega_c} = 224 \text{ kips} \quad [\text{ASD}]$$

Determine the cross-sectional area of the reinforced member.

$$\begin{aligned} A_{\text{reinforced}} &= A_{\text{column}} + n_{\text{plates}} A_{\text{plate}} \\ &= 14.4 \text{ in}^2 + (2)\big((0.5 \text{ in})(11 \text{ in})\big) \\ &= 25.4 \text{ in}^2 \end{aligned}$$

Determine the moment of inertia for the reinforced member about each axis. For the x-axis,

$$\begin{aligned} I_{x,\text{reinforced}} &= I_{x,\text{column}} + n_{\text{plates}} I_{x,\text{plate}} \\ &= I_{x,\text{column}} + n_{\text{plates}} \left(\frac{tw^3}{12} \right) \\ &= 272 \text{ in}^4 + (2) \left(\frac{(0.5 \text{ in})(11 \text{ in})^3}{12} \right) \\ &= 383 \text{ in}^4 \end{aligned}$$

For the y-axis,

$$\begin{aligned} I_{y,\text{reinforced}} &= I_{y,\text{column}} + n_{\text{plates}} I_{y,\text{plate}} \\ &= I_{y,\text{column}} + n_{\text{plates}} \left(\frac{wt^3}{12} + Ad^2 \right) \\ &= 93.4 \text{ in}^4 + (2) \left(\frac{(11 \text{ in})(0.5 \text{ in})^3}{12} + (5.5 \text{ in}^2)(5.25 \text{ in})^2 \right) \\ &= 397 \text{ in}^4 \end{aligned}$$

Determine the radius of gyration of the reinforced member about each axis.

$$r_x = \sqrt{\frac{I_{x,\text{reinforced}}}{A_{\text{reinforced}}}} = \sqrt{\frac{383 \text{ in}^4}{25.4 \text{ in}^2}} = 3.88 \text{ in}$$

$$r_y = \sqrt{\frac{I_{y,\text{reinforced}}}{A_{\text{reinforced}}}} = \sqrt{\frac{397 \text{ in}^4}{25.4 \text{ in}^2}} = 3.95 \text{ in}$$

Use Eq. 7.2 to determine the nominal strength of the reinforced section.

$$P_n = F_{\text{cr}} A_g$$

The gross area is

$$A_g = A_{\text{reinforced}} = 25.4 \text{ in}^2$$

The column's effective length, KL, is 20 ft. Check the slenderness ratio, KL/r, to determine the applicable formula for F_{cr}.

$$4.71\sqrt{\frac{E}{F_y}} = 4.71\sqrt{\frac{29{,}000 \ \frac{\text{kips}}{\text{in}^2}}{50 \ \frac{\text{kips}}{\text{in}^2}}} = 113.43$$

$$\frac{KL}{r} = \frac{(20 \text{ ft})\left(12 \ \frac{\text{in}}{\text{ft}}\right)}{3.88 \text{ in}} = 61.86 \quad \left[< 4.71\sqrt{E/F_y}, \text{ so use Eq. 7.6}\right]$$

From Eq. 7.8, the elastic critical buckling stress is

$$\begin{aligned} F_e &= \frac{\pi^2 E}{\left(\frac{KL}{r}\right)^2} \\ &= \frac{\pi^2 \left(29{,}000 \ \frac{\text{kips}}{\text{in}^2}\right)}{(61.86)^2} \\ &= 74.80 \text{ ksi} \end{aligned}$$

From Eq. 7.6, the flexural buckling stress is

$$\begin{aligned} F_{\text{cr}} &= 0.658^{F_y/F_e} F_y \\ &= \left(0.658^{50 \ \frac{\text{kips}}{\text{in}^2} / 74.80 \ \frac{\text{kips}}{\text{in}^2}}\right)\left(50 \ \frac{\text{kips}}{\text{in}^2}\right) \\ &= 37.8 \text{ ksi} \end{aligned}$$

The nominal strength, then, is

$$P_n = F_{cr}A_g = \left(37.8\ \frac{\text{kips}}{\text{in}^2}\right)\left(25.4\ \text{in}^2\right) = 960\ \text{kips}$$

Determine the new design strength (LRFD) or available strength (ASD) and the magnitude of increase.

LRFD	ASD
$\phi_c P_n = (0.90)(960\ \text{kips}) = 864\ \text{kips}$	$\frac{P_n}{\Omega_c} = \frac{960\ \text{kips}}{1.67} = 575\ \text{kips}$
$\frac{(\phi_c P_n)_{\text{reinforced}}}{(\phi_c P_n)_{\text{unreinforced}}} = \frac{864\ \text{kips}}{337\ \text{kips}} = 2.56 \quad (2.6)$	$\frac{\left(\frac{P_n}{\Omega_c}\right)_{\text{reinforced}}}{\left(\frac{P_n}{\Omega_c}\right)_{\text{unreinforced}}} = \frac{575\ \text{kips}}{224\ \text{kips}} = 2.56 \quad (2.6)$

The answer is (C).

PRACTICE PROBLEM 2

In the truss shown, the load at each of points A, C, D, and E consists of a dead load of 3 kips and a live load of 9 kips. The load at point B consists of a dead load of 6 kips and a live load of 18 kips. All panel points act as pinned connections and $F_y = 36$ ksi.

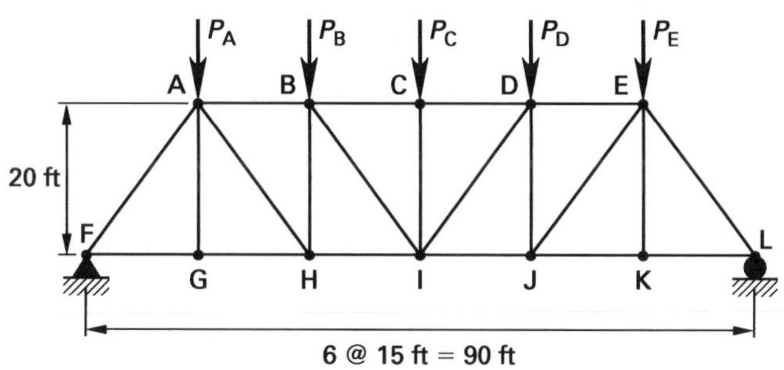

The gross cross-sectional area of steel required for member A-H is most nearly

(A) 1.5 in^2

(B) 1.8 in^2

(C) 2.0 in^2

(D) 2.3 in^2

Solution

Analyze the truss to determine the reaction at panel point F and then determine the force acting on member A-G. Break P_B into two loads, P_{B1} and P_{B2}, each consisting of a dead load of 3 kips and a live load of 9 kips. The six loads P_A, P_{B1}, P_{B2}, P_C, P_D, and P_E are then equal. Calculate the required load for each.

LRFD	ASD
$P_u = 1.2P_D + 1.6P_L$ $= (1.2)(3\text{ kips}) + (1.6)(9\text{ kips})$ $= 18\text{ kips}$	$P_a = P_D + P_L$ $= 3\text{ kips} + 9\text{ kips}$ $= 12\text{ kips}$

Calculate the reaction at point F. Treat the five equally spaced loads P_A, P_{B1}, P_C, P_D, and P_E as a single load of $5P_C$ at point C.

LRFD	ASD
$R_{u,F} = \frac{1}{2}(5P_C) + \frac{2}{3}P_{B2}$ $= \frac{5P_u}{2} + \frac{2P_u}{3}$ $= \frac{(5)(18\text{ kips})}{2} + \frac{(2)(18\text{ kips})}{3}$ $= 57\text{ kips}$	$R_{u,F} = \frac{1}{2}(5P_C) + \frac{2}{3}P_{B2}$ $= \frac{5P_u}{2} + \frac{2P_u}{3}$ $= \frac{(5)(12\text{ kips})}{2} + \frac{(2)(12\text{ kips})}{3}$ $= 38\text{ kips}$

Because F-G and G-H are collinear, A-G is a zero-force member. Therefore, the vertical component of member A-H is equal to the vertical reaction at F less the downward load of P_A at panel point A.

LRFD	ASD
$P_{AH,vert} = R_{u,F} - P_A$ $= 57\text{ kips} - 18\text{ kips}$ $= 39\text{ kips}$	$P_{AH,vert} = R_{u,F} - P_A$ $= 38\text{ kips} - 12\text{ kips}$ $= 26\text{ kips}$

ΔBAH is a 3-4-5 triangle, so L_{AH} is 25 ft. The force in A-H is

$$P_{AH} = \frac{P_{AH,vert}}{\sin\angle BAH} = \frac{P_{AH,vert}}{\frac{L_{BH}}{L_{AH}}} = \frac{P_{AH,vert}}{\frac{20\text{ ft}}{25\text{ ft}}}$$

$$= \frac{5P_{AH,vert}}{4}$$

LRFD	ASD
$P_{AH} = \frac{(5)(39\ \text{kips})}{4} = 48.75\ \text{kips}\quad [\text{in tension}]$	$P_{AH} = \frac{(5)(26\ \text{kips})}{4} = 32.5\ \text{kips}\quad [\text{in tension}]$

Use Eq. 4.15 (LRFD) or Eq. 4.22 (ASD) to determine the minimum gross area of steel for the required strength.

LRFD	ASD
$A_{g,AH} \geq \frac{R_{u,AH}}{\phi_t F_y} \geq \frac{48.75\ \text{kips}}{(0.90)\left(36\ \frac{\text{kips}}{\text{in}^2}\right)} \geq 1.51\ \text{in}^2\quad (1.5\ \text{in}^2)$	$A_g \geq \frac{\Omega_t R_{a,AH}}{F_y} \geq \frac{(1.67)(32.5\ \text{kips})}{36\ \frac{\text{kips}}{\text{in}^2}} \geq 1.51\ \text{in}^2\quad (1.5\ \text{in}^2)$

The answer is (A).

PRACTICE PROBLEM 3

For the truss in Prob. 2, select the lightest pair of 6 × 4 in angles that meets the available strength requirements for member A-F.

(A) 2L6 × 4 × $^9/_{16}$ in, long legs back to back

(B) 2L6 × 4 × $^5/_8$ in, long legs back to back

(C) 2L6 × 4 × $^3/_4$ in, long legs back to back

(D) 2L6 × 4 × $^7/_8$ in, long legs back to back

Solution

From the beginning of the solution to Prob. 2, the reaction at point F is

LRFD	ASD
$R_{u,F} = 57\ \text{kips}$	$R_{u,F} = 38\ \text{kips}$

ΔAFG is a 3-4-5 triangle, so L_{AF} is 25 ft. The force in member A-F is

$$P_{AF} = \frac{R_{u,F}}{\sin \angle AFG} = \frac{R_{u,F}}{\frac{L_{AG}}{L_{AF}}} = \frac{R_{u,F}}{\frac{20\ \text{ft}}{25\ \text{ft}}} = \frac{5R_{u,F}}{4}$$

LRFD	ASD
$P_{AF} = \frac{(5)(57\ \text{kips})}{4} = 71.25\ \text{kips}$	$P_{AF} = \frac{(5)(38\ \text{kips})}{4} = 47.5\ \text{kips}$

Examining *AISC Manual* Table 4-9 shows that the *Y-Y* axis is controlling, because the load capacity is always greater about the *X-X* axis for the same length. Select the lightest pair of 6 × 4 angles that meets the following strength requirement with the *Y-Y* axis controlling and using an effective length of 26 ft. The required strength is

LRFD	ASD
$\phi_c P_n \geq R_u = P_{AF} = 71.25\ \text{kips}$	$\frac{P_n}{\Omega_c} \geq R_a = P_{AF} = 47.5\ \text{kips}$

From *AISC Manual* Table 4-9, for 2L6 × 4 × $^5/_8$,

LRFD	ASD
$\phi_c P_n = 66.9\ \text{kips}\quad [< R_u,\ \text{not enough}]$	$\frac{P_n}{\Omega_c} = 44.5\ \text{kips}\quad [< R_u,\ \text{not enough}]$

For 2L6 × 4 × $^3/_4$,

LRFD	ASD
$\phi_c P_n = 81.5\ \text{kips}\quad [> R_u,\ \text{so OK}]$	$\frac{P_n}{\Omega_c} = 54.2\ \text{kips}\quad [> R_u,\ \text{so OK}]$

The answer is (C).

PRACTICE PROBLEM 4

The anchor rods shown are subject to a horizontal shear force and an overturning moment. The anchor rods are spaced at 12 in centers on both axes. The threads on the anchor rods are excluded from the shear plane. The length of anchor rod is sufficient to develop full available strength.

Anchor rods

$^7/_8$ in diameter

$A = 0.601\ \text{in}^2$

Material properties

ASTM F1554, grade 36

$F_y = 36$ ksi

$F_u = 58$ ksi

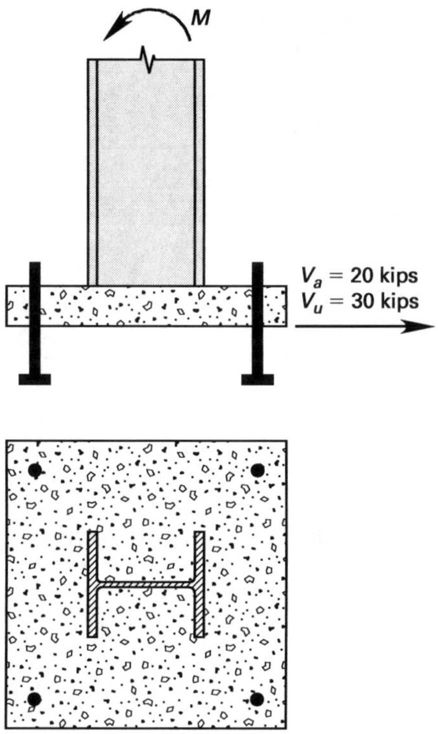

What is most nearly the available moment-resisting capacity? (LRFD options are in parentheses.)

(A) 14 ft-kips (20 ft-kips)

(B) 19 ft-kips (29 ft-kips)

(C) 25 ft-kips (38 ft-kips)

(D) 30 ft-kips (44 ft-kips)

Solution

Determine the nominal tension resistance capacity of the bolts. From *AISC Specification* Table J3.2,

$$\begin{aligned} F_{nt} &= 0.75F_u \\ &= (0.75)\left(58\ \frac{\text{kips}}{\text{in}^2}\right) \\ &= 43.5\ \text{ksi} \\ R_{nt} &= F_{nt}A_b \\ &= \left(43.5\ \frac{\text{kips}}{\text{in}^2}\right)\left(0.601\ \frac{\text{in}^2}{\text{bolt}}\right) \\ &= 26.14\ \text{kips/bolt} \end{aligned}$$

Determine the nominal shear resistance capacity of the bolts. From *AISC Specification* Table J3.2,

$$\begin{aligned} F_{nv} &= 0.563F_u \\ &= (0.563)\left(58\ \frac{\text{kips}}{\text{in}^2}\right) \\ &= 32.65\ \text{ksi} \\ R_{nv} &= F_{nv}A_b \\ &= \left(32.65\ \frac{\text{kips}}{\text{in}^2}\right)\left(0.601\ \frac{\text{in}^2}{\text{bolt}}\right) \\ &= 19.62\ \text{kips/bolt} \end{aligned}$$

Determine the design strength (LRFD) and the allowable strength (ASD) for the bolts. For tension,

LRFD	ASD
$\phi_t R_{nt} = (0.75)\left(26.14\ \frac{\text{kips}}{\text{bolt}}\right)$ $= 19.61\ \text{kips/bolt}$	$\frac{R_{nt}}{\Omega_t} = \frac{26.14\ \frac{\text{kips}}{\text{bolt}}}{2.0}$ $= 13.07\ \text{kips/bolt}$

For shear,

LRFD	ASD
$\phi_v R_{nv} = (0.75)\left(19.62\frac{\text{kips}}{\text{bolt}}\right)$ $= 14.72\ \text{kips/bolt}$	$\frac{R_{nv}}{\Omega_v} = \frac{19.62\ \frac{\text{kips}}{\text{bolt}}}{2.0}$ $= 9.81\ \text{kips/bolt}$

Check the ratio of shear required to shear available in order to determine whether the available tensile strength must be reduced (see Sec. 9.9).

LRFD	ASD
$\frac{R_{uv}}{\phi_v R_{nv}} = \frac{\frac{V_u}{n_{\text{bolts}}}}{\phi_v R_{nv}} = \frac{\frac{30\ \text{kips}}{4\ \text{bolts}}}{14.72\ \frac{\text{kips}}{\text{bolt}}}$ $= 0.51\quad [> 0.20]$	$\frac{R_{av}}{\frac{R_{nv}}{\Omega_v}} = \frac{\frac{V_a}{n_{\text{bolts}}}}{\frac{R_{nv}}{\Omega_v}} = \frac{\frac{20\ \text{kips}}{4\ \text{bolts}}}{\frac{19.62\ \frac{\text{kips}}{\text{bolt}}}{2.0}}$ $= 0.51\quad [> 0.20]$

Because the ratio exceeds 0.20, the available tensile strength must be reduced.

LRFD	ASD
$F'_{nt} = 1.3F_{nt} - \left(\frac{F_{nt}}{\phi F_{nv}}\right) f_v \le F_{nt}$ $= (1.3)\left(43.5\ \frac{\text{kips}}{\text{in}^2}\right) - \left(\frac{43.5\ \frac{\text{kips}}{\text{in}^2}}{(0.75)\left(32.65\ \frac{\text{kips}}{\text{in}^2}\right)}\right) \times \left(\frac{7.5\ \text{kips}}{0.601\ \text{in}^2}\right)$ $= 34.38\ \text{ksi}$	$F'_{nt} = 1.3F_{nt} - \left(\frac{\Omega F_{nt}}{F_{nv}}\right) f_v \le F_{nt}$ $= (1.3)\left(43.5\ \frac{\text{kips}}{\text{in}^2}\right) - \left(\frac{(2.00)\left(43.5\ \frac{\text{kips}}{\text{in}^2}\right)}{32.65\ \frac{\text{kips}}{\text{in}^2}}\right) \times \left(\frac{5.0\ \text{kips}}{0.601\ \text{in}^2}\right)$ $= 34.38\ \text{ksi}$

Calculate the reduced strength of the bolts for combined shear and tension. From Eq. 9.2,

$$R_{nt} = F'_{nt}A_b = \left(34.38\ \frac{\text{kips}}{\text{in}^2}\right)\left(0.601\ \frac{\text{in}^2}{\text{bolt}}\right) = 20.66\ \text{kips/bolt}$$

LRFD	ASD
$\phi R_{nt} = (0.75)\left(20.66\ \frac{\text{kips}}{\text{bolt}}\right)$ $= 15.50\ \text{kips/bolt}$	$\frac{R_{nt}}{\Omega} = \frac{20.66\ \frac{\text{kips}}{\text{bolt}}}{2.0}$ $= 10.33\ \text{kips/bolt}$

The available moment-resisting capacity is equal to the available tensile strength times the moment arm.

LRFD	ASD
$M_u = Td = n_{\text{bolts}}(\phi R_{nt})d$ $= (2\ \text{bolts})\left(15.50\ \frac{\text{kips}}{\text{bolt}}\right)(1\ \text{ft})$ $= 31.0\ \text{ft-kips}\quad (29\ \text{ft-kips})$	$M_a = Td = n_{\text{bolts}}\left(\frac{R_{nt}}{\Omega}\right)d$ $= (2\ \text{bolts})\left(10.33\ \frac{\text{kips}}{\text{bolt}}\right)(1\ \text{ft})$ $= 20.66\ \text{ft-kips}$

The answer is (B).

PRACTICE PROBLEM 5

The center-to-center span of an HSS18 × 6 × $^3/_8$ spandrel beam is 40 ft with the strong axis vertical. The beam is braced laterally only at the ends of the beam. The bending load imparted to the weak axis from wind load is 0.23 kips/ft; based on ASCE 7 no bending load is imparted to the weak axis from dead load.

Section properties		Material properties
$t = 0.349$ in	$S_x = 66.9\text{ in}^3$	ASTM A500, Grade B
weight = 58.10 lbf/ft	$r_x = 6.15$ in	$F_y = 46$ ksi
$A = 16.0\text{ in}^2$	$Z_x = 86.4\text{ in}^3$	$F_u = 58$ ksi
$b/t = 14.2$	$I_y = 106\text{ in}^4$	
$h/t = 48.6$	$S_y = 35.5\text{ in}^3$	
$I_x = 602\text{ in}^4$	$r_y = 2.58$ in	
	$Z_y = 39.5\text{ in}^3$	

What is most nearly the load per foot that can be imparted to the x-axis of the beam? (LRFD options are in parentheses.)

(A) 0.28 kips/ft (0.52 kips/ft)

(B) 0.38 kips/ft (0.62 kips/ft)

(C) 0.48 kips/ft (0.72 kips/ft)

(D) 0.58 kips/ft (0.82 kips/ft)

Solution

Both the flanges and webs are compact, so only the limit state of yielding applies. Determine the required strength for the y-axis.

LRFD	ASD
$w_u = 1.2w_D + 1.0w_W = (1.2)\left(0\ \frac{\text{kip}}{\text{ft}}\right) + (1.0)\left(0.23\ \frac{\text{kip}}{\text{ft}}\right) = 0.23\ \text{kip/ft}$	$w_a = w_D + 0.60w_W = 0.00\ \frac{\text{kip}}{\text{ft}} + (0.60)\left(0.23\ \frac{\text{kip}}{\text{ft}}\right) = 0.14\ \text{kip/ft}$
$M_{uy} = \frac{w_uL^2}{8} = \frac{\left(0.23\ \frac{\text{kip}}{\text{ft}}\right)(40\text{ ft})^2}{8} = 46\text{ ft-kips}$	$M_{ay} = \frac{w_aL^2}{8} = \frac{\left(0.14\ \frac{\text{kip}}{\text{ft}}\right)(40\text{ ft})^2}{8} = 28\text{ ft-kips}$

Determine the design flexural strength (LRFD) and allowable strength (ASD) for each of the axes. For the x-axis, from Eq. 5.23,

$$M_{nx} = M_{px} = F_y Z_x = \frac{\left(46\ \frac{\text{kips}}{\text{in}^2}\right)\left(86.4\ \text{in}^3\right)}{12\ \frac{\text{in}}{\text{ft}}} = 331.2\ \text{ft-kips}$$

LRFD	ASD
$M_{cx} = \phi_b M_{nx} = 0.90 M_{nx}$ $= (0.90)(331.2\ \text{ft-kips})$ $= 298\ \text{ft-kips}$	$M_{cx} = \frac{M_{nx}}{\Omega_b} = \frac{331.2\ \text{ft-kips}}{1.67}$ $= 198\ \text{ft-kips}$

Alternatively, *AISC Manual* Table 3-12 gives the same values. For the y-axis, *AISC Manual* Table 3-12 gives

LRFD	ASD
$M_{cy} = \phi_b M_{ny} = 102\ \text{ft-kips}$	$M_{cy} = \frac{M_{ny}}{\Omega_b} = 68\ \text{ft-kips}$

Determine the applicable interaction equation to use for calculating available flexural stress on the x-axis. Because $P_r = 0$ lbf, $P_r/P_c < 0.2$. Use Eq. 8.2.

$$\frac{P_r}{2P_c} + \left(\frac{M_{rx}}{M_{cx}} + \frac{M_{ry}}{M_{cy}}\right) \le 1.0$$

As there is no axial load, the first term is zero.

LRFD	ASD
$\frac{M_{rx}}{M_{cx}} + \frac{M_{ry}}{M_{cy}} \le 1.0$ $M_{rx} \le \left(1.0 - \frac{M_{ry}}{M_{cy}}\right) M_{cx}$ $\le \left(1.0 - \frac{46\ \text{ft-kips}}{102\ \text{ft-kips}}\right)$ $\times (298\ \text{ft-kips})$ $\le 163.61\ \text{ft-kips}$	$\frac{M_{rx}}{M_{cx}} + \frac{M_{ry}}{M_{cy}} \le 1.0$ $M_{rx} \le \left(1.0 - \frac{M_{ry}}{M_{cy}}\right) M_{cx}$ $\le \left(1.0 - \frac{28\ \text{ft-kips}}{68\ \text{ft-kips}}\right)$ $\times (198\ \text{ft-kips})$ $M_{rx} \le 116.47\ \text{ft-kips}$

Determine the load per foot that will use the available strength in the x-axis.

LRFD	ASD
$M_{ux} = M_{rx} = 163.61$ ft-kips	$M_{ax} = M_{rx} = 116.47$ ft-kips
$w_u = \frac{8M_{ux}}{L^2} = \frac{(8)(163.61 \text{ ft-kips})}{(40 \text{ ft})^2} = 0.82 \text{ kips/ft}$	$w_a = \frac{8M_{ax}}{L^2} = \frac{(8)(116.47 \text{ ft-kips})}{(40 \text{ ft})^2} = 0.58 \text{ kips/ft}$

The answer is (D).

PRACTICE PROBLEM 6

A W10 × 60 composite steel column is encased in 16 in × 16 in of concrete. The concrete has a compressive strength of 8 ksi. The effective length of the column about both axes is 20 ft with pinned ends. The concrete section is reinforced with four no. 14 reinforcing bars spaced at 12 in centers.

Steel section properties

W10 × 60

$A_s = 17.6 \text{ in}^2$

$d = 10.2$ in

$t_w = 0.402$ in

$b_f = 10.1$ in

$t_f = 0.680$

$b_f/2t_f = 7.41$

$h/t_w = 18.7$

$I_x = 341 \text{ in}^4$

$S_x = 66.7 \text{ in}^3$

$r_x = 4.39$ in

$Z_x = 74.6 \text{ in}^3$

$I_y = 116 \text{ in}^4$

$S_y = 23.0 \text{ in}^3$

$r_y = 2.57$ in

$Z_y = 35.0 \text{ in}^3$

Steel material properties

ASTM A992

$F_y = 50$ ksi

$F_u = 65$ ksi

Concrete section properties

$A_g = 256 \text{ in}^2$

$w_c = 150 \text{ lbf/ft}^3$

$A_{sr} = 2.25 \text{ in}^2\text{/bar}$

Concrete material properties

$f_c' = 8$ ksi

$F_{y,sr} = 60$ ksi

What is most nearly the design strength (LRFD) or the allowable strength (ASD)? (LRFD options are in parentheses.)

(A) 780 kips (1200 kips)

(B) 910 kips (1400 kips)

(C) 1100 kips (1700 kips)

(D) 1200 kips (1800 kips)

Solution

Determine the area of concrete.

$$A_c = A_g - A_s - A_{sr} = (16\ \text{in})^2 - 17.6\ \text{in}^2 - (4\ \text{bars})\left(2.25\ \frac{\text{in}^2}{\text{bar}}\right) = 229.4\ \text{in}^2$$

Use Eq. 12.6 to determine P_{no}.

$$\begin{aligned} P_{no} &= F_y A_s + F_{y,sr} A_{sr} + 0.85 f'_c A_c \\ &= (17.6\ \text{in}^2)\left(50\ \frac{\text{kips}}{\text{in}^2}\right) + \left(60\ \frac{\text{kips}}{\text{in}^2}\right)\left((4\ \text{bars})\left(2.25\ \frac{\text{in}^2}{\text{bar}}\right)\right) \\ &\quad + (0.85)\left(8\ \frac{\text{kips}}{\text{in}^2}\right)(229.4\ \text{in}^2) \\ &= 2980\ \text{kips} \end{aligned}$$

Use Eq. 12.9 to determine the coefficient, C_1.

$$\begin{aligned} C_1 &= 0.1 + 2\left(\frac{A_s}{A_c + A_s}\right) \le 0.3 \\ &= 0.1 + (2)\left(\frac{17.6\ \text{in}^2}{229.4\ \text{in}^2 + 17.6\ \text{in}^2}\right) \\ &= 0.24 \end{aligned}$$

Use Eq. 12.10 to determine the modulus of elasticity for the concrete.

$$\begin{aligned} E_c &= 33 w_c^{1.5} \sqrt{f'_c} \\ &= \frac{(33)\left(150\ \frac{\text{lbf}}{\text{ft}^3}\right)^{1.5} \sqrt{8000\ \frac{\text{lbf}}{\text{in}^2}}}{1000\ \frac{\text{lbf}}{\text{kip}}} \\ &= 5422\ \text{ksi} \end{aligned}$$

Determine the moment of inertia for the steel reinforcement.

$$\begin{aligned} I_{sr} &= n_{bars} I_x + A_{sr} d^2 = n_{bars}\left(\frac{\pi r^4}{4}\right) + A_{sr} d^2 \\ &= (4)\left(\frac{\pi (0.85\ \text{in})^4}{4}\right) + \left((4\ \text{bars})\left(2.25\ \frac{\text{in}^2}{\text{bar}}\right)\right)(6\ \text{in})^2 \\ &= 325.64\ \text{in}^4 \end{aligned}$$

Determine the moment of inertia for the weak axis slenderness check.

$$\begin{aligned} I_c &= I_g - I_{y,\text{steel}} - I_{\text{sr}} = \frac{bh^3}{12} - I_{y,\text{steel}} - I_{\text{sr}} \\ &= \frac{(16 \text{ in})(16 \text{ in})^3}{12} - 116 \text{ in}^4 - 325.64 \text{ in}^4 \\ &= 5020 \text{ in}^4 \end{aligned}$$

Use Eq. 12.8 to determine the effective stiffness of the composite section.

$$\begin{aligned} (EI)_{\text{eff}} &= E_s I_s + 0.5 E_s I_{\text{sr}} + C_1 E_c I_c \\ &= \left(29{,}000 \ \frac{\text{kips}}{\text{in}^2}\right)(116 \text{ in}^4) + (0.5)\left(29{,}000 \ \frac{\text{kips}}{\text{in}^2}\right)(325.64 \text{ in}^4) \\ &\quad + (0.24)\left(5422 \ \frac{\text{kips}}{\text{in}^2}\right)(5020 \text{ in}^4) \\ &= 14{,}618{,}206 \text{ in}^2\text{-kips} \end{aligned}$$

Use Eq. 12.7 to determine the elastic buckling load.

$$P_e = \frac{\pi^2 (EI)_{\text{eff}}}{(KL)^2} = \frac{\pi^2 (14{,}618{,}206 \text{ in}^2\text{-kips})}{(20 \text{ ft})^2 \left(12 \ \frac{\text{in}}{\text{ft}}\right)^2} = 2505 \text{ kips}$$

Determine the applicable interaction formula.

$$\frac{P_{no}}{P_e} = \frac{2980 \text{ kips}}{2505 \text{ kips}} = 1.19 \quad [< 2.25, \text{ so use Eq. 12.4}]$$

From Eq. 12.4,

$$\begin{aligned} P_n &= 0.658^{P_{no}/P_e} P_{no} \\ &= (0.658)^{2980 \text{ kips}/2505 \text{ kips}} (2980 \text{ kips}) \\ &= 1811 \text{ kips} \end{aligned}$$

Determine the design strength (LRFD) or allowable strength (ASD).

LRFD	ASD
$\phi_c P_n = (0.75)(1811 \text{ kips}) = 1358 \text{ kips} \quad (1400 \text{ kips})$	$\frac{P_n}{\Omega_c} = \frac{1811 \text{ kips}}{2} = 905.5 \text{ kips} \quad (910 \text{ kips})$

The answer is (B).

PRACTICE PROBLEM 7

The column shown is located in a braced frame with both ends fixed and an effective length of 30 ft.

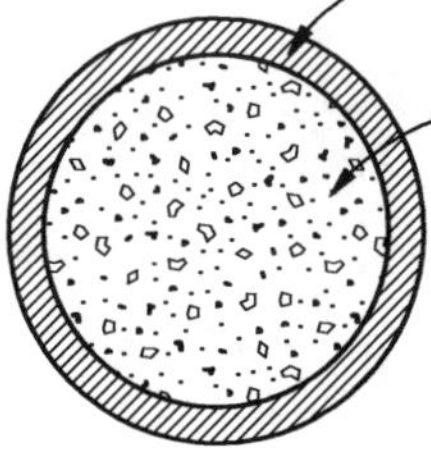

HSS section properties	HSS material properties	Concrete properties
$t_w = 0.465$ in	ASTM A500, grade B	$f_c' = 8$ ksi
$A = 19.8\ \text{in}^2$	$F_y = 42$ ksi	$w_c = 145\ \text{lbf/ft}^3$
$D/t = 30.1$	$F_u = 58$ ksi	$E_c = 5422$ ksi
$I = 453\ \text{in}^4$	$E_s = 29{,}000$ ksi	
$S = 64.8\ \text{in}^3$		
$r = 4.79$ in		
$Z = 85.2\ \text{in}^3$		

What is most nearly the design strength (LRFD) or the allowable strength (ASD)? (LRFD options are in parentheses.)

(A) 490 kips (730 kips)

(B) 520 kips (780 kips)

(C) 550 kips (830 kips)

(D) 660 kips (990 kips)

Solution

Determine the gross area of the filled composite column.

$$A_g = \frac{\pi d^2}{4} = \frac{\pi (14\ \text{in})^2}{4} = 153.94\ \text{in}^2$$

Determine the area of the concrete.

$$\begin{aligned} A_c &= A_g - A_s = 153.94\ \text{in}^2 - 19.8\ \text{in}^2 \\ &= 134.14\ \text{in}^2 \end{aligned}$$

Check the requirements for filled composite columns (see Sec. 12.6). Determine whether the requirement for minimum steel area is met.

$$\begin{aligned} A_{s,\min} &= 0.01A_g = (0.01)(153.94 \text{ in}^2) \\ &= 1.54 \text{ in}^2 \quad [\leq 19.8 \text{ in}^2, \text{ so OK}] \end{aligned}$$

Use Eq. 12.16 and Eq. 12.17 to determine whether the column is compact, noncompact, or slender.

$$\lambda_p = \frac{0.15E}{F_y} = \frac{(0.15)\left(29{,}000 \ \frac{\text{kips}}{\text{in}^2}\right)}{42 \ \frac{\text{kips}}{\text{in}^2}} = 103.6$$

$$\lambda_r = \frac{0.19E}{F_y} = \frac{(0.19)\left(29{,}000 \ \frac{\text{kips}}{\text{in}^2}\right)}{42 \ \frac{\text{kips}}{\text{in}^2}} = 131.2$$

$D/t < \lambda_p$, so the section is compact. Verify that the width-to-thickness ratio is acceptable. According to *AISC Specification* Table I1.1A, for a compact, round HSS element used in a composite member subject to axial compression, the limiting ratio is

$$\frac{D}{t} \leq \frac{0.15E}{F_y}$$

$$30.1 \leq \frac{(0.15)\left(29{,}000 \ \frac{\text{kips}}{\text{in}^2}\right)}{42 \ \frac{\text{kips}}{\text{in}^2}} = 103.6 \quad [\text{so OK}]$$

Use Eq. 12.18 and Eq. 12.19 to determine P_{no}. C_2 is 0.95 for a round section.

$$\begin{aligned} P_{no} &= F_y A_s + C_2 f'_c \left(A_c + A_{sr} \left(\frac{E_s}{E_c} \right) \right) \\ &= \left(42 \ \frac{\text{kips}}{\text{in}^2}\right)(19.8 \text{ in}^2) + (0.95)\left(8 \ \frac{\text{kips}}{\text{in}^2}\right) \\ &\quad \times \left(134.14 \text{ in}^2 + (0 \text{ in}^2) \left(\frac{29{,}000 \ \frac{\text{kips}}{\text{in}^2}}{5422 \ \frac{\text{kips}}{\text{in}^2}} \right) \right) \\ &= 1851 \text{ kips} \end{aligned}$$

Calculate the effective stiffness of the column. First, use Eq. 12.26 to determine the coefficient C_3.

$$\begin{aligned} C_3 &= 0.6+2\left(\frac{A_s}{A_c+A_s}\right) \le 0.9 \\ &= 0.6+(2)\left(\frac{19.8\ \text{in}^2}{134.14\ \text{in}^2+19.8\ \text{in}^2}\right) \\ &= 0.86 \quad [\le 0.9,\ \text{so OK}] \end{aligned}$$

Use Eq. 12.10 to determine the modulus of elasticity for the concrete.

$$E_c = 33w_c^{1.5}\sqrt{f'_c} = \frac{(33)\left(150\ \frac{\text{lbf}}{\text{ft}^3}\right)^{1.5}\sqrt{8000\ \frac{\text{lbf}}{\text{ft}^3}}}{1000\ \frac{\text{lbf}}{\text{kip}}} = 5422\ \text{ksi}$$

Determine the moment of inertia of the concrete.

$$I_c = \frac{\pi D_c^4}{64} = \frac{\pi\left(14\ \text{in}-(2)(0.465\ \text{in})\right)^4}{64} = 1432\ \text{in}^4$$

Use Eq. 12.25 to determine effective stiffness of the composite column, $(EI)_{\text{eff}}$.

$$\begin{aligned} (EI)_{\text{eff}} &= E_sI_s + E_sI_{\text{sr}} + C_3E_cI_c \\ &= \left(29{,}000\ \frac{\text{kips}}{\text{in}^2}\right)\left(453\ \text{in}^4\right)+\left(0\ \frac{\text{kips}}{\text{in}^2}\right)\left(0\ \text{in}^4\right) \\ &\quad +(0.86)\left(5422\ \frac{\text{kips}}{\text{in}^2}\right)\left(1432\ \text{in}^4\right) \\ &= 19{,}814{,}301\ \text{in}^2\text{-kips} \end{aligned}$$

Use this value and Eq. 12.7 to determine P_e.

$$P_e = \frac{\pi^2(EI)_{\text{eff}}}{(KL)^2} = \frac{\pi^2\left(19{,}814{,}301\ \text{in}^2\text{-kips}\right)}{(30\ \text{ft})^2\left(12\ \frac{\text{in}}{\text{ft}}\right)^2} = 1509\ \text{kips}$$

To determine which formula to use to determine P_n, check the ratio P_e/P_{no}.

$$\frac{P_e}{P_{no}} = \frac{1509\ \text{kips}}{1851\ \text{kips}} = 0.82 \quad [> 0.44,\ \text{so use Eq. 12.4}]$$

Use Eq. 12.4 to determine the nominal compressive strength, P_n, of the composite column.

$$\begin{aligned} P_n &= \left(0.658^{P_{no}/P_e}\right)P_{no} \\ &= (0.658)^{1851 \text{ kips}/1509 \text{ kips}}(1851 \text{ kips}) \\ &= 1108 \text{ kips} \end{aligned}$$

Determine the design strength (LRFD) or the allowable strength (ASD).

LRFD	ASD
$\begin{aligned}\phi_c P_n &= (0.75)(1108 \text{ kips}) \\ &= 831 \text{ kips} \quad (830 \text{ kips})\end{aligned}$	$\begin{aligned}\frac{P_n}{\Omega_c} &= \frac{1108 \text{ kips}}{2} \\ &= 554 \text{ kips} \quad (550 \text{ kips})\end{aligned}$

The answer is (C).

PRACTICE PROBLEM 8

The framing plan is shown for a corner bay of an office building that utilizes composite steel construction. The concrete slab consists of a 3 in, 16 gage composite formed steel deck with a total depth of 5 in of lightweight concrete above the top of the steel beams. Live load deflection is limited to span/360.

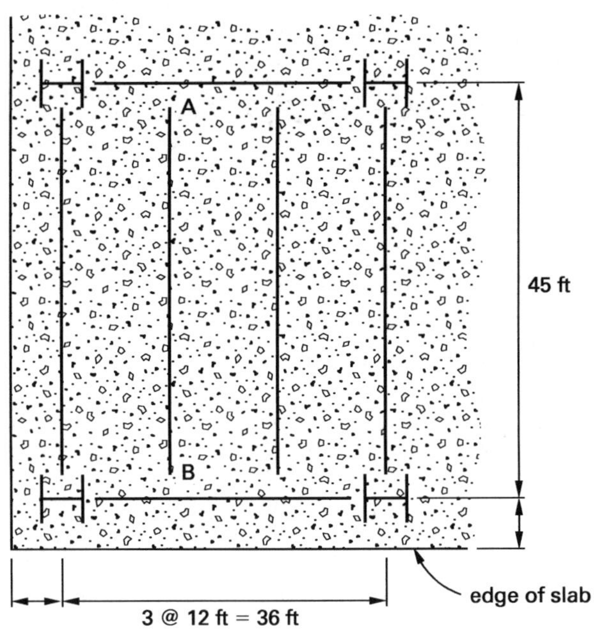

Steel properties

ASTM A992

$F_y = 50$ ksi

$F_u = 65$ ksi

$E_s = 29{,}000$ ksi

Concrete properties

$f_c' = 4$ ksi

$w_c = 115\ \text{lbf/ft}^3$

$E_c = 2573$ ksi

$A_c = 21.2\ \text{in}^2/\text{ft}$

weight = 32 lbf/ft^2

Design loads before concrete sets

concrete + formed steel deck = 35 lbf/ft

working live load = 20 lbf/ft^2

estimated beam weight = 80 lbf/ft

Design loads after concrete sets

concrete + formed steel deck = 35 lbf/ft

design live load = 100 lbf/ft^2

estimated beam weight = 80 lbf/ft

mechanical, electrical, and plumbing + finishes = 10 lbf/ft^2

Select a beam size to satisfy design loads for beam A-B.

(A) W18 × 55

(B) W18 × 60

(C) W21 × 55

(D) W21 × 57

Solution

Before the concrete sets, the dead load per foot of beam, $w_{D,\text{before}}$, is the weight of a 12 ft wide portion of the formed steel deck and the concrete, plus the weight of the beam itself.

$$w_{D,\text{before}} = \left(35\ \frac{\text{lbf}}{\text{ft}^2}\right)(12\ \text{ft}) + 80\ \frac{\text{lbf}}{\text{ft}} = 500\ \text{lbf/ft}$$

After the concrete sets, the dead load also includes the weight of the mechanical, electrical, and plumbing systems and the finishes.

$$w_{D,\text{after}} = \left(35\ \frac{\text{lbf}}{\text{ft}^2} + 10\ \frac{\text{lbf}}{\text{ft}^2}\right)(12\ \text{ft}) + 80\ \frac{\text{lbf}}{\text{ft}} = 620\ \text{lbf/ft}$$

The live load per foot of beam before the concrete sets is

$$w_{L,\text{before}} = \left(20\ \frac{\text{lbf}}{\text{ft}^2}\right)(12\ \text{ft}) = 240\ \text{lbf/ft}$$

After the concrete sets, it is

$$w_{L,\text{after}} = \left(100\ \frac{\text{lbf}}{\text{ft}^2}\right)(12\ \text{ft}) = 1200\ \text{lbf/ft}$$

Determine the design loads for before and after the concrete sets.

LRFD	ASD
$w_{u,\text{before}} = 1.2w_{D,\text{before}} + 1.6w_{L,\text{before}}$ $= \dfrac{(1.2)\left(500\ \frac{\text{lbf}}{\text{ft}}\right) + (1.6)\left(240\ \frac{\text{lbf}}{\text{ft}}\right)}{1000\ \frac{\text{lbf}}{\text{kip}}}$ $= 0.984\ \text{kips/ft}$	$w_{a,\text{before}} = w_{D,\text{before}} + w_{L,\text{before}}$ $= \dfrac{500\ \frac{\text{lbf}}{\text{ft}} + 240\ \frac{\text{lbf}}{\text{ft}}}{1000\ \frac{\text{lbf}}{\text{kip}}}$ $= 0.74\ \text{kips/ft}$
$w_{u,\text{after}} = 1.2w_{D,\text{after}} + 1.6w_{L,\text{after}}$ $= \dfrac{(1.2)\left(620\ \frac{\text{lbf}}{\text{ft}}\right) + (1.6)\left(1200\ \frac{\text{lbf}}{\text{ft}}\right)}{1000\ \frac{\text{lbf}}{\text{kip}}}$ $= 2.66\ \text{kips/ft}$	$w_{a,\text{after}} = w_{D,\text{after}} + w_{L,\text{after}}$ $= \dfrac{620\ \frac{\text{lbf}}{\text{ft}} + 1200\ \frac{\text{lbf}}{\text{ft}}}{1000\ \frac{\text{lbf}}{\text{kip}}}$ $= 1.82\ \text{kips/ft}$

Calculate the required moments before and after the concrete sets.

LRFD	ASD
$M_{u,\text{before}} = \dfrac{w_{u,\text{before}}L^2}{8}$ $= \dfrac{\left(0.984\ \frac{\text{kips}}{\text{ft}}\right)(45\ \text{ft})^2}{8}$ $= 249\ \text{ft-kips}$	$M_{a,\text{before}} = \dfrac{w_{a,\text{before}}L^2}{8}$ $= \dfrac{\left(0.74\ \frac{\text{kips}}{\text{ft}}\right)(45\ \text{ft})^2}{8}$ $= 187\ \text{ft-kips}$
$M_{u,\text{after}} = \dfrac{w_{u,\text{after}}L^2}{8}$ $= \dfrac{\left(2.66\ \frac{\text{kips}}{\text{ft}}\right)(45\ \text{ft})^2}{8}$ $= 673\ \text{ft-kips}$	$M_{a,\text{after}} = \dfrac{w_{a,\text{after}}L^2}{8}$ $= \dfrac{\left(1.82\ \frac{\text{kips}}{\text{ft}}\right)(45\ \text{ft})^2}{8}$ $= 461\ \text{ft-kips}$

Determine the effective width of the concrete slab (see Sec. 12.8).

$$b_{e,\text{half}} \le \begin{cases} \dfrac{L}{8} = \dfrac{45 \text{ ft}}{8} \\ \quad = 5.625 \text{ ft} \quad [\text{controls}] \\ \dfrac{s}{2} = \dfrac{12 \text{ ft}}{2} \\ \quad = 6 \text{ ft} \end{cases}$$

$$b_e = 2b_{e,\text{half}} = (2)(5.625 \text{ ft}) = 11.25 \text{ ft}$$

Use Eq. 12.42 to calculate the moment arm distance for the concrete, $Y2$, making the assumption that the depth of the concrete in compression, a, is 1.0 in.

$$\begin{aligned} Y2 &= t_{\text{slab}} - \frac{a}{2} = 5 \text{ in} - \frac{1 \text{ in}}{2} \\ &= 4.5 \text{ in} \end{aligned}$$

Make a trial selection for the composite beam from *AISC Manual* Table 3-19. Enter the table with M_u = 673 ft-kips (LRFD) or M_a = 461 ft-kips (ASD), $Y2$ = 4.5 in, and an assumption that the plastic neutral axis (PNA) will be located at the bottom of the top flange (BFL), location 5. Try a W18 × 60.

LRFD	ASD
$\phi_b M_p = 710$ ft-kips $[> M_u = 673 \text{ ft-kips, so OK}]$	$M_p/\Omega_b = 472$ ft-kips $[> M_a = 461 \text{ ft-kips, so OK}]$

From the same table, the horizontal shear capacity of the beam is ΣQ_n = 357 kips. From *AISC Manual* Table 1-1, for a W18 × 60, I_x = 984 in^4. Check beam deflection under the load of the concrete weight, neglecting the temporary live load of 20 lbf/ft^2.

$$w = \frac{\left(35 \ \dfrac{\text{lbf}}{\text{ft}^2}\right)(12 \text{ ft}) + 80 \ \dfrac{\text{lbf}}{\text{ft}}}{1000 \ \dfrac{\text{lbf}}{\text{kip}}} = 0.50 \ \text{kips/ft}$$

$$\begin{aligned} \Delta &= \frac{5wL^4}{384EI_x} \\ &= \frac{(5)\left(0.50 \ \dfrac{\text{kips}}{\text{ft}}\right)(45 \text{ ft})^4 \left(12 \ \dfrac{\text{in}}{\text{ft}}\right)^3}{(384)\left(29{,}000 \ \dfrac{\text{kips}}{\text{in}^2}\right)(984 \text{ in}^4)} \\ &= 1.62 \text{ in} \quad [\text{recommend 1.5 in or 1.75 in camber}] \end{aligned}$$

Check steel strength for unshored construction loads, assuming that deck welds provide adequate lateral support. From *AISC Manual* Table 3-2, for a W18 × 60,

LRFD	ASD
$\phi_b M_{px} = 461 \text{ ft-kips}$ $\left[\geq M_{u,\text{before}} = 249 \text{ ft-kips, so OK}\right]$	$M_{px}/\Omega_b = 307 \text{ ft-kips}$ $\left[\geq M_{a,\text{before}} = 187 \text{ ft-kips, so OK}\right]$

Use Eq. 12.43 to check the depth of compression concrete.

$$\begin{aligned} a &= \frac{\sum Q_n}{0.85 f'_c b_e} \\ &= \frac{355 \text{ kips}}{(0.85)\left(4 \ \frac{\text{kips}}{\text{in}^2}\right)(11.25 \text{ ft})\left(12 \ \frac{\text{in}}{\text{ft}}\right)} \\ &= 0.77 \text{ in} \quad [\leq 1 \text{ in assumed, so OK}] \end{aligned}$$

Check that the live load deflection is less than $L/360$. From *AISC Manual* Table 3-20, with the PNA at BFL and $Y2$ = 4.5 in, the lower bound moment of inertia is I_{LB} = 1920 in^4. Calculate the deflection.

$$\begin{aligned} w_L &= \left(w_{L,\text{after}} + w_{\text{MEP}}\right) b_e \\ &= \frac{\left(100 \ \frac{\text{lbf}}{\text{ft}^2} + 10 \ \frac{\text{lbf}}{\text{ft}^2}\right)(12 \text{ ft})}{1000 \ \frac{\text{lbf}}{\text{kip}}} \\ &= 1.32 \text{ kips/ft} \end{aligned}$$

$$\begin{aligned} \Delta_L &= \frac{5 w_L L^4}{384 E I_{\text{LB}}} \\ &= \frac{(5)\left(1.32 \ \frac{\text{kips}}{\text{ft}}\right)(45 \text{ ft})^4 \left(12 \ \frac{\text{in}}{\text{ft}}\right)^3}{(384)\left(29{,}000 \ \frac{\text{kips}}{\text{in}^2}\right)(1920 \text{ in}^4)} \\ &= 2.19 \text{ in} \end{aligned}$$

The maximum deflection is

$$\Delta_{L,\max} = \frac{L}{360} = \frac{(45 \text{ ft})\left(12 \ \frac{\text{in}}{\text{ft}}\right)}{360} = 1.5 \text{ in}$$

The deflection of 2.18 exceeds the maximum, so a heavier or deeper section should be tried.

Determine the lower bound moment of inertia required to limit deflection to $L/360$.

$$\begin{aligned} I_{\text{LB,req}} &= \left(\frac{\Delta_{\text{current}}}{\Delta_{\text{req}}}\right) I_{\text{LB,current}} \\ &= \left(\frac{2.19\ \text{in}}{1.5\ \text{in}}\right)\left(1920\ \text{in}^4\right) \\ &= 2803\ \text{in}^4 \end{aligned}$$

From *AISC Manual* Table 3-20, try a W21 × 57 with PNA at location 2 and $Y2$ = 4.5 in, giving I_{LB} = 2930 in^4. Use Eq. 12.43 to check the depth of compression concrete. From *AISC Manual* Table 3.19, ΣQ_n = 728kips.

$$\begin{aligned} a &= \frac{\sum Q_n}{0.85 f'_c b_e} \\ &= \frac{728\ \text{kips}}{(0.85)\left(4\ \dfrac{\text{kips}}{\text{in}^2}\right)(11.25\ \text{ft})\left(12\ \dfrac{\text{in}}{\text{ft}}\right)} \\ &= 1.6\ \text{in} \quad [>1\ \text{in assumed}] \end{aligned}$$

The depth of compression concrete remains above the flute of the metal form deck, so this is satisfactory. Use Eq. 12.42 to calculate $Y2$.

$$Y2 = t_{\text{slab}} - \frac{a}{2} = 5\ \text{in} - \frac{1.6\ \text{in}}{2} = 4.2\ \text{in}$$

Calculate the lower bound moment of inertia for $Y2$ = 4.2 in by interpolating in *AISC Manual* Table 3-20 between the tabulated values for $Y2$ = 4 in and $Y2$ = 4.5 in (2820 in^4 and 2930 in^4, respectively).

$$\begin{aligned} I_{\text{LB}} &= 2820\ \text{in}^4 + \left(\frac{0.2}{0.5}\right)\left(2930\ \text{in}^4 - 2820\ \text{in}^4\right) \\ &= 2864\ \text{in}^4 \quad \left[>2805\ \text{in}^4,\ \text{so OK}\right] \end{aligned}$$

The lower bound moment is sufficient, so use a W21 × 57. Use Eq. 12.44 to determine the number of studs required for placement in the strong position. From *AISC Manual* Table 3-21, for $^3/_4$ in studs placed in the strong position, Q_n = 21.2 kips/stud.

$$\begin{aligned} n_{\text{half}} &= \frac{\sum Q_n}{Q_n} = \frac{728\ \text{kips}}{21.2\ \dfrac{\text{kips}}{\text{stud}}} \\ &= 34.33\ \text{studs} \quad \left[\begin{array}{l}\text{on each side of point} \\ \text{of maximum moment}\end{array}\right] \end{aligned}$$

Check the stud diameter requirements.

$$\begin{aligned} d_{\text{stud}} &\leq 2.5t_f \\ 0.75\text{ in} &\leq (2.5)(0.65\text{ in}) \\ &\leq 1.625\text{ in} \quad [\text{so OK}] \end{aligned}$$

The $^3/_4$ in diameter studs are good.

The answer is (D).

PRACTICE PROBLEM 9

A plant engineer wants to increase the load capacity of the W30 × 108 beam shown by installing a new column at the midpoint of the existing span. Assume that the effective height of the column will be 20 ft and that the beam is laterally braced for the full length. After the new column is installed, it is desired that the beam have the capacity to support as great a load as is possible without any modification or reinforcement of the existing columns.

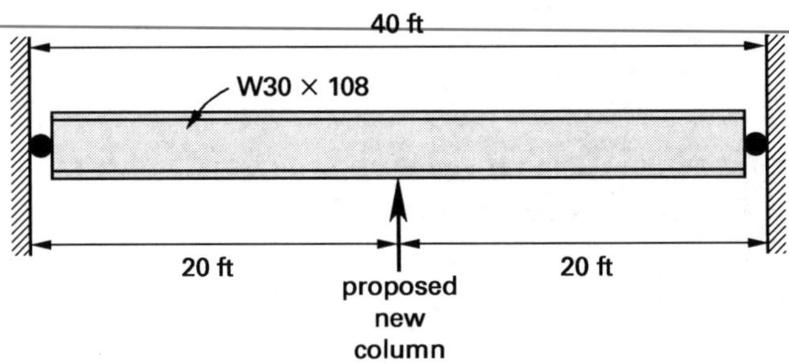

Section properties		Material properties
$A = 31.7\text{ in}^2$	$I_x = 4470\text{ in}^4$	ASTM A992
$d = 29.8\text{ in}$	$S_x = 299\text{ in}^3$	$F_y = 50\text{ ksi}$
$t_w = 0.545\text{ in}$	$r_x = 11.9\text{ in}$	$F_u = 65\text{ ksi}$
$b_f = 10.5\text{ in}$	$Z_x = 346\text{ in}^3$	$E_s = 29{,}000\text{ ksi}$
$t_f = 0.760\text{ in}$	$I_y = 146\text{ in}^4$	
$b_f/2t_f = 6.89$	$S_y = 27.9\text{ in}^3$	
$h/t_w = 49.6$	$r_y = 2.15\text{ in}$	
	$Z_y = 43.9\text{ in}^3$	

The lightest W10 member that can be used for the new column is a

(A) W10 × 33

(B) W10 × 45

(C) W10 × 60

(D) W10 × 68

Solution

Determine the load capacity before the new column is installed. From *AISC Manual* Table 3-2, for a W30 × 108 beam,

LRFD	ASD
$\phi_b M_{px} = 1300 \text{ ft-kips}$	$\dfrac{M_{px}}{\Omega_b} = 863 \text{ ft-kips}$
$w_{u,\text{old}} = \dfrac{8(\phi_b M_{px})}{L^2} = \dfrac{(8)(1300 \text{ ft-kips})}{(40 \text{ ft})^2} = 6.50 \text{ kips/ft}$	$w_{a,\text{old}} = \dfrac{8\left(\dfrac{M_{px}}{\Omega_b}\right)}{L^2} = \dfrac{(8)(863 \text{ ft-kips})}{(40 \text{ ft})^2} = 4.32 \text{ kips/ft}$
$R_{u,\text{old ext}} = \dfrac{w_u L}{2} = \dfrac{\left(6.50 \ \dfrac{\text{kips}}{\text{ft}}\right)(40 \text{ ft})}{2} = 130 \text{ kips}$	$R_{a,\text{old ext}} = \dfrac{w_a L}{2} = \dfrac{\left(4.32 \ \dfrac{\text{kips}}{\text{ft}}\right)(40 \text{ ft})}{2} = 86.4 \text{ kips}$

Determine the load capacity after the new column is installed. For a uniformly loaded continuous beam with two equal spans, the governing moment is a negative moment of $-wL^2/8$ at the interior support; the reaction at the midpoint is $10wL/8$.

LRFD	ASD
$w_{u,\text{new}} = \dfrac{8(\phi_b M_{px})}{L^2} = \dfrac{(8)(1300 \text{ ft-kips})}{(20 \text{ ft})^2} = 26.0 \text{ kips/ft}$	$w_{a,\text{new}} = \dfrac{8\left(\dfrac{M_{px}}{\Omega_b}\right)}{L^2} = \dfrac{(8)(863 \text{ ft-kips})}{(20 \text{ ft})^2} = 17.3 \text{ kips/ft}$
$R_{u,\text{new int}} = \dfrac{10 w_u L}{8} = \dfrac{(10)\left(26.0 \ \dfrac{\text{kips}}{\text{ft}}\right)(20 \text{ ft})}{8} = 650 \text{ kips} \quad [\text{at new column}]$	$R_{a,\text{new int}} = \dfrac{10 w_a L}{8} = \dfrac{(10)\left(17.3 \ \dfrac{\text{kips}}{\text{ft}}\right)(20 \text{ ft})}{8} = 433 \text{ kips} \quad [\text{at new column}]$

$R_{u,\text{new ext}} = \dfrac{3w_uL}{8} = \dfrac{(3)\left(26.0\ \dfrac{\text{kips}}{\text{ft}}\right)(20\ \text{ft})}{8} = 195\ \text{kips}$ [at original columns]	$R_{u,\text{new ext}} = \dfrac{3w_uL}{8} = \dfrac{(3)\left(17.3\ \dfrac{\text{kips}}{\text{ft}}\right)(20\ \text{ft})}{8} = 130\ \text{kips}$ [at original columns]

With the added column, the beam itself will be capable of supporting a greater uniform load; however, this maximum load would produce a greater reaction at the original columns than there was before. If the original columns and footings are not to require reinforcement, this reaction must be no greater than it was before. The load capacity, then, must be reduced so that it produces the same reaction at the original columns.

LRFD	ASD
$w_{u,\text{reduced}} = w_{u,\text{new}}\left(\dfrac{R_{u,\text{old ext}}}{R_{u,\text{new ext}}}\right) = \left(26.0\ \dfrac{\text{kips}}{\text{ft}}\right)\left(\dfrac{130\ \text{kips}}{195\ \text{kips}}\right) = 17.33\ \text{kips/ft}$	$w_{a,\text{reduced}} = w_{a,\text{new}}\left(\dfrac{R_{a,\text{old ext}}}{R_{a,\text{new ext}}}\right) = \left(17.3\ \dfrac{\text{kips}}{\text{ft}}\right)\left(\dfrac{86.4\ \text{kips}}{130\ \text{kips}}\right) = 11.5\ \text{kips/ft}$

Determine what the reduced load will be on the new interior column.

LRFD	ASD
$P_{u,\text{reduced int}} = R_{u,\text{new int}}\left(\dfrac{R_{u,\text{old ext}}}{R_{u,\text{new ext}}}\right) = (650\ \text{kips})\left(\dfrac{130\ \text{kips}}{195\ \text{kips}}\right) = 433\ \text{kips}$	$P_{a,\text{reduced int}} = R_{a,\text{new int}}\left(\dfrac{R_{a,\text{old ext}}}{R_{a,\text{new ext}}}\right) = (433\ \text{kips})\left(\dfrac{86.4\ \text{kips}}{130\ \text{kips}}\right) = 287\ \text{kips}$

From *AISC Manual* Table 4-1, entering the table with an effective length of 20 ft, select the lightest W10 that has the available strength to support the reduced load on the new column (433 kips for LRFD, 287 kips for ASD). This is a W10 × 68 with

LRFD	ASD
$\phi_cP_n = 478\ \text{kips}$	$\dfrac{P_n}{\Omega_c} = 318\ \text{kips}$

The answer is (D).

PRACTICE PROBLEM 10

The 12 in standard steel pipe shown is part of an aboveground plant water distribution system. The pipe is not subject to freezing or seismic forces. It is subject to a lateral wind load of 30 lbf/ft. Based on ASCE 7, the supports are uniformly spaced along its length at the maximum spacing that will limit the deflection to span/240. ASCE 7 wind loads are ultimate, with a load factor of 1.0; for ASD the load factor is 0.60. For well-defined fluid loads, *F*, use the same load factor used for dead loads.

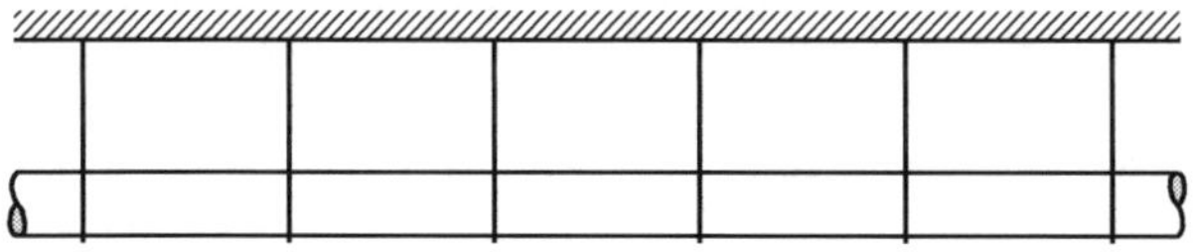

Section properties		Material properties
$w_{pipe} = 49.6$ lbf/ft	$A = 13.6$ in^2	ASTM A53, grade B
$D = 12.8$ in	$D/t = 36.5$	$F_y = 35$ ksi
$d = 12.0$ in	$I = 262$ in^4	$F_u = 60$ ksi
$t = 0.375$ in	$S = 41$ in^3	$E_s = 29{,}000$ ksi
$t_{des} = 0.349$ in	$r = 4.39$ in	a

What is most nearly the required hanger strength? (LRFD options are given in parentheses.)

(A) 6.5 kips (8.0 kips)

(B) 8.0 kips (9.0 kips)

(C) 12 kips (15 kips)

(D) 15 kips (18 kips)

Solution

The cross-sectional area of flow inside the pipe is

$$A_{in} = \frac{\pi d^2}{4} = \frac{\pi(12 \text{ in})^2}{4} = 113 \text{ in}^2$$

The weight of the water in the pipe is

$$\begin{aligned} w_{water} &= A_{in}\gamma_{water} \\ &= \left(\frac{113 \text{ in}^2}{\left(12 \frac{\text{in}}{\text{ft}}\right)^2}\right)\left(62.4 \frac{\text{lbf}}{\text{ft}^3}\right) \\ &= 49.0 \text{ lbf/ft} \end{aligned}$$

Determine the effects of the loads on the pipe. Without wind,

LRFD	ASD
$w_{u,\text{no wind}} = 1.2D + 1.2F$ $= 1.2\left(w_{\text{pipe}} + w_{\text{water}}\right)$ $= (1.2)\left(\left(49.6\ \frac{\text{lbf}}{\text{ft}}\right) + \left(49.0\ \frac{\text{lbf}}{\text{ft}}\right)\right)$ $= 118.32\ \text{lbf/ft}$	$w_{a,\text{no wind}} = D + F$ $= w_{\text{pipe}} + w_{\text{water}}$ $= 49.6\ \frac{\text{lbf}}{\text{ft}} + 49.0\ \frac{\text{lbf}}{\text{ft}}$ $= 98.6\ \text{lbf/ft}$

With wind,

LRFD	ASD
$w_u = \sqrt{w_{u,\text{no wind}}^2 + (1.0w_{\text{wind}})^2}$ $= \sqrt{\left(118.32\ \frac{\text{lbf}}{\text{ft}}\right)^2 + \left((1.0)\left(30\ \frac{\text{lbf}}{\text{ft}}\right)\right)^2}$ $= 122\ \text{lbf/ft}\ \ (0.122\ \text{kip/ft})$ [controls]	$w_a = \sqrt{w_{a,\text{no wind}}^2 + 0.6w_{\text{wind}}^2}$ $= \sqrt{\left(98.6\ \frac{\text{lbf}}{\text{ft}}\right)^2 + (0.6)\left(30\ \frac{\text{lbf}}{\text{ft}}\right)^2}$ $= 101.30\ \text{lbf/ft}\ \ (0.101\ \text{kip/ft})$ [controls]

The combined loading governs. Determine the design strength (LRFD) or allowable strength (ASD) of the pipe. From *AISC Manual* Table 3-15,

LRFD	ASD
$\phi_b M_n = 141$ ft-kips	$\frac{M_n}{\Omega_b} = 93.8$ ft-kips

Determine the maximum length of pipe between hangers. From *AISC Manual* Table 3-23, case 42, the maximum positive and negative moments are

$$M_{\text{max}}^{+} = 0.0772wL^2$$

$$M_{\text{max}}^{-} = -0.107wL^2 \quad \text{[governs]}$$

Rearranging,

$$L = \sqrt{\frac{M_{\text{max}}^{-}}{-0.107w}} = \sqrt{\frac{M_{\text{max}}}{0.107w}}$$

The maximum deflection is

$$\Delta_{max} = \frac{0.0065wL^4}{EI}$$

The maximum length is limited both because the moment must not exceed the design strength and because the deflection must not exceed span/240.

The maximum length based on the moment not exceeding the design strength is

LRFD	ASD
$L = \sqrt{\frac{M_{max}}{0.107w_u}} = \sqrt{\frac{141 \text{ ft-kips}}{(0.107)\left(0.122 \frac{\text{kips}}{\text{ft}}\right)}} = 103.92 \text{ ft}$	$L = \sqrt{\frac{M_{max}}{0.107w_a}} = \sqrt{\frac{93.8 \text{ ft-kips}}{(0.107)\left(0.101 \frac{\text{kips}}{\text{ft}}\right)}} = 93.16\text{ft}$

The maximum length based on the deflection not exceeding $L/240$ is

LRFD	ASD
$\Delta_{max} = \frac{L}{240} = \frac{0.0065w_aL^4}{EI}$ $L = \sqrt[3]{\frac{EI}{(240)(0.0065w_a)}}$ $= \sqrt[3]{\frac{\left(29{,}000 \frac{\text{kips}}{\text{in}^2}\right)(262 \text{ in}^4)}{(240)(0.0065) \times \left(0.122 \frac{\text{kips}}{\text{ft}}\right) \times \left(12 \frac{\text{in}}{\text{ft}}\right)^2}}$ $= 65.21 \text{ ft} \quad [\text{controls}]$	$\Delta_{max} = \frac{L}{240} = \frac{0.0065w_aL^4}{EI}$ $L = \sqrt[3]{\frac{EI}{(240)(0.0065w_a)}}$ $= \sqrt[3]{\frac{\left(29{,}000 \frac{\text{kips}}{\text{in}^2}\right)(262 \text{ in}^4)}{(240)(0.0065) \times \left(0.101 \frac{\text{kips}}{\text{ft}}\right) \times \left(12 \frac{\text{in}}{\text{ft}}\right)^2}}$ $= 69.44 \text{ ft} \quad [\text{controls}]$

Determine the maximum reaction. From *AISC Manual* Table 3-23, case 42,

$$R_{max} = 1.14wL$$

LRFD	ASD
$R_{u,max} = (1.14)\left(0.122\ \frac{\text{kips}}{\text{ft}}\right)(65.21\ \text{ft}) = 9.07\ \text{kips}$	$R_{a,max} = (1.14)\left(0.101\ \frac{\text{kips}}{\text{ft}}\right)(69.44\ \text{ft}) = 8.00\ \text{kips}$

Use Eq. 5.51 to check the shear capacity of pipe.

$$V_n = \frac{F_{cr}A_g}{2}$$

In this equation, F_{cr} is the larger of the values given by Eq. 5.49 and Eq. 5.50, but no larger than $0.6F_y$. From Eq. 5.49,

LRFD	ASD
$F_{cr} = \frac{1.60E}{\sqrt{\frac{L_v}{D}}\left(\frac{D}{t}\right)^{5/4}} = \frac{(1.6)\left(29{,}000\ \frac{\text{kips}}{\text{in}^2}\right)}{\sqrt{\frac{\left(\frac{65.21\ \text{ft}}{2}\right)\left(12\ \frac{\text{in}}{\text{ft}}\right)}{12.8\ \text{in}}}(36.5)^{5/4}} = 93.5\ \text{ksi}$	$F_{cr} = \frac{1.60E}{\sqrt{\frac{L_v}{D}}\left(\frac{D}{t}\right)^{5/4}} = \frac{(1.6)\left(29{,}000\ \frac{\text{kips}}{\text{in}^2}\right)}{\sqrt{\frac{\left(\frac{69.44\ \text{ft}}{2}\right)\left(12\ \frac{\text{in}}{\text{ft}}\right)}{12.8\ \text{in}}}(36.5)^{5/4}} = 90.7\ \text{ksi}$

From Eq. 5.53,

$$F_{cr} = \frac{0.78E}{\left(\frac{D}{t}\right)^{3/2}} = \frac{(0.78)\left(29{,}000\ \frac{\text{kips}}{\text{in}^2}\right)}{(36.5)^{3/2}} = 103\ \text{ksi}$$

The upper limit for F_{cr} is

$$\begin{aligned} F_{cr} &\le 0.6F_y \\ &\le (0.6)\left(35\ \frac{\text{kips}}{\text{in}^2}\right) \\ &\le 21\ \text{ksi}\quad \text{[controls]} \end{aligned}$$

From Eq. 5.48, the shear capacity of the pipe is

$$V_n = \frac{F_{cr}A_g}{2} = \frac{\left(21\ \frac{\text{kips}}{\text{in}^2}\right)(13.6\ \text{in}^2)}{2} = 142.8\ \text{kips}\quad [> R_{max},\ \text{so OK}]$$

This is more than enough, so the required hanger strength is

LRFD	ASD
$R_{u,max} = 9.0\ \text{kips}$	$R_{a,max} = 7.9\ \text{kips}\quad (8.0\ \text{kips})$

The answer is (B).

PRACTICE PROBLEM 11

The plate shown is welded to the flange and centered on the web of the W-column. Assume that the flange thickness is sufficient for the weld.

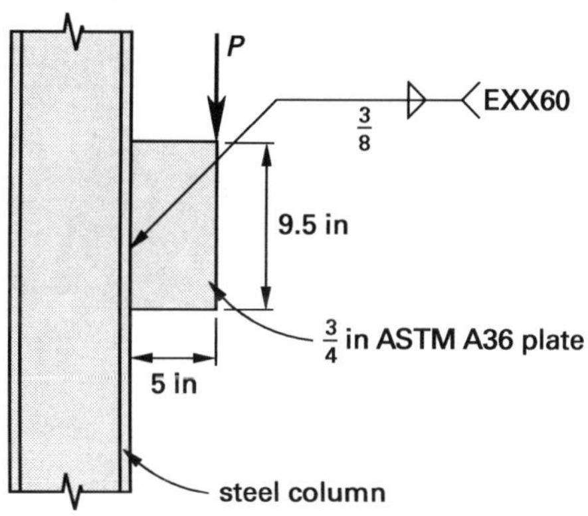

Material properties for plate and column

ASTM A36

$F_y = 36$ ksi

$F_u = 58$ ksi

$E_s = 29{,}000$ ksi

What is most nearly the maximum load that can be applied to the plate? (LRFD options are in parentheses.)

(A) 40 kips (60 kips)

(B) 50 kips (80 kips)

(C) 60 kips (90 kips)

(D) 70 kips (110 kips)

Solution

Find the electrode strength coefficient, C_1, in *AISC Manual* Table 8-3. For E60 electrodes, $C_1 = 0.857$. Determine the coefficient C from *AISC Manual* Table 8-4. First determine a.

$$a = \frac{e_x}{l} = \frac{5 \text{ in}}{9.5 \text{ in}} = 0.526$$

According to the diagram accompanying the table, when the load is not in the plane of the weld group, take k as zero. Enter *AISC Manual* Table 8-4 with angle = 0° and $k = 0$, and interpolate between the values for $a = 0.500$ and $a = 0.600$. For $a = 0.526$, $C = 2.215$.

Use Eq. 10.9 or Eq. 10.10 to determine the design load, P_u (LRFD) or the allowable load, P_a (ASD).[1]

LRFD	ASD
$P_u = \phi C C_1 D l = (0.75)(2.215)(0.857)(6) \times \left(9.5 \ \frac{\text{in}}{\text{weld}}\right) = 81.15 \text{ kips} \quad (80 \text{ kips})$	$P_a = \frac{C C_1 D l}{\Omega} = \frac{(2.215)(0.857)(6)\left(9.5 \ \frac{\text{in}}{\text{weld}}\right)}{2.00} = 54.10 \text{ kips} \quad (50 \text{ kips})$

The answer is (B).

PRACTICE PROBLEM 12

The welded plate girder shown is fabricated from ASTM A572 steel. The distance from the point where the loads are concentrated to the support on either side is greater than the depth d.

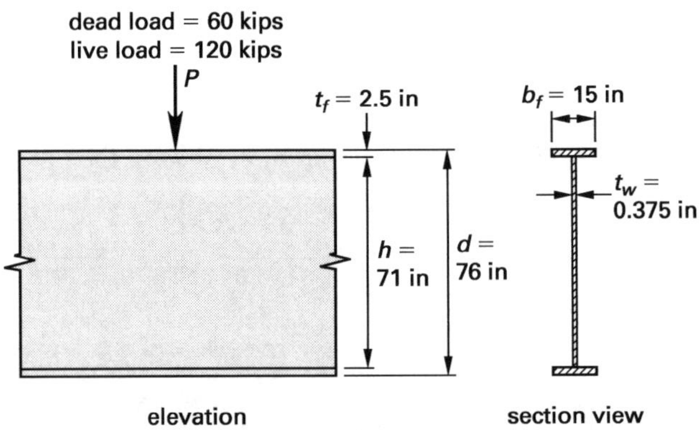

[1]The *AISC Manual* procedure is based on two welds, one on each side of the plate. It is incorrect to multiply by two welds here.

Material properties

ASTM A572, grade 50

$F_y = 50$ ksi

$F_u = 65$ ksi

$E_s = 29{,}000$ ksi

In calculations, make the conservative assumption that the length of bearing is zero. Select the smallest bearing stiffener that will meet the load requirements.

(A) $^3/_8$ in × 4 in

(B) $^3/_8$ in × 5 in

(C) $^1/_2$ in × 6 in

(D) $^5/_8$ in × 7 in

Solution

Determine the design loads.

LRFD	ASD
$R_u = 1.2D + 1.6L$ $= (1.2)(60 \text{ kips}) + (1.6)(120 \text{ kips})$ $= 264 \text{ kips}$	$R_a = D + L$ $= 60 \text{ kips} + 120 \text{ kips}$ $= 180 \text{ kips}$

Use Eq. 6.4 to check the limit state for web local yielding. Assume that the length of bearing, l_b, is zero.

$$\begin{aligned} R_n &= F_{yw} t_w (5k + l_b) \\ &= \left(50\ \frac{\text{kips}}{\text{in}^2}\right)(0.375 \text{ in})\big((5)(2.5 \text{ in}) + 0 \text{ in}\big) \\ &= 234 \text{ kips} \end{aligned}$$

LRFD	ASD
$\phi R_n = (1.0)(234 \text{ kips})$ $= 234 \text{ kips}$ $[< R_u = 264 \text{ kips, not OK}]$	$\frac{R_n}{\Omega} = \frac{234 \text{ kips}}{1.5}$ $= 156 \text{ kips}$ $[< R_a = 180 \text{ kips, not OK}]$

Use Eq. 6.6 to check the limit state of web crippling.

$$
\begin{aligned}
R_n &= 0.80t_w^2\left(1+3\left(\frac{l_b}{d}\right)\left(\frac{t_w}{t_f}\right)^{1.5}\right)\sqrt{\frac{EF_{yw}t_f}{t_w}} \\
&= (0.80)(0.375\text{ in})^2\left(1+(3)\left(\frac{0\text{ in}}{76\text{ in}}\right)\left(\frac{0.375\text{ in}}{2.5\text{ in}}\right)^{1.5}\right) \\
&\quad \times\sqrt{\frac{\left(29{,}000\ \frac{\text{kips}}{\text{in}^2}\right)\left(50\ \frac{\text{kips}}{\text{in}^2}\right)(2.5\text{ in})}{0.375\text{ in}}} \\
&= 350\text{ kips}
\end{aligned}
$$

LRFD	ASD
$\phi R_n = (0.75)(350\text{ kips})$ $= 263\text{ kips}$ $[< R_u = 264\text{ kips, not OK}]$	$\frac{R_n}{\Omega} = \frac{350\text{ kips}}{2}$ $= 175\text{ kips}$ $[< R_a = 180\text{ kips, not OK}]$

A stiffener is required. Use Eq. 6.50 to determine the maximum stiffener width.

$$
b_{\text{st,max}} = \frac{b_f - t_w}{2} = \frac{15\text{ in} - 0.375\text{ in}}{2} = 7.3125\text{ in}
$$

Determine the limiting width-thickness ratio from *AISC Specification* Table B4.1, case 3.

$$
\begin{aligned}
\frac{b_{\text{st}}}{t_{\text{st}}} &\le 0.56\sqrt{\frac{E}{F_y}} \\
&\le 0.56\sqrt{\frac{29{,}000\ \frac{\text{kips}}{\text{in}^2}}{50\ \frac{\text{kips}}{\text{in}^2}}} \\
&\le 13.49 \\
t_{\text{st}} &\ge \frac{b_{\text{st}}}{13.49}
\end{aligned}
$$

For a 7 in stiffener, the thickness must be

$$
t_{\text{st}} \ge \frac{7\text{ in}}{13.49} = 0.52\text{ in}
$$

For a 6 in stiffener,

$$t_{st} \geq \frac{6 \text{ in}}{13.49} = 0.44 \text{ in}$$

For a 5 in stiffener,

$$t_{st} \geq \frac{5 \text{ in}}{13.49} = 0.37 \text{ in}$$

For a 4 in stiffener,

$$t_{st} \geq \frac{4 \text{ in}}{13.49} = 0.30 \text{ in}$$

Try stiffener plates 5 in wide and $^3/_8$ in thick. The stiffeners are not at the ends of the member, so use Eq. 6.56 to calculate the gross area of the cross-shaped column formed by the beam web and stiffeners.

$$\begin{aligned} A_{st} &= n_{st} b_{st} t_{st} = (2)(5 \text{ in})(0.375 \text{ in}) \\ &= 3.75 \text{ in}^2 \\ A_{g,\text{cross}} &= A_{st} + 25t_w^2 \\ &= 3.75 \text{ in}^2 + (25)(0.375 \text{ in})^2 \\ &= 7.27 \text{ in}^2 \end{aligned}$$

From Eq. 6.48, the effective web length is

$$L_{w,\text{eff}} = 25t_w = (25)(0.375 \text{ in}) = 9.375 \text{ in}$$

Calculate the moment of inertia of the cross-shaped column about the centerline of the beam web.

$$\begin{aligned} I_{\text{cross}} &= I_{st} + I_w = \frac{(bd^3)_{st}}{12} + \frac{(bd^3)_w}{12} \\ &= \frac{t_{st}(t_w + 2b_{st})^3}{12} + \frac{(L_{w,\text{eff}} - t_{st})t_w^3}{12} \\ &= \frac{(0.375 \text{ in})(0.375 \text{ in} + (2)(5 \text{ in}))^3}{12} + \frac{(9.375 \text{ in} - 0.375 \text{ in})(0.375 \text{ in})^3}{12} \\ &= 34.94 \text{ in}^4 \end{aligned}$$

The radius of gyration for the cross-shaped column is

$$r_{\text{cross}} = \sqrt{\frac{I_{\text{cross}}}{A_{g,\text{cross}}}} = \sqrt{\frac{34.94 \text{ in}^4}{7.27 \text{ in}^2}} = 2.19 \text{ in}$$

Determine the effective slenderness ratio for the column, using an effective length factor of $K = 0.75$.

$$\frac{KL}{r} = \frac{Kh_{\text{cross}}}{r_{\text{cross}}} = \frac{(0.75)(71\ \text{in})}{2.19\ \text{in}} = 24.32$$

Determine the correct formula to use for calculating the critical flexural buckling stress.

$$4.71\sqrt{\frac{E}{F_y}} = 4.71\sqrt{\frac{29{,}000\ \frac{\text{kips}}{\text{in}^2}}{50\ \frac{\text{kips}}{\text{in}^2}}} = 113.43 \quad [\geq KL/r,\ \text{so use Eq. 6.58}]$$

From Eq. 6.60, the elastic critical buckling stress is

$$F_e = \frac{\pi^2 E}{\left(\frac{KL}{r}\right)^2} = \frac{\pi^2\left(29{,}000\ \frac{\text{kips}}{\text{in}^2}\right)}{(24.32)^2} = 484\ \text{ksi}$$

Calculate the critical flexural buckling stress using Eq. 6.58.

$$F_{\text{cr}} = 0.658^{F_y/F_e}F_y = (0.658)^{50\ \frac{\text{kips}}{\text{in}^2}/484\ \frac{\text{kips}}{\text{in}^2}}\left(50\ \frac{\text{kips}}{\text{in}^2}\right) = 47.88\ \text{ksi}$$

From Eq. 6.57, the nominal axial compression load capacity for the cross-shaped stiffener column is

$$P_n = F_{\text{cr}}A_{g,\text{cross}} = \left(47.88\ \frac{\text{kips}}{\text{in}^2}\right)(7.27\ \text{in}^2) = 348\ \text{kips}$$

Determine design strength (LRFD) or allowable strength (ASD) of the cross-shaped column.

LRFD	ASD
$P_{u,\mathrm{st}} = \phi_c P_n = (0.90)(348 \text{ kips})$ $= 313 \text{ kips}$ $[> R_u = 264 \text{ kips, so OK}]$	$P_{a,\mathrm{st}} = \dfrac{P_n}{\Omega_c} = \dfrac{348 \text{ kips}}{1.67}$ $= 208 \text{ kips}$ $[> R_a = 180 \text{ kips, so OK}]$

The answer is (B).

PRACTICE PROBLEM 13

The W10 × 88 column shown bears on a concrete pedestal of the same size as the square column base plate. The column load is $P_u = 600$ kips (LRFD) or $P_a = 400$ kips (ASD). The compressive strength of the concrete is 5 ksi.

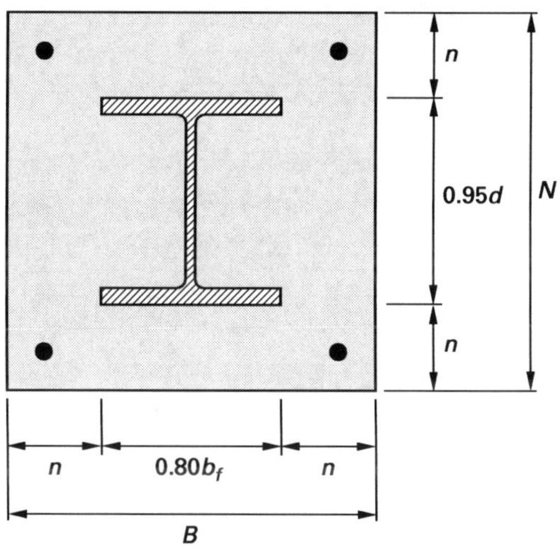

Section properties

$A = 25.9 \text{ in}^2$

$d = 10.8$ in

$t_w = 0.605$ in

$b_f = 10.3$ in

$t_f = 0.99$ in

$b_f/2t_f = 5.18$

$h/t_w = 13.0$

Material properties

W10 × 88

ASTM A992

$F_y = 50$ ksi

$F_u = 65$ ksi

plate

ASTM A572, grade 50

$F_y = 50$ ksi

$F_u = 65$ ksi

The smallest of the following base plates that meets the design criteria is

(A) 15 in × 15 in × 1 in

(B) 15 in × 15 in × 1.25 in

(C) 16 in × 16 in × 1 in

(D) 16 in × 16 in × 1.25 in

Solution

Determine the required area of the base plate in accordance with ACI 318 Sec. 10.17.1 (see Sec. 7.7 in this book). For the ASD method, only the total load is given and no allocation is made between live and dead load, so multiply the total column load by the average load factor of 1.5.

$$P_u = 1.5P = (1.5)(400 \text{ kips}) = 600 \text{ kips}$$

This modified load for the ASD method is equal to the column load for the LRFD method. Calculate the required area of the base plate.

$$\begin{aligned} P_u &\le \phi(0.85 f'_c) A_1 \\ A_1 &\ge \frac{P_u}{\phi(0.85 f'_c)} \\ &\ge \frac{600 \text{ kips}}{(0.65)(0.85)\left(5 \ \frac{\text{kips}}{\text{in}^2}\right)} \\ &\ge 217.19 \text{ in}^2 \end{aligned}$$

For a square base plate,

$$\begin{aligned} BN &= A_1 \\ B = N &= \sqrt{A_1} \\ &= \sqrt{217.19 \text{ in}^2} \\ &= 14.73 \text{ in} \quad [\text{use 15 in} \times 15 \text{ in}] \end{aligned}$$

Determine the governing cantilever projection (taking λ conservatively as 1.0).

$$m = \frac{N - 0.95d}{2} = \frac{15 \text{ in} - (0.95)(10.8 \text{ in})}{2} = 2.37 \text{ in}$$

$$n = \frac{B - 0.80b_f}{2} = \frac{15 \text{ in} - (0.80)(10.3 \text{ in})}{2} = 3.38 \text{ in} \quad [\text{controls}]$$

$$\lambda n' = \tfrac{1}{4}\lambda\sqrt{db_f} = \left(\frac{1}{4}\right)(1)\sqrt{(10.8 \text{ in})(10.3 \text{ in})} = 2.64 \text{ in}$$

Determine the bearing stress.

LRFD	ASD
$f_u = \dfrac{P_u}{A_{\text{plate}}} = \dfrac{600 \text{ kips}}{(15 \text{ in})^2} = 2.66 \text{ ksi}$	$f_a = \dfrac{P_a}{A_{\text{plate}}} = \dfrac{400 \text{ kips}}{(15 \text{ in})^2} = 1.77 \text{ ksi}$

Use Eq. 7.40 (LRFD) or Eq. 7.41 (ASD) to calculate the required thickness of the base plate.

$$l = n = 3.38 \text{ in} \quad [\text{controlling value}]$$

LRFD	ASD
$t_{p,\min} = l\sqrt{\dfrac{2P_u}{0.9F_yBN}} = (3.38 \text{ in}) \times \sqrt{\dfrac{(2)(600 \text{ kips})}{(0.9)\left(50 \dfrac{\text{kips}}{\text{in}^2}\right) \times (16 \text{ in})(16 \text{ in})}} = 1.09 \text{ in}$	$t_{p,\min} = l\sqrt{\dfrac{3.33P_a}{F_yBN}} = (3.38 \text{ in}) \times \sqrt{\dfrac{(3.33)(400 \text{ kips})}{\left(50 \dfrac{\text{kips}}{\text{in}^2}\right)(16 \text{ in})(16 \text{ in})}} = 1.09 \text{ in}$

Use a base plate 15 in × 15 in × 1.25 in.

The answer is (B).

PRACTICE PROBLEM 14

The rolled steel angle shown is to span an opening of 8 ft. It will be uniformly loaded through the y-axis.

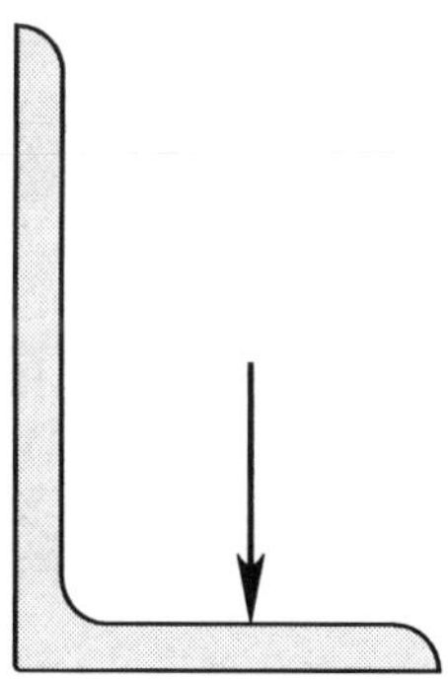

Section properties		Material properties
L5 × 3 × $^3/_8$ in	S_y = 0.874 in^3	ASTM A36
A = 2.86 in^2	r_y = 0.838 in	F_y = 36 ksi
I_x = 7.35 in^4	$\overline{x}$ = 0.698 in	F_u = 58 ksi
S_x = 2.22 in^3	Z_y = 1.57 in^3	
r_x = 1.60 in	I_z = 1.20 in^4	
$\overline{y}$ = 1.69 in	S_z = 0.726 in^3	
Z_x = 3.93 in^3	r_z = 0.646 in	
I_y = 2.01 in^4	β_w = 2.40 in [from *AISC Commentary* Table C-F10.1]	

What is most nearly the design strength (LRFD) or allowable strength (ASD) of the angle? (LRFD options are in parentheses.)

(A) 4.4 ft-kips (6.6 ft-kips)

(B) 5.2 ft-kips (7.8 ft-kips)

(C) 6.2 ft-kips (9.3 ft-kips)

(D) 6.8 ft-kips (10 ft-kips)

Solution

The nominal flexural strength of a single angle is governed by the limit states of lateral-torsional buckling and the yield moment about the axis of bending. First calculate the yield moment.

$$\begin{aligned} M_y &= S_x F_y \\ &= \frac{(2.22 \text{ in}^3)\left(36 \frac{\text{kips}}{\text{in}^2}\right)}{12 \frac{\text{in}}{\text{ft}}} \\ &= 6.66 \text{ ft-kips} \end{aligned}$$

For the limit state of yielding, from *AISC Specification* Eq. F10-1, the nominal flexural strength is

$$M_n = 1.5 M_y = (1.5)(6.66 \text{ ft-kips}) = 9.99 \text{ ft-kips}$$

For the limit state of lateral-torsional buckling, without continuous lateral-torsional restraint and with the maximum compression in the toe, start with *AISC Specification* Eq. F10-5, taking C_b conservatively as 1.0.

$$M_e = \left(\frac{4.9EI_zC_b}{L^2}\right)\left(\sqrt{\beta_w^2 + 0.052\left(\frac{Lt}{r_z}\right)^2} + \beta_w\right)$$

$$= \left(\frac{(4.9)\left(29{,}000\ \frac{\text{kips}}{\text{in}^2}\right)\left(1.20\ \text{in}^4\right)(1.0)}{(8\ \text{ft})^2\left(12\ \frac{\text{in}}{\text{ft}}\right)^2}\right)$$

$$\times\left(\sqrt{\left(\frac{2.40\ \text{in}}{12\ \frac{\text{in}}{\text{ft}}}\right)^2 + (0.052)\left(\frac{(8\ \text{ft})(0.375\ \text{in})}{0.646\ \text{in}}\right)^2} + \frac{2.40\ \text{in}}{12\ \frac{\text{in}}{\text{ft}}}\right)$$

$$= 23.64\ \text{ft-kips} \quad [> M_y]$$

M_e is greater than M_y, so use *AISC Specification* Eq. F10-3 to determine M_n for lateral-torsional buckling.

$$M_n \le \begin{cases} \left(1.92 - 1.17\sqrt{\dfrac{M_y}{M_e}}\right)M_y \\ \quad = \left(1.92 - 1.17\sqrt{\dfrac{6.66\ \text{ft-kips}}{23.64\ \text{ft-kips}}}\right)(6.66\ \text{ft-kips}) \\ \quad = 8.65\ \text{ft-kips} \quad [\text{controls}] \\ 1.5M_y = 9.99\ \text{ft-kips} \end{cases}$$

Determine the design strength (LRFD) or available strength (ASD) of the angle.

LRFD	ASD
$M_u \le \phi_b M_n$ $\le (0.90)(8.65\ \text{ft-kips})$ $\le 7.79\ \text{ft-kips} \quad (7.8\ \text{ft-kips})$	$M_a \le \dfrac{M_n}{\Omega_b}$ $\le \dfrac{8.65\ \text{ft-kips}}{1.67}$ $\le 5.18\ \text{ft-kips} \quad (5.2\ \text{ft-kips})$

The answer is (B).

PRACTICE PROBLEM 15

A steel column 14 ft long is part of a braced frame system and has pinned connections at both ends. The column is subjected to the following loads as determined by the direct analysis method.

LRFD	ASD
$P_u = 400$ kips $M_{ux} = 250$ ft-kips $M_{uy} = ?$	$P_a = 267$ kips $M_{ax} = 167$ ft-kips $M_{ay} = ?$

Section properties		Material properties
W14 × 99	$I_x = 1110\ \text{in}^4$	ASTM A992
$A = 29.1\ \text{in}^2$	$S_x = 157\ \text{in}^3$	$F_y = 50$ ksi
$d = 14.2$ in	$r_x = 6.17$ in	$F_u = 65$ ksi
$t_w = 0.485$ in	$Z_x = 173\ \text{in}^3$	
$b_f = 14.6$ in	$I_y = 402\ \text{in}^4$	
$t_f = 0.780$ in	$S_y = 55.2\ \text{in}^3$	
$b_f/2t_f = 9.34$	$r_y = 3.71$ in	
$t/h_w = 23.5$	$Z_y = 83.6\ \text{in}^3$	

Determine the available design strength (LRFD) or the allowable strength (ASD) about the y-axis. (LRFD options are in parentheses.)

(A) 56 ft-kips (84 ft-kips)

(B) 63 ft-kips (95 ft-kips)

(C) 69 ft-kips (105 ft-kips)

(D) 78 ft-kips (117 ft-kips)

Solution

The loads are determined by the direct analysis method, so $K = 1.0$. (See Sec. 7.1.) The column's effective length is

$$KL_x = (1.00)(14\ \text{ft}) = 14\ \text{ft}$$

From *AISC Manual* Table 4-1, for a W14 × 99 with $KL_x = 14$ ft,

LRFD	ASD
$\phi_c P_n = 1130$ kips	$\dfrac{P_n}{\Omega_c} = 750$ kips

From *AISC Manual* Table 6-1, the combined stress coefficients for a W14 × 99 with $KL_x = 14$ ft are

LRFD	ASD
$p = \dfrac{0.887}{10^3 \text{ kips}}$ $b_x = \dfrac{1.38}{10^3 \text{ ft-kips}}$ $b_y = \dfrac{2.85}{10^3 \text{ ft-kips}}$	$p = \dfrac{1.33}{10^3 \text{ kips}}$ $b_x = \dfrac{2.08}{10^3 \text{ ft-kips}}$ $b_y = \dfrac{4.29}{10^3 \text{ ft-kips}}$

Determine the moment capacity of the member's y-axis. Check the ratio of required axial strength to available axial strength to determine which interaction formula to use.

LRFD	ASD
$\dfrac{P_r}{P_c} = \dfrac{P_u}{\phi_c P_n} = \dfrac{400 \text{ kips}}{1130 \text{ kips}} = 0.354 \quad [> 0.2, \text{ so use Eq. 8.5}]$	$\dfrac{P_r}{P_c} = \dfrac{P_a}{\dfrac{P_n}{\Omega_c}} = \dfrac{267 \text{ kips}}{750 \text{ kips}} = 0.356 \quad [> 0.2, \text{ so use Eq. 8.5}]$

$P_r/P_c > 0.2$, so use Eq. 8.5.

$$pP_u + b_x M_{ux} + b_y M_{uy} \le 1.0$$

$$M_{uy} = \frac{1.0 - pP_u - b_x M_{ux}}{b_y}$$

LRFD	ASD
$M_{uy} = \dfrac{1.0 - \left(\dfrac{0.886}{10^3 \text{ kips}}\right)(400 \text{ kips}) - \left(\dfrac{1.38}{10^3 \text{ ft-kips}}\right) \times (250 \text{ ft-kips})}{\dfrac{2.85}{10^3 \text{ ft-kips}}}$ $= 105.47 \text{ ft-kips} \quad (110 \text{ ft-kips})$	$M_{uy} = \dfrac{1.0 - \left(\dfrac{1.33}{10^3 \text{ kips}}\right)(267 \text{ kips}) - \left(\dfrac{2.08}{10^3 \text{ ft-kips}}\right) \times (167 \text{ ft-kips})}{\dfrac{4.29}{10^3 \text{ ft-kips}}}$ $= 69.35 \text{ ft-kips} \quad (69 \text{ ft-kips})$

The answer is (C).

PRACTICE PROBLEM 16

The HSS16 × 4 × $^1/_4$ member shown is a flexural member with an unbraced length of 20 ft about both axes. There is no axial load on the member. The bending moment about the x-axis is M_{ux} = 70 ft-kips (LRFD) or M_{ax} = 47 ft-kips (ASD).

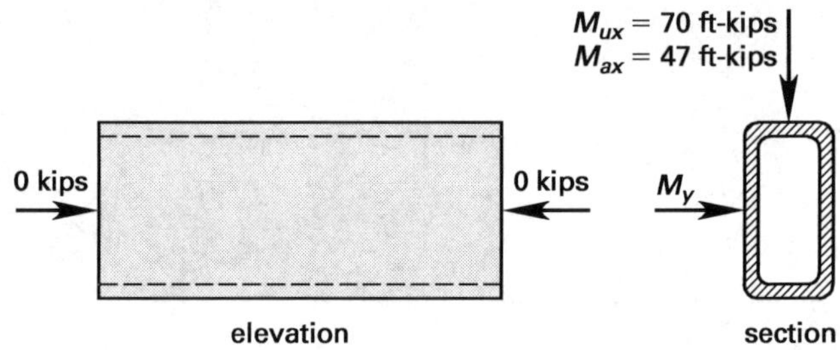

Section properties		Material properties
$A = 8.96\text{ in}^2$	$r_x = 5.31\text{ in}$	ASTM A500, grade B
$t_{des} = 0.233\text{ in}$	$Z_x = 41.7\text{ in}^3$	$F_y = 46\text{ ksi}$
$b/t = 14.2$	$I_y = 27.7\text{ in}^4$	$F_u = 58\text{ ksi}$
$h/t = 65.7$	$S_y = 13.8\text{ in}^4$	$E_s = 29{,}000\text{ ksi}$
$I_x = 253\text{ in}^4$	$r_y = 1.76\text{ in}$	
$S_x = 31.6\text{ in}^3$	$Z_y = 15.2\text{ in}^3$	

What is most nearly the load that can be applied to the y-axis? (LRFD options are in parentheses.)

(A) 0.08 kips/ft (0.11 kips/ft)

(B) 0.15 kips/ft (0.22 kips/ft)

(C) 0.22 kips/ft (0.33 kips/ft)

(D) 0.29 kips/ft (0.44 kips/ft)

Solution

From *AISC Manual* Table 3-12, for an HSS16 × 4 × $^1/_4$ member,

LRFD	ASD
$\phi M_{nx} = 142$ ft-kips $\phi M_{ny} = 32.8$ ft-kips	$\dfrac{M_{nx}}{\Omega} = 94.3$ ft-kips $\dfrac{M_{ny}}{\Omega} = 21.8$ ft-kips

The axial force is zero, so use Eq. 8.2 with the first term zero.

$$\frac{M_{rx}}{M_{cx}}+\frac{M_{ry}}{M_{cy}}\leq 1.0$$

$$M_{ry}\leq M_{cy}\left(1.0-\frac{M_{rx}}{M_{cx}}\right)$$

LRFD	ASD
$M_{uy}\leq(32.8\text{ ft-kips})\times\left(1.0-\frac{70\text{ ft-kips}}{142\text{ ft-kips}}\right)\leq 16.63\text{ ft-kips}$	$M_{ay}\leq(21.8\text{ ft-kips})\times\left(1.0-\frac{47\text{ ft-kips}}{94.3\text{ ft-kips}}\right)\leq 10.93\text{ ft-kips}$

Determine the uniform load per linear foot that can be applied to the y-axis based on the available design strength or allowable strength.

LRFD	ASD
$w_u=\frac{8M_{uy}}{L^2}=\frac{(8)(16.63\text{ ft-kips})}{(20\text{ ft})^2}=0.33\text{ kips/ft}$	$w_u=\frac{8M_{ay}}{L^2}=\frac{(8)(10.93\text{ ft-kips})}{(20\text{ ft})^2}=0.22\text{ kips/ft}$

The answer is (C).

PRACTICE PROBLEM 17

The W8 × 35 shown is rigidly attached to the overhead structure and is 15 ft long. A concentric axial load is suspended from the bottom of the member, and horizontal loads are applied to the bottom of the member in both the x- and y-axes. The vertical and y-axis loads are known.

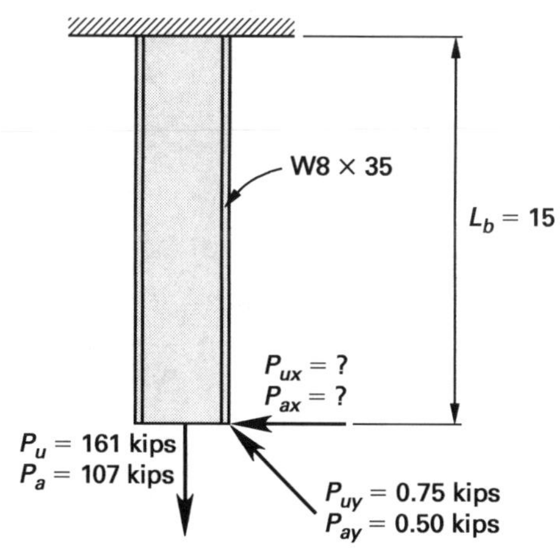

Section properties		Material properties
$A = 10.3 \text{ in}^2$	$I_x = 127 \text{ in}^4$	ASTM A992
$d = 8.12 \text{ in}$	$S_x = 31.2 \text{ in}^3$	$F_y = 50 \text{ ksi}$
$t_w = 0.310 \text{ in}$	$r_x = 3.51 \text{ in}$	$F_u = 65 \text{ ksi}$
$b_f = 8.02 \text{ in}$	$Z_x = 34.7 \text{ in}^3$	$E_s = 29{,}000 \text{ ksi}$
$t_f = 0.495 \text{ in}$	$I_y = 42.6 \text{ in}^4$	
$b_f/2t_f = 8.10$	$S_y = 10.6 \text{ in}^3$	
$h/t_w = 20.5$	$r_y = 2.03 \text{ in}$	
	$Z_y = 16.1 \text{ in}^3$	

Which of the following is most nearly the load that can be applied to the x-axis without exceeding the available strength of the member? (LRFD options are in parentheses.)

(A) 2.7 kips (4.0 kips)

(B) 3.2 kips (4.7 kips)

(C) 3.5 kips (5.3 kips)

(D) 3.9 kips (5.9 kips)

Solution

For a W8 × 35 member, from *AISC Manual* Table 3-2,

$$L_p = 7.17 \text{ ft}$$
$$L_r = 27 \text{ ft}$$

LRFD	ASD
$\phi_b M_{px} = 130 \text{ ft-kips}$ $\phi_b M_{rx} = 81.9 \text{ ft-kips}$ $\text{BF} = 2.43 \text{ kips}$	$\frac{M_{px}}{\Omega} = 86.6 \text{ ft-kips}$ $\frac{M_{rx}}{\Omega} = 54.5 \text{ ft-kips}$ $\text{BF} = 1.62 \text{ kips}$

From *AISC Manual* Table 3-4,

LRFD	ASD
$\phi_b M_{py} = 60.4 \text{ ft-kips}$	$\frac{M_{py}}{\Omega_b} = 40.2 \text{ ft-kips}$

From *AISC Manual* Table 5-1, for tension, $A_g = 10.3\ \text{in}^2$.

$$A_e = 0.75A_g = (0.75)(10.3\ \text{in}^2) = 7.73\ \text{in}^2$$

For yielding,

LRFD	ASD
$\phi_t P_n = 464\ \text{kips}$	$P_n/\Omega_t = 308\ \text{kips}$

For rupture,

LRFD	ASD
$\phi_t P_n = 377\ \text{kips}$	$P_n/\Omega_t = 251\ \text{kips}$

Rupture does not control because there are no holes in the W8 × 35 hanger. Check the ratio of required axial strength to available axial strength to determine which interaction formula applies, Eq. 8.1 or Eq. 8.2.

LRFD	ASD
$\frac{P_r}{P_c} = \frac{P_u}{\phi_t P_n} = \frac{161\ \text{kips}}{464\ \text{kips}} = 0.35$	$\frac{P_r}{P_c} = \frac{P_a}{\frac{P_n}{\Omega_t}} = \frac{107\ \text{kips}}{308\ \text{kips}} = 0.35$

$P_r/P_c \geq 0.2$, so Eq. 8.1 applies. Calculate M_{ry}.

LRFD	ASD
$M_{ry} = P_{uy}L_b = (0.75\ \text{kips})(15\ \text{ft}) = 11.25\ \text{ft-kips}$	$M_{ry} = P_{ay}L_b = (0.50\ \text{kips})(15\ \text{ft}) = 7.5\ \text{ft-kips}$

Use Eq. 5.10 (LRFD) or Eq. 5.11 (ASD) to determine the maximum available moment capacity that will not exceed the lateral torsional buckling limit state. For cantilevers or overhangs where the free end is unbraced, $C_b = 1.0$ (per *AISC Specification* Sec. F1).

LRFD	ASD
$\phi_b M_n = C_b\left(\phi_b M_{px} - (\text{BF}) \times (L_b - L_p)\right) \leq \phi_b M_{px}$ $= (1)\left(130\ \text{ft-kips} - (2.43\ \text{kips}) \times (15\ \text{ft} - 7.17\ \text{ft})\right)$ $= 110.97\ \text{ft-kips}\quad [\leq 130\ \text{ft-kips}]$	$\frac{M_n}{\Omega_b} = C_b\left(\frac{M_{px}}{\Omega_b} - (\text{BF}) \times (L_b - L_p)\right) \leq \frac{M_{px}}{\Omega_b}$ $= (1)\left(86.6\ \text{ft-kips} - (1.62\ \text{kips}) \times (15\ \text{ft} - 7.17\ \text{ft})\right)$ $= 73.92\ \text{ft-kips}\quad [\leq 86.6\ \text{ft-kips}]$

Use Eq. 8.1 to determine the available flexural strength in the x-axis.

$$\frac{P_r}{P_c}+\left(\frac{8}{9}\right)\left(\frac{M_{rx}}{M_{cx}}+\frac{M_{ry}}{M_{cy}}\right)\leq 1.0$$

$$M_{rx}\leq M_{cx}\left(\left(\frac{9}{8}\right)\left(1.0-\frac{P_r}{P_c}\right)-\frac{M_{ry}}{M_{cy}}\right)$$

LRFD	ASD
$M_{rx}\leq(110.97\text{ ft-kips})\times\left(\left(\frac{9}{8}\right)\left(1.0-\frac{161\text{ kips}}{464\text{ kips}}\right)-\frac{11.25\text{ ft-kips}}{60.4\text{ ft-kips}}\right)$ $=60.85\text{ ft-kips}$	$M_{rx}\leq(73.92\text{ ft-kips})\times\left(\left(\frac{9}{8}\right)\left(1.0-\frac{107\text{ kips}}{308\text{ kips}}\right)-\frac{7.5\text{ ft-kips}}{40.2\text{ ft-kips}}\right)$ $=40.48\text{ ft-kips}$

Determine the load that develops the design strength (LRFD) or the allowable strength (ASD) about the x-axis.

LRFD	ASD
$P_x=\frac{M_{rx}}{L}=\frac{60.85\text{ ft-kips}}{15\text{ ft}}=4.06\text{ kips}\quad(4.0\text{ kips})$	$P_x=\frac{M_{rx}}{L}=\frac{40.48\text{ ft-kips}}{15\text{ ft}}=2.70\text{ kips}\quad(2.7\text{ kips})$

Use Eq. 8.1 to check the solution.

LRFD	ASD
$\frac{P_r}{P_c}+\left(\frac{8}{9}\right)\left(\frac{M_{rx}}{M_{cx}}+\frac{M_{ry}}{M_{cy}}\right)$ $=0.35+\left(\frac{8}{9}\right)\left(\frac{60.85\text{ ft-kips}}{110.97\text{ ft-kips}}+\frac{11.25\text{ ft-kips}}{60.4\text{ ft-kips}}\right)$ $=1.00\quad[\text{OK}]$	$\frac{P_r}{P_c}+\left(\frac{8}{9}\right)\left(\frac{M_{rx}}{M_{cx}}+\frac{M_{ry}}{M_{cy}}\right)$ $=0.35+\left(\frac{8}{9}\right)\left(\frac{40.48\text{ ft-kips}}{73.92\text{ ft-kips}}+\frac{7.5\text{ ft-kips}}{40.2\text{ ft-kips}}\right)$ $=1.00\quad[\text{OK}]$

The answer is (A).

PRACTICE PROBLEM 18

The structural section shown is fabricated from three plates that are welded together in an H shape. The welds are sufficient to develop full section strength.

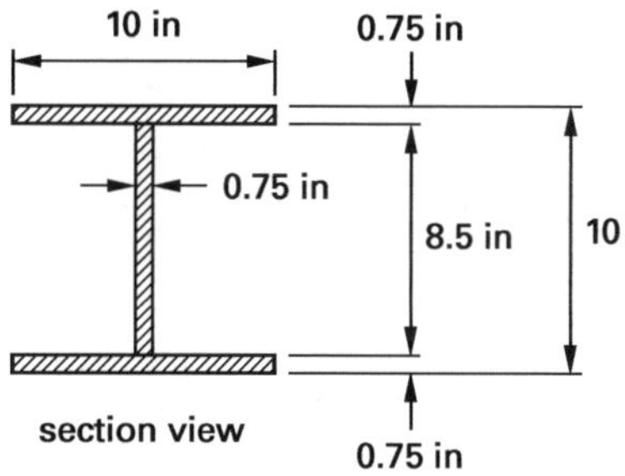

Material properties

ASTM A572, grade B

$F_y = 50$ ksi

$F_u = 65$ ksi

Which of the following is most nearly the available flexural strength about the weak axis? (LRFD options are in parentheses.)

(A) 87 ft-kips (130 ft-kips)

(B) 97 ft-kips (150 ft-kips)

(C) 110 ft-kips (170 ft-kips)

(D) 130 ft-kips (190 ft-kips)

Solution

Determine the elastic section modulus for the weak axis, S_y. For each flange alone,

$$S_{yf} = \frac{bd^2}{6} = \frac{(0.75 \text{ in})(10 \text{ in})^2}{6} = 12.5 \text{ in}^3$$

For the web,

$$S_{yw} = \frac{bd^2}{6} = \frac{(8.5 \text{ in})(0.75 \text{ in})^2}{6} = 0.80 \text{ in}^3$$

For the entire section (both flanges and the web),

$$S_y = 2S_{yf} + S_{yw} = (2)(12.5 \text{ in}^3) + 0.80 \text{ in}^3 = 25.8 \text{ in}^3$$

Determine the plastic section modulus for the weak axis, Z_y. For each flange alone,

$$Z_{yf} = \frac{bd^2}{4} = \frac{(0.75\text{ in})(10\text{ in})^2}{4} = 18.75\text{ in}^3$$

For the web,

$$Z_{yw} = \frac{bd^2}{4} = \frac{(8.5\text{ in})(0.75\text{ in})^2}{4} = 1.20\text{ in}^3$$

For the entire section (both flanges and the web),

$$\begin{aligned} Z_y &= 2Z_{yf} + Z_{yw} \\ &= (2)(18.75\text{ in}^3) + 1.20\text{ in}^3 \\ &= 38.7\text{ in}^3 \end{aligned}$$

Determine whether the flanges are compact, in which case the limit state of yielding controls. From *AISC Specification* Table B4.1b, case 11,

$$\begin{aligned} \lambda_p &= 0.38\sqrt{\frac{E}{F_y}} \\ &= 0.38\sqrt{\frac{29{,}000\ \frac{\text{kips}}{\text{in}^2}}{50\ \frac{\text{kips}}{\text{in}^2}}} \\ &= 9.15 \end{aligned}$$

$$\frac{b}{t} = \frac{5\text{ in}}{0.75\text{ in}} = 6.67 \quad [< 9.15\text{, so compact}]$$

The flanges are compact and the limit state of yielding controls. From Eq. 5.19, determine the nominal flexural strength, M_n, based on the limit state of yielding.

$$M_n = M_p \le \begin{cases} F_y Z_y = \dfrac{\left(50\ \dfrac{\text{kips}}{\text{in}^2}\right)(38.7\text{ in}^3)}{12\ \dfrac{\text{in}}{\text{ft}}} \\ \quad = 161.25\text{ ft-kips} \quad [\text{controls}] \\ 1.6F_y S_y = \dfrac{(1.6)\left(50\ \dfrac{\text{kips}}{\text{in}^2}\right)(25.8\text{ in}^3)}{12\ \dfrac{\text{in}}{\text{ft}}} \\ \quad = 172\text{ ft-kips} \end{cases}$$

Determine the design strength (LRFD) or allowable strength (ASD).

LRFD	ASD
$M_u \le \phi M_n$ $\le (0.90)(161.25\text{ ft-kips})$ $\le 145.13\text{ ft-kips} \quad (150\text{ ft-kips})$	$M_a \le \dfrac{M_n}{\Omega}$ $\le \dfrac{161.25\text{ ft-kips}}{1.67}$ $\le 96.56\text{ ft-kips} \quad (97\text{ ft-kips})$

The answer is (B).

PRACTICE PROBLEM 19

An HSS8 × 4 × $^3/_8$ is to be used as a beam.

Section properties

$t = 0.349$ in

$A = 7.58\text{ in}^2$

$b/t = 8.46$

$h/t = 19.9$

$I_x = 58.7\text{ in}^4$

$S_x = 14.7\text{ in}^3$

$r_x = 2.78$ in

$Z_x = 18.8\text{ in}^3$

$I_y = 19.6\text{ in}^4$

$S_y = 9.8\text{ in}^3$

$r_y = 1.61$ in

$Z_y = 11.5\text{ in}^3$

Material properties

ASTM A500, grade B

$F_y = 46$ ksi

$F_u = 58$ ksi

Which of the following is most nearly the available shear strength about the strong axis? (LRFD options are in parentheses.)

(A) 80 kips (120 kips)

(B) 93 kips (140 kips)

(C) 100 kips (150 kips)

(D) 110 kips (170 kips)

Solution

Determine the effective web height for shear.

$$h_{\text{eff}} = h - 3t = 8\text{ in} - (3)(0.349\text{ in}) = 6.95\text{ in}$$

The web area is

$$\begin{aligned} A_w &= 2h_{\text{eff}}t = (2)(6.95\text{ in})(0.349\text{ in}) \\ &= 4.85\text{ in}^2 \end{aligned}$$

Determine which formula to use for the web shear coefficient, C_v. The height-thickness ratio, h/t, is less than 260, so the web plate buckling coefficient is $k_v = 5.0$ (see Sec. 11.4).

$$1.10\sqrt{\frac{k_v E}{F_y}} = 1.10\sqrt{\frac{(5)\left(29{,}000\ \frac{\text{kips}}{\text{in}^2}\right)}{46\ \frac{\text{kips}}{\text{in}^2}}} = 61.76\quad [> h/t = 19.2]$$

From Eq. 11.24, then, $C_v = 1.0$. Use Eq. 11.23 to determine the nominal shear capacity.

$$\begin{aligned} V_n &= 0.6F_y A_w C_v \\ &= (0.60)\left(46\ \frac{\text{kips}}{\text{in}^2}\right)(4.85\ \text{in}^2)(1.0) \\ &= 134\ \text{kips} \end{aligned}$$

Determine the design strength (LRFD) or allowable strength (ASD).

LRFD	ASD
$V_u \le \phi_v V_n$ $\le (0.90)(134\ \text{kips})$ $\le 120.6\ \text{kips}\quad (120\ \text{kips})$	$V_a \le \frac{V_n}{\Omega_v}$ $\le \frac{134\ \text{kips}}{1.67}$ $\le 80.24\ \text{kips}\quad (80\ \text{kips})$

The answer is (A).

PRACTICE PROBLEM 20

The connection plate shown is $^1/_2$ in thick and is punched to receive $^3/_4$ in diameter bolts.

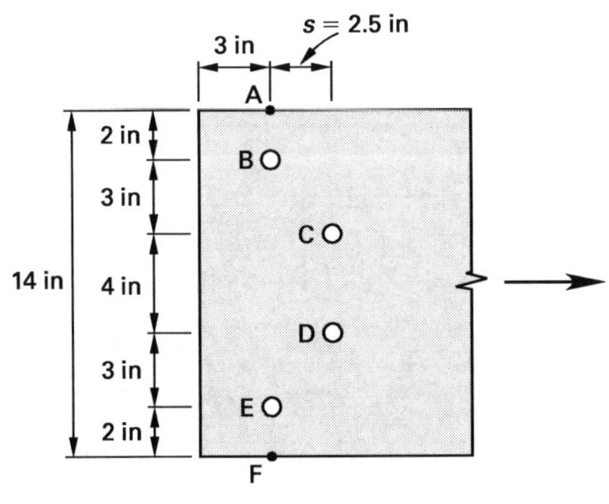

Material properties

ASTM A36

$F_y = 36$ ksi

$F_u = 58$ ksi

Which of the following is most nearly the available tensile strength? (LRFD options are in parentheses.)

(A) 150 kips (230 kips)

(B) 170 kips (250 kips)

(C) 190 kips (290 kips)

(D) 220 kips (330 kips)

Solution

Determine the gross area of the plate.

$$A_g = bt = (14 \text{ in})(0.50 \text{ in}) = 7.0 \text{ in}^2$$

Determine the net effective width of the plate. The effective diameter of a hole is $^1/_8$ in larger than the nominal diameter of its bolt, so

$$d_{\text{hole}} = 0.75 \text{ in} + 0.125 \text{ in} = 0.875 \text{ in}$$

To compute the effective net width for a chain of holes, use a variant of Eq. 4.9, dividing each term by the plate thickness.

$$\frac{A_n}{t} = \frac{A_g}{t} - \frac{\sum d_{\text{hole}} t}{t} + \frac{\sum\left(\frac{s^2}{4g}\right)t}{t}$$

$$b_n = b - \sum d_{\text{hole}} + \sum \frac{s^2}{4g}$$

For chain A-B-E-F,

$$\begin{aligned} b_n &= 14 \text{ in} - (2)(0.875 \text{ in}) + 0 \text{ in} \\ &= 12.3 \text{ in} \end{aligned}$$

For chain A-B-C-D-E-F,

$$b_n = 14 \text{ in} - (4)(0.875 \text{ in}) + \left(\frac{(2.50 \text{ in})^2}{(4)(3 \text{ in})} + \frac{(2.50 \text{ in})^2}{(4)(3 \text{ in})}\right)$$

$$= 11.54 \text{ in}$$

For chain A-B-D-E-F,

$$b_n = 14\text{ in} - (3)(0.875\text{ in}) + \left(\frac{(2.50\text{ in})^2}{(4)(7\text{ in})} + \frac{(2.50\text{ in})^2}{(4)(3\text{ in})} \right)$$
$$= 12.12\text{ in}$$

Chain A-B-C-D-E-F has the least effective net width and therefore is controlling. All elements of the plate are in contact, so the shear lag factor is $U = 1.0$. From Eq. 4.10,

$$\begin{aligned} A_e &= UA_n = Ub_nt \\ &= (1.0)(11.54\text{ in})(0.5\text{ in}) \\ &= 5.77\text{ in}^2 \end{aligned}$$

Use Eq. 4.2 to calculate the nominal strength based on the limit state of yielding on the gross area.

$$P_n = F_yA_g = \left(36\ \frac{\text{kips}}{\text{in}^2}\right)(7.0\text{ in}^2) = 252\text{ kips}$$

Use Eq. 4.3 to calculate the nominal strength based on the limit state of rupture on the net section.

$$P_n = F_uA_e = \left(58\ \frac{\text{kips}}{\text{in}^2}\right)(5.77\text{ in}^2) = 334.7\text{ kips}$$

Yielding on the gross section is smaller and governs. Calculate the design strength (LRFD) or the allowable strength (ASD).

LRFD	ASD
$P_u \le \phi_t P_n$ $\le (0.90)(252\text{ kips})$ $\le 226.8\text{ kips}\quad(230\text{ kips})$	$P_a \le \dfrac{P_n}{\Omega_t}$ $\le \dfrac{252\text{ kips}}{1.67}$ $\le 150.9\text{ kips}\quad(150\text{ kips})$

The answer is (A).

PRACTICE PROBLEM 21

An HSS6 × 6 × $^3/_8$ member is secured to two tension tabs with $^7/_8$ in diameter through-bolts as shown.

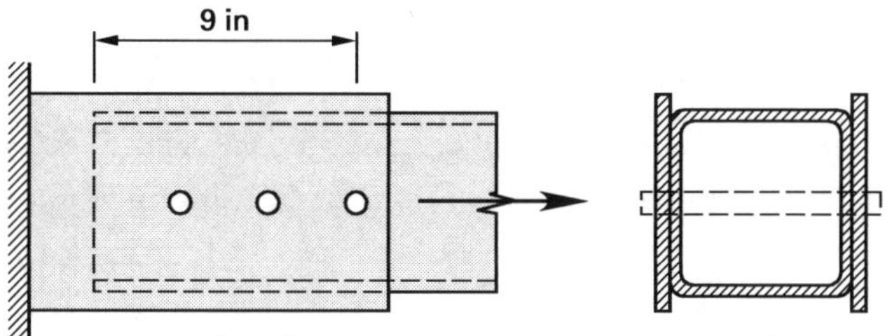

Section properties

		Material properties
$t = 0.349$ in	$I = 39.5$ in^4	ASTM A500, grade B
$A = 7.58$ in^2	$S = 13.2$ in^3	$F_y = 46$ ksi
$b/t = 14.2$	$r = 2.28$ in	$F_u = 58$ ksi
$h/t = 14.2$	$Z = 15.8$ in^3	

What is most nearly the available strength of the HSS member? (LRFD options are in parentheses.)

(A) 170 kips (250 kips)

(B) 180 kips (270 kips)

(C) 220 kips (320 kips)

(D) 240 kips (360 kips)

Solution

From *AISC Manual* Table 5-5, for an HSS6 × 6 × $^3/_8$ member,

$$A_g = 7.58 \text{ in}^2$$
$$0.75A_g = 5.69 \text{ in}^2$$

From the same table, for yielding on the gross section ($\phi_t = 0.90$, $\Omega_t = 1.67$),

LRFD	ASD
$\phi_t P_n = 314$ kips	$\dfrac{P_n}{\Omega_t} = 209$ kips

For rupture on the net section ($\phi_t = 0.75$, $\Omega_t = 2.00$),

LRFD	ASD
$\phi_t P_n = 248 \text{ kips}$	$\dfrac{P_n}{\Omega_t} = 165 \text{ kips}$

Determine the net area of the HSS member at the holes. The effective diameter of a hole is $^1/_8$ in larger than nominal diameter of its bolt, so from Eq. 4.6 and Eq. 4.7,

$$\begin{aligned} d_{\text{hole}} &= 0.875 \text{ in} + 0.125 \text{ in} \\ &= 1 \text{ in} \\ A_h &= n_{\text{holes}} t d_{\text{hole}} \\ &= (2)(0.349 \text{ in})(1 \text{ in}) \\ &= 0.698 \text{ in}^2 \\ A_n &= A_g - A_h \\ &= 7.58 \text{ in}^2 - 0.698 \text{ in}^2 \\ &= 6.88 \text{ in}^2 \end{aligned}$$

Using Eq. 4.10, determine the net effective area of the HSS member at the holes.

$$\begin{aligned} A_e &= U A_n \\ &= (0.90)(6.88 \text{ in}^2) \\ &= 6.19 \text{ in}^2 \quad [> 5.69 \text{ in}^2] \end{aligned}$$

Using Eq. 4.3, determine the nominal resistance to rupture.

$$\begin{aligned} P_n &= F_u A_e \\ &= \left(58 \ \frac{\text{kips}}{\text{in}^2}\right)(6.19 \text{ in}^2) \\ &= 359 \text{ kips} \end{aligned}$$

For rupture on the net section ($\phi_t = 0.75$, $\Omega_t = 2.00$),

LRFD	ASD
$\phi_t P_n = (0.75)(359 \text{ kips})$ $= 269.25 \text{ kips} \quad (270 \text{ kips})$	$\dfrac{P_n}{\Omega_t} = \dfrac{359 \text{ kips}}{2.0}$ $= 179.5 \text{ kips} \quad (180 \text{ kips})$

This is less than the yielding strength, so the rupture strength controls.

The answer is (B).

PRACTICE PROBLEM 22

An HSS12 × 6 × 1/2 member has a length of 20 ft.

Section properties		Material properties
$t = 0.465$ in	$Z_x = 57.4$ in^3	ASTM A500, grade B
$A = 15.3$ in^2	$I_y = 19.1$ in^4	$F_y = 46$ ksi
$b/t = 9.9$	$S_y = 30.4$ in^3	$F_u = 58$ ksi
$h/t = 22.8$	$r_y = 2.44$ in	
$I_x = 271$ in^4	$Z = 35.2$ in^3	
$S_x = 45.2$ in^3	$J = 227$ in^4	
$r_x = 4.21$ in	$C = 59.0$ in^3	

What is most nearly the available torsional strength of the HSS member? (LRFD options are in parentheses.)

(A) 60 ft-kips (90 ft-kips)

(B) 67 ft-kips (100 ft-kips)

(C) 74 ft-kips (110 ft-kips)

(D) 81 ft-kips (120 ft-kips)

Solution

Determine which equation to use to calculate the critical stress.

$$\begin{aligned} 2.45\sqrt{\frac{E}{F_y}} &= 2.45\sqrt{\frac{29{,}000\ \frac{\text{kips}}{\text{in}^2}}{46\ \frac{\text{kips}}{\text{in}^2}}} \\ &= 61.52 \quad [\geq h/t\text{, so use Eq. 8.12}] \end{aligned}$$

Using Eq. 8.12, the critical stress is

$$\begin{aligned} F_{\text{cr}} &= 0.6F_y \\ &= (0.6)\left(46\ \frac{\text{kips}}{\text{in}^2}\right) \\ &= 27.6\ \text{ksi} \end{aligned}$$

Use Eq. 8.8 to calculate the nominal torsional resistance.

$$\begin{aligned} T_n &= F_{cr}C \\ &= \frac{\left(27.6\ \frac{\text{kips}}{\text{in}^2}\right)\left(59.0\ \text{in}^3\right)}{12\ \frac{\text{in}}{\text{ft}}} \\ &= 135.7\ \text{ft-kips} \end{aligned}$$

Calculate the design torsional strength (LRFD) or allowable torsional strength (ASD).

LRFD	ASD
$T_u \le \phi_T T_n$ $\le (0.90)(135.7\ \text{ft-kips})$ $\le 122.13\ \text{ft-kips}\quad (120\ \text{ft-kips})$	$T_a \le \frac{T_n}{\Omega_T} \le \frac{135.7\ \text{ft-kips}}{1.67}$ $\le 81.26\ \text{ft-kips}\quad (81\ \text{ft-kips})$

The answer is (D).

PRACTICE PROBLEM 23

A gusset plate is to be connected to a beam with $^5/_{16}$ in E70XX fillet welds on each side of the plate as shown. The beam is sufficiently stiff not to control design. The plate will be subjected to loads at an angle as shown.

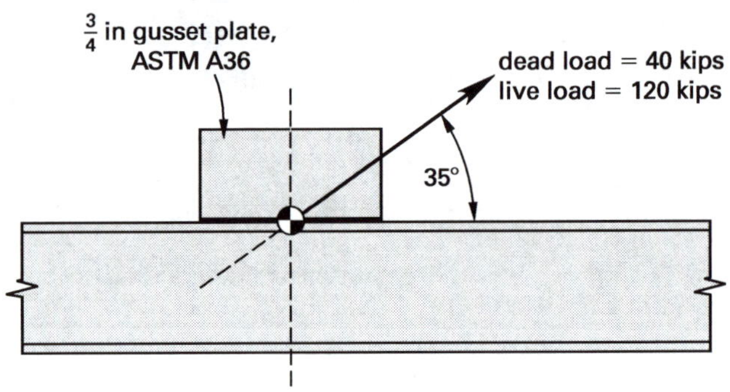

To resist the loads, the required length of the welds is most nearly

(A) 15 in

(B) 17 in

(C) 19 in

(D) 21 in

Solution

Calculate the required strength.

LRFD	ASD
$P_u = 1.2D + 1.6L$ $= (1.2)(40 \text{ kips}) + (1.6)(120 \text{ kips})$ $= 240 \text{ kips}$	$P_a = D + L$ $= 40 \text{ kips} + 120 \text{ kips}$ $= 160 \text{ kips}$

Calculate the shear strength of a $^5/_{16}$ in fillet weld. (D is the number of sixteenths of an inch in the weld length. See Sec. 10.7.)

LRFD	ASD
$r_n = D\left(1.392 \ \frac{\text{kips}}{\text{in}}\right) \quad [\text{per 1/16 in}]$ $= (5)\left(1.392 \ \frac{\text{kips}}{\text{in}}\right)$ $= 6.96 \text{ kips/in}$	$r_n = D\left(0.928 \ \frac{\text{kips}}{\text{in}}\right) \quad [\text{per 1/16 in}]$ $= (5)\left(0.928 \ \frac{\text{kips}}{\text{in}}\right)$ $= 4.64 \text{ kips/in}$

Use *AISC Specification* Eq. J2-5 to calculate the allowable increase in weld capacity due to the angle of load.

$$\begin{aligned} F_w &= 0.60F_{\text{EXX}}\left(1.0 + 0.50\sin^{1.5}\theta\right) \\ &= (0.60)\left(70 \ \frac{\text{kips}}{\text{in}^2}\right)\left(1 + (0.50)(\sin 35°)^{1.5}\right) \\ &= 51.12 \text{ ksi} \end{aligned}$$

Use Eq. 10.4 to determine the nominal resistance capacity of each $^5/_{16}$ in weld.

$$\begin{aligned} R_{n,\text{per weld}} &= F_w A_w = F_w(0.707w) \\ &= \left(51.12 \ \frac{\text{kips}}{\text{in}^2}\right)(0.707)\left(\frac{5}{16} \text{ in}\right) \\ &= 11.29 \text{ kips/in} \quad [\text{per weld}] \end{aligned}$$

For two $^5/_{16}$ welds, R_n = (2)(11.29 kips/in) = 22.59 kips/in.

Determine the length of weld required.

LRFD	ASD
$L = \dfrac{P_u}{\phi R_n} = \dfrac{240 \text{ kips}}{(0.75)\left(22.59 \dfrac{\text{kips}}{\text{in}}\right)}$ $= 14.17 \text{ in} \quad (14 \text{ in})$	$L = \dfrac{P_a}{\dfrac{R_n}{\Omega}} = \dfrac{160 \text{ kips}}{\dfrac{22.59 \dfrac{\text{kips}}{\text{in}}}{2.0}}$ $= 14.17 \text{ in} \quad (14 \text{ in})$

The answer is (A).

PRACTICE PROBLEM 24

An HSS10 × 6 × $^3/_8$ column is filled with concrete that has a specified compressive strength of 5 ksi.

Section properties

$t = 0.349$ in

$A = 10.4 \text{ in}^2$

$b/t = 14.2$

$h/t = 25.7$

$I_x = 137 \text{ in}^4$

$S_x = 27.4 \text{ in}^3$

$r_x = 3.63$ in

$Z_x = 33.8 \text{ in}^3$

$I_y = 61.8 \text{ in}^4$

$S_y = 20.6 \text{ in}^3$

$r_y = 2.44$ in

$Z_y = 23.7 \text{ in}^3$

Material properties

ASTM A500, grade B

$F_y = 46$ ksi

$F_u = 58$ ksi

Which of the following is most nearly the available shear strength about the strong axis of the column? (LRFD options are in parentheses.)

(A) 75 kips (110 kips)

(B) 92 kips (140 kips)

(C) 100 kips (160 kips)

(D) 124 kips (190 kips)

Solution

According to *AISC Specification* Sec. I2.1d, the available shear strength for a filled concrete column is the shear strength of the steel section alone or that of the concrete section alone, whichever is greater. Calculate the shear strength of the steel section alone. The effective height is

$$\begin{aligned} h_{\text{eff}} &= d - 3t \\ &= 10 \text{ in} - (3)(0.349 \text{ in}) \\ &= 8.95 \text{ in} \end{aligned}$$

The shear area is

$$A_w = 2h_{\text{eff}}t = (2)(8.95 \text{ in})(0.349 \text{ in}) = 6.25 \text{ in}^2$$

The shear strength is found from Eq. 5.44. From the criteria at the end of *AISC Manual* Table 1-12, the web shear coefficient, C_v, is 1.0.

$$\begin{aligned} V_n &= 0.6F_yA_wC_v \\ &= (0.6)\left(46 \ \frac{\text{kips}}{\text{in}^2}\right)(6.25 \text{ in}^2)(1.0) \\ &= 173 \text{ kips} \end{aligned}$$

Calculate the shear strength of the concrete alone, neglecting reduction of the concrete area due to the rounded corners. The concrete area is

$$\begin{aligned} A_c &= b_cd_c = (b-2t)(d-2t) \\ &= (6 \text{ in} - (2)(0.349 \text{ in}))(10 \text{ in} - (2)(0.349 \text{ in})) \\ &= 49.32 \text{ in}^2 \end{aligned}$$

From ACI 318 Eq. 11-3, the shear strength is

$$\begin{aligned} V_c &= 2\lambda\sqrt{f_c'}b_wd = 2\lambda\sqrt{f_c'}A_c \\ &= \frac{(2)(1.0)\sqrt{5000 \ \frac{\text{lbf}}{\text{in}^2}}(49.32 \text{ in}^2)}{1000 \ \frac{\text{lbf}}{\text{kip}}} \\ &= 6.97 \text{ kips} \end{aligned}$$

The factor λ is 1.0 for normal weight concrete. The nominal shear strength of steel section is greater and controls. Determine the design shear strength (LRFD) or allowable shear strength (ASD).

LRFD	ASD
$V_u \leq \phi_vV_n$ $\leq (0.90)(173 \text{ kips})$ $\leq 156 \text{ kips} \quad (160 \text{ kips})$	$V_a \leq \frac{V_n}{\Omega_v}$ $\leq \frac{173 \text{ kips}}{1.67}$ $\leq 104 \text{ kips} \quad (100 \text{ kips})$

The answer is (C).

PRACTICE PROBLEM 25

Two $^5/_{16}$ in fillet welds are needed for the tension brace shown. Use E70 electrodes.

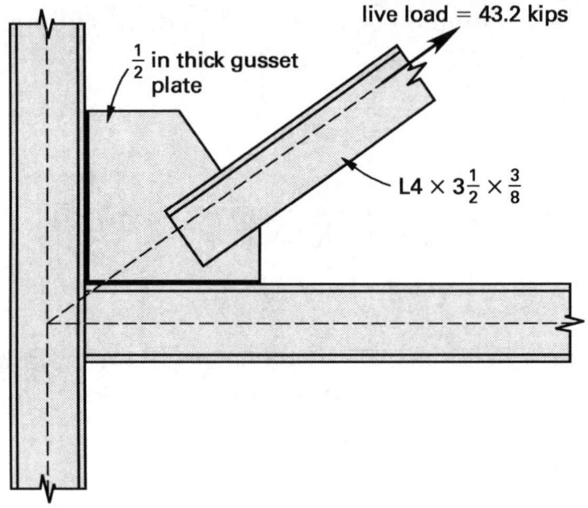

Section properties

$A = 2.687\ \text{in}^2$	$S_y = 1.16\ \text{in}^3$
$I_x = 4.15\ \text{in}^4$	$r_y = 1.05\ \text{in}$
$S_x = 1.48\ \text{in}^3$	$Z_y = 2.06\ \text{in}^3$
$r_x = 1.25\ \text{in}$	$I_z = 1.38\ \text{in}^4$
$Z_x = 2.66\ \text{in}^3$	$S_z = 0.938\ \text{in}^3$
$I_y = 2.96\ \text{in}^4$	$r_z = 0.719\ \text{in}$

Material properties for angle and gusset plate

ASTM A36

$F_y = 36$ ksi

$F_u = 58$ ksi

The required length of each weld is most nearly

(A) $5^1/_2$ in

(B) 6 in

(C) $6^1/_2$ in

(D) 7 in

Solution

Calculate the required design strengths.

LRFD	ASD
$P_u = 1.2D + 1.6L$ $= (1.2)(14.4\ \text{kips})$ $+ (1.6)(43.2\ \text{kips})$ $= 86.4\ \text{kips}$	$P_a = D + L$ $= 14.4\ \text{kips} + 43.2\ \text{kips}$ $= 57.6\ \text{kips}$

The minimum weld size, based on *AISC Specification* Table J2.4, is $^3/_{16}$ in. The maximum weld size at a rolled edge is the nominal edge thickness (in this case, $^3/_8$ in) less $^1/_{16}$ in. Therefore, the maximum weld size that can be used is $^5/_{16}$ in, and this will produce the shortest weld length.

Calculate the required length of weld. D is the number of sixteenths of an inch in the weld size. (See Sec. 10.7.)

LRFD	ASD
$\phi R_n = D\left(1.392\ \frac{\text{kips}}{\text{in}}\right)\ \text{[per 1/16 in]}$ $= (5)\left(1.392\ \frac{\text{kips}}{\text{in}}\right)$ $= 6.96\ \text{kips/in}$	$\frac{R_n}{\Omega} = D\left(0.928\ \frac{\text{kips}}{\text{in}}\right)\ \text{[per 1/16 in]}$ $= (5)\left(0.928\ \frac{\text{kips}}{\text{in}}\right)$ $= 4.64\ \text{kips/in}$
$L = \frac{P_u}{\phi R_n} = \frac{86.4\ \text{kips}}{6.96\ \frac{\text{kips}}{\text{in}}} = 12.4\ \text{in}$	$L = \frac{P_a}{\frac{R_n}{\Omega}} = \frac{57.6\ \text{kips}}{4.64\ \frac{\text{kips}}{\text{in}}} = 12.4\ \text{in}$

The length of the weld is distributed equally to the toe and heel of the angle.

$$L' = \frac{L}{2} = \frac{12.4\ \text{in}}{2} = 6.2\ \text{in}$$

Use 6.5 in. Check the minimum weld length. From *AISC Specification* Sec. J2.2b, this is four times the weld size.

$$L'_{\text{min}} = 4w = (4)\left(\frac{5}{16}\ \text{in}\right) = 1.25\ \text{in}$$

Two 6.5 in welds are OK.

The answer is (C).

PRACTICE PROBLEM 26

The tension brace shown in Prob. 25 is to be connected instead with Group A bolts with threads excluded from the shear plane. Bolt holes are of standard size and spacing is 3 in.

Section properties

$A = 2.67\ \text{in}^2$ $\quad S_y = 1.16\ \text{in}^3$

$I_x = 4.15\ \text{in}^4$ $\quad r_y = 1.05\ \text{in}$

$S_x = 1.48\ \text{in}^3$ $\quad Z_y = 2.06\ \text{in}^3$

$r_x = 1.25\ \text{in}$ $\quad I_z = 1.38\ \text{in}^4$

$Z_x = 2.66\ \text{in}^3$ $\quad S_z = 0.938\ \text{in}^3$

$I_y = 2.96\ \text{in}^4$ $\quad r_z = 0.719\ \text{in}$

Material properties for angle and gusset plate

ASTM A36 steel

$F_y = 36\ \text{ksi}$

$F_u = 58\ \text{ksi}$

How many $^3/_4$ in Group A bolts with threads excluded from the shear plane are required for the tension brace?

(A) two

(B) three

(C) four

(D) five

Solution

Calculate the required design strengths.

LRFD	ASD
$P_u = 1.2D + 1.6L$ $= (1.2)(14.4\ \text{kips}) + (1.6)(43.2\ \text{kips})$ $= 86.4\ \text{kips}$	$P_a = D + L$ $= 14.4\ \text{kips} + 43.2\ \text{kips}$ $= 57.6\ \text{kips}$

The applicable limit states are single shear on the bolt and the bolt bearing on the $^3/_8$ in thick angle leg. Determine the number of bolts required based on single shear on the bolt. From *AISC Manual* Table 7-1,

LRFD	ASD
$\phi_v r_n = 22.5\ \text{kips/bolt}$	$\dfrac{r_n}{\Omega_v} = 15.0\ \text{kips/bolt}$
$n = \dfrac{P_u}{\phi_v r_n} = \dfrac{86.4\ \text{kips}}{22.5\ \dfrac{\text{kips}}{\text{bolt}}}$ $= 3.84\ \text{bolts}\quad [4\ \text{bolts}]$	$n = \dfrac{P_a}{\dfrac{r_n}{\Omega_v}} = \dfrac{57.6\ \text{kips}}{15.0\ \dfrac{\text{kips}}{\text{bolt}}}$ $= 3.84\ \text{bolts}\quad [4\ \text{bolts}]$

Check bearing on the $^3/_8$ in thick angle leg. From *AISC Manual* Table 7-5,

LRFD	ASD
$\phi_v r_n = \left(78.3\ \frac{\text{kips}}{\text{in thickness}}\right)t$ $= \left(78.3\ \frac{\text{kips}}{\text{in thickness}}\right)\left(\frac{3}{8}\ \text{in}\right)$ $= 29.4\ \text{kips}\quad [\text{per bolt}]$	$\frac{r_n}{\Omega_v} = \left(52.2\ \frac{\text{kips}}{\text{in thickness}}\right)t$ $= \left(52.2\ \frac{\text{kips}}{\text{in thickness}}\right)\left(\frac{3}{8}\ \text{in}\right)$ $= 19.6\ \text{kips}\quad [\text{per bolt}]$
$n = \frac{P_u}{\phi_v r_n} = \frac{86.4\ \text{kips}}{29.3\ \frac{\text{kips}}{\text{bolt}}}$ $= 2.95\ \text{bolts}\quad [3\ \text{bolts}]$	$n = \frac{P_a}{\frac{r_n}{\Omega_v}} = \frac{57.6\ \text{kips}}{19.6\ \frac{\text{kips}}{\text{bolt}}}$ $= 2.94\ \text{bolts}\quad [3\ \text{bolts}]$

The limit state of single shear controls, and four bolts are required.

The answer is (C).

PRACTICE PROBLEM 27

The compression flange of the W16 × 26 steel beam shown is braced at 3.5 ft centers. The tension flange contains holes for $^3/_4$ in diameter bolts in pairs at 2 ft centers to support a movable partition.

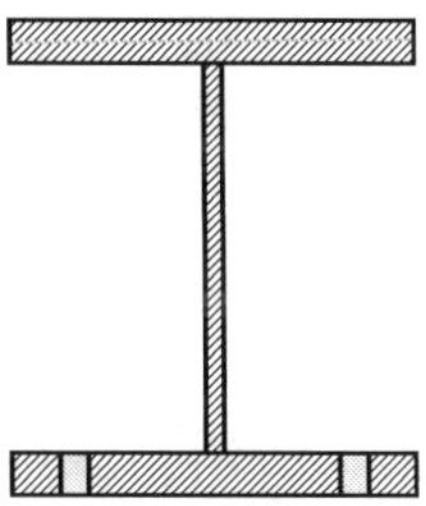

section view

Section properties

$A = 7.68\ \text{in}^2$

$d = 15.7\ \text{in}$

$t_w = 0.25\ \text{in}$

$b_f = 5.50\ \text{in}$

$t_f = 0.345\ \text{in}$

$b_f/2t_f = 7.97$

$h/t_w = 56.8\ \text{in}$

$I_x = 301\ \text{in}^4$

$S_x = 38.4\ \text{in}^3$

$r_x = 6.26\ \text{in}$

$Z_x = 44.2\ \text{in}^3$

$I_y = 9.59\ \text{in}^4$

$S_y = 3.49\ \text{in}^3$

$r_y = 1.12\ \text{in}$

$Z_y = 5.48\ \text{in}^3$

Material properties

ASTM A992

$F_y = 50\ \text{ksi}$

$F_u = 65\ \text{ksi}$

Which of the following is most nearly the available flexural strength of the steel beam? (LRFD options are in parentheses.)

(A) 57 ft-kips (85 ft-kips)

(B) 67 ft-kips (100 ft-kips)

(C) 85 ft-kips (130 ft-kips)

(D) 110 ft-kips (170 ft-kips)

Solution

The actual unbraced length is 3.5 ft, which is less than L_p = 3.96 ft, so the compression flange is capable of reaching its full plastic moment. Determine whether the available flexural strength has to be reduced as a result of the holes in tension flange.

$$\begin{aligned} A_{fg} &= b_f t_f = (5.50 \text{ in})(0.345 \text{ in}) \\ &= 1.90 \text{ in}^2 \\ d_{\text{hole}} &= 0.75 \text{ in} + 0.125 \text{ in} \\ &= 0.875 \text{ in} \\ A_h &= n_{\text{holes}} t_f d_{\text{hole}} = (2)(0.345 \text{ in})(0.875 \text{ in}) \\ &= 0.604 \text{ in}^2 \\ A_{fn} &= A_{fg} - A_h = 1.90 \text{ in}^2 - 0.604 \text{ in}^2 \\ &= 1.30 \text{ in}^2 \end{aligned}$$

Use Eq. 5.39 to check whether the limit state of tensile rupture applies.

$$\begin{aligned} F_u A_{fn} &= \left(65 \ \frac{\text{kips}}{\text{in}^2}\right)(1.30 \text{ in}^2) = 84.5 \text{ kips} \\ Y_t F_y A_g &= (1)\left(50 \ \frac{\text{kips}}{\text{in}^2}\right)(1.90 \text{ in}^2) = 95.0 \text{ kips} \\ F_u A_{fn} &< Y_t F_y A_g \quad [\text{tensile rupture does apply}] \end{aligned}$$

From Eq. 5.43, the nominal flexural strength is limited by

$$\begin{aligned} M_n &\le \left(\frac{F_u A_{fn}}{A_{fg}}\right) S_x \\ &\le \left(\frac{\left(65 \ \frac{\text{kips}}{\text{in}^2}\right)(1.30 \text{ in}^2)}{(1.90 \text{ in}^2)\left(12 \ \frac{\text{in}}{\text{ft}}\right)}\right)(38.4 \text{ in}^3) \\ &\le 142 \text{ ft-kips} \end{aligned}$$

Determine the design strength (LRFD) or allowable strength (ASD).

LRFD	ASD
$M_u \le \phi_b M_n$ $\le (0.90)(142 \text{ ft-kips})$ $\le 128 \text{ ft-kips} \quad (130 \text{ ft-kips})$	$M_a \le \frac{M_n}{\Omega_b} = \frac{142 \text{ ft-kips}}{1.67}$ $\le 85.0 \text{ ft-kips} \quad (85 \text{ ft-kips})$

The answer is (C).

PRACTICE PROBLEM 28

A fully composite steel beam is shown. The beam's plastic neutral axis is at the bottom of the top flange. A 4 in thick concrete slab is placed directly on top of the beam (no formed steel deck). The shear studs are $^3/_4$ in in diameter by 3.5 in long. The concrete has a design compressive strength of 4 ksi and a unit weight of 145 lbf/ft^3.

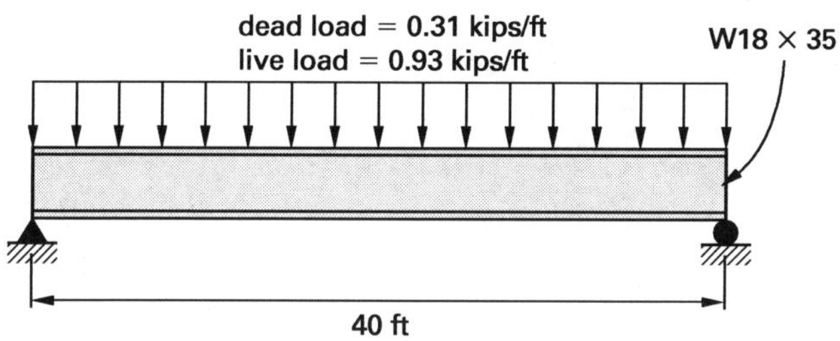

How many shear studs are required?

(A) 20 studs

(B) 26 studs

(C) 30 studs

(D) 34 studs

Solution

From *AISC Manual* Table 3-19, with the plastic neutral axis (PNA) located at the bottom of the flange, ΣQ_n = 260 kips. From *AISC Manual* Table 3-21, for $^3/_4$ in diameter studs with no deck, $Q_n = 21.5$ kips/stud. The number of studs required on each side of the point of maximum moment is

$$n = \frac{\sum Q_n}{Q_n} = \frac{260 \text{ kips}}{21.5 \ \frac{\text{kips}}{\text{stud}}} = 12 \text{ studs}$$

For two sides, a total of 24 studs are required.

The answer is (B).

PRACTICE PROBLEM 29

The column shown is part of a braced frame and is pinned at both ends. The moments about the weak axis are $M_{y,D} = 4$ ft-kips and $M_{y,L} = 12$ ft-kips.

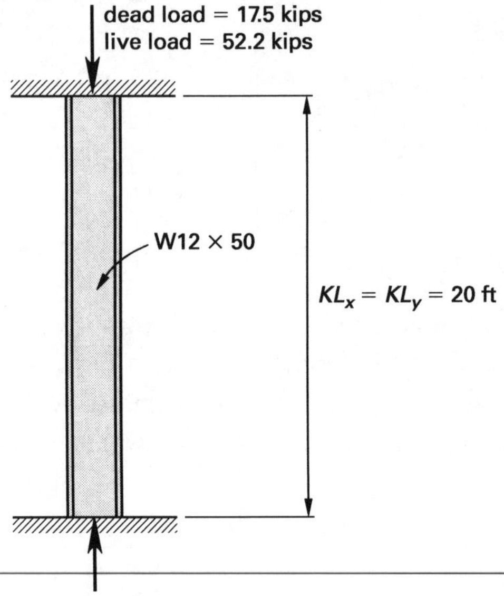

Section properties		Material properties
$A = 14.6$ in^2	$I_x = 391$ in^4	ASTM A992
$d = 12.2$ in	$S_x = 64.2$ in^3	$F_y = 50$ ksi
$t_w = 0.370$	$r_x = 5.18$ in	$F_u = 65$ ksi
$b_f = 8.08$ in	$Z_x = 71.9$ in^3	
$t_f = 0.640$ in	$I_y = 56.3$ in^4	
$b_f/2t_f = 6.31$	$S_y = 13.9$ in^3	
$h/t_w = 26.8$	$r_y = 1.96$ in	
	$Z_y = 21.3$ in^3	

What is most nearly the available flexural strength about the strong axis? (LRFD options are in parentheses.)

(A) 24 ft-kips (36 ft-kips)

(B) 30 ft-kips (44 ft-kips)

(C) 37 ft-kips (55 ft-kips)

(D) 46 ft-kips (68 ft-kips)

Solution

Determine the required strengths.

LRFD	ASD
$P_u = 1.2D + 1.6L$ $= (1.2)(17.50 \text{ kips}) + (1.6)(52.20 \text{ kips})$ $= 104.52 \text{ kips}$	$P_a = D + L$ $= 17.50 \text{ kips} + 52.20 \text{ kips}$ $= 69.70 \text{ kips}$
$M_{uy} = 1.2M_{yD} + 1.6M_{yL}$ $= (1.2)(4 \text{ ft-kips}) + (1.6)(12 \text{ ft-kips})$ $= 24.0 \text{ ft-kips}$	$M_{ay} = M_{yD} + M_{yL}$ $= 4 \text{ ft-kips} + 12 \text{ ft-kips}$ $= 16.0 \text{ ft-kips}$

Determine the ratio of required axial load to available axial load. From *AISC Manual* Table 4-1,

LRFD	ASD
$\phi_c P_n = 220 \text{ kips}$ $\dfrac{P_u}{\phi_c P_n} = \dfrac{104.52 \text{ kips}}{220 \text{ kips}}$ $= 0.48 \quad [> 0.20]$	$\dfrac{P_n}{\Omega_c} = 146 \text{ kips}$ $\dfrac{P_a}{\dfrac{P_n}{\Omega_c}} = \dfrac{69.70 \text{ kips}}{146 \text{ kips}}$ $= 0.48 \quad [> 0.20]$

The ratio of required axial load to available axial load exceeds 0.20, so use Eq. 8.5 to determine the effects of the combined loads.

LRFD	ASD
$pP_u + b_x M_{ux} + b_y M_{uy} \le 1.0$	$pP_u + b_x M_{ax} + b_y M_{ay} \le 1.0$

Determine the combined stress coefficients from *AISC Manual* Table 6-1.

LRFD	ASD
$p \times 10^3 = 4.55 \text{ kips}^{-1}$ $b_x \times 10^3 = 4.64 \ (\text{ft-kips})^{-1}$ $b_y \times 10^3 = 11.1 \ (\text{ft-kips})^{-1}$	$p \times 10^3 = 6.85 \text{ kips}^{-1}$ $b_x \times 10^3 = 6.98 \ (\text{ft-kips})^{-1}$ $b_y \times 10^3 = 16.7 \ (\text{ft-kips})^{-1}$

Solve for the flexural strength around the strong axis.

LRFD	ASD
$M_{ux} \leq \dfrac{1.0 - pP_u - b_y M_{uy}}{b_x}$	$M_{ax} \leq \dfrac{1.0 - pP_a - b_y M_{ay}}{b_x}$
$\leq \dfrac{1.0 - \left(\dfrac{4.55}{10^3 \text{ kips}}\right) \times (104.52 \text{ kips}) - \left(\dfrac{11.1}{10^3 \text{ ft-kips}}\right) \times (24.0 \text{ ft-kips})}{\dfrac{4.64}{10^3 \text{ ft-kips}}}$	$\leq \dfrac{1.0 - \left(\dfrac{6.83}{10^3 \text{ kips}}\right) \times (69.70 \text{ kips}) - \left(\dfrac{16.7}{10^3 \text{ ft-kips}}\right) \times (16.0 \text{ ft-kips})}{\dfrac{6.98}{10^3 \text{ ft-kips}}}$
$= 55.6 \text{ ft-kips} \quad (55 \text{ ft-kips})$	$= 36.7 \text{ ft-kips} \quad (37 \text{ ft-kips})$

The answer is (C).

PRACTICE PROBLEM 30

The I-shaped section shown is fabricated from plate steel and is welded together with fillet welds. The maximum shear due to a uniformly distributed load is $V_D = 14$ kips and $V_L = 42$ kips.

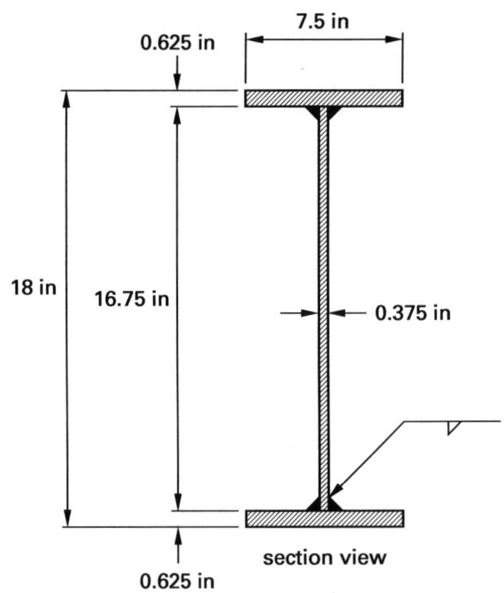

Material properties

ASTM A572, grade B

$F_y = 50$ ksi

$F_u = 65$ ksi

Determine the required size of fillet weld if E70 electrodes will be used.

(A) $^1/_{16}$ in

(B) $^1/_8$ in

(C) $^3/_{16}$ in

(D) $^1/_4$ in

Solution

The distance in the *y*-direction from the centroid of the section to the centroid of one flange is

$$\begin{aligned}\bar{y} &= \frac{h_w}{2} + \frac{t_f}{2} \\ &= \frac{16.75 \text{ in}}{2} + \frac{0.625 \text{ in}}{2} \\ &= 8.69 \text{ in}\end{aligned}$$

Determine the required section properties.

$$A_f = b_f t_f = (7.5 \text{ in})(0.625 \text{ in}) = 4.69 \text{ in}^2$$

$$A_w = d_w t_w = (16.75 \text{ in})(0.375 \text{ in}) = 6.28 \text{ in}^2$$

$$\begin{aligned}I_f &= I_c + A\bar{y}^2 = \frac{b_f t_f^3}{12} + A\bar{y}^2 \\ &= \frac{(7.5 \text{ in})(0.625 \text{ in})^3}{12} + (4.69 \text{ in}^2)(8.69 \text{ in})^2 \\ &= 354 \text{ in}^4\end{aligned}$$

$$\begin{aligned}I_w &= \frac{t_w d_w^3}{12} = \frac{(0.375 \text{ in})(16.75 \text{ in})^3}{12} \\ &= 147 \text{ in}^4\end{aligned}$$

$$\begin{aligned}I &= I_w + 2I_f = 147 \text{ in}^4 + (2)(354 \text{ in}^4) \\ &= 855 \text{ in}^4\end{aligned}$$

Compute the static moment of the flange about the neutral axis of the member.

$$Q_f = A_f\bar{y} = (4.69 \text{ in}^2)(8.69 \text{ in}) = 40.8 \text{ in}^3$$

Determine the required shear resistance.

LRFD	ASD
$V_u = 1.2V_D + 1.6V_L$ $= (1.2)(14 \text{ kips}) + (1.6)(42 \text{ kips})$ $= 84 \text{ kips}$	$V_a = V_D + V_L$ $= 14 \text{ kips} + 42 \text{ kips}$ $= 56 \text{ kips}$

Determine the horizontal shear stress at the flange-web interface.

LRFD	ASD
$\tau_h = \dfrac{V_u Q_f}{I t_w} = \dfrac{(84 \text{ kips})(40.8 \text{ in}^3)}{(855 \text{ in}^4)(0.375 \text{ in})}$ $= 10.7 \text{ ksi}$	$\tau_h = \dfrac{V_a Q_f}{I t_w} = \dfrac{(56 \text{ kips})(40.8 \text{ in}^3)}{(855 \text{ in}^4)(0.375 \text{ in})}$ $= 7.13 \text{ ksi}$

Determine the horizontal force per inch.

LRFD	ASD
$\tau_{h,\text{per inch}} = \tau_h t_w$ $= \left(10.7 \ \dfrac{\text{kips}}{\text{in}^2}\right)(0.375 \text{ in})$ $= 4.01 \text{ kips/in}$	$\tau_{h,\text{per inch}} = \tau_h t_w$ $= \left(7.13 \ \dfrac{\text{kips}}{\text{in}^2}\right)(0.375 \text{ in})$ $= 2.67 \text{ kips/in}$

Find the required length of weld to resist the horizontal force. *D* is the number of sixteenths of an inch in the weld size. (See Sec. 10.7.) There are two welds, one on each side of the web, so *D* is multiplied by two.

LRFD	ASD
$r_n \le \tau_{h,\text{per inch}}$ $r_n = 2D\left(1.392 \ \dfrac{\text{kips}}{\text{in}}\right)$ $D = \dfrac{r_n}{(2)\left(1.392 \ \dfrac{\text{kips}}{\text{in}}\right)} \le \dfrac{4.01 \ \dfrac{\text{kips}}{\text{in}}}{(2)\left(1.392 \ \dfrac{\text{kips}}{\text{in}}\right)}$ $\le 1.44 \quad [\text{use at least 1/8 in weld}]$	$r_n \le \tau_{h,\text{per inch}}$ $r_n = 2D\left(0.928 \ \dfrac{\text{kips}}{\text{in}}\right)$ $D = \dfrac{r_n}{(2)\left(0.928 \ \dfrac{\text{kips}}{\text{in}}\right)} \le \dfrac{2.67 \ \dfrac{\text{kips}}{\text{in}}}{(2)\left(0.928 \ \dfrac{\text{kips}}{\text{in}}\right)}$ $\le 1.44 \quad [\text{use at least 1/8 in weld}]$

Check *AISC Specification* Eq. J2.4 for the minimum weld size. The thinner part being joined is 0.375 in, which is between ¼ in and ½ in, so the minimum weld size is $^3/_{16}$ in.

Use *AISC Specification* Eq. J4-3 to check the minimum required web thickness for shear yielding for two $^3/_{16}$ in fillet welds.

LRFD	ASD
$\phi R_n = \phi(0.60F_yA_g)$ $= (1.0)(0.60)\left(50\ \frac{\text{kips}}{\text{in}^2}\right)$ $\times(0.375\ \text{in}^2)$ $= 11.25$ ksi $[>10.7$ ksi required, so OK$]$	$\frac{R_n}{\Omega} = \frac{0.60F_yA_g}{\Omega}$ $= \frac{(0.60)\left(50\ \frac{\text{kips}}{\text{in}^2}\right)(0.375\ \text{in}^2)}{1.5}$ $= 7.50$ ksi $[>7.13$ ksi required, so OK$]$

Use two $^3/_{16}$ in fillet welds.

The answer is (C).

PRACTICE PROBLEM 31

The angle brace shown is 8 ft long and is connected to gusset plates at each end through the same leg with a minimum of two bolts.

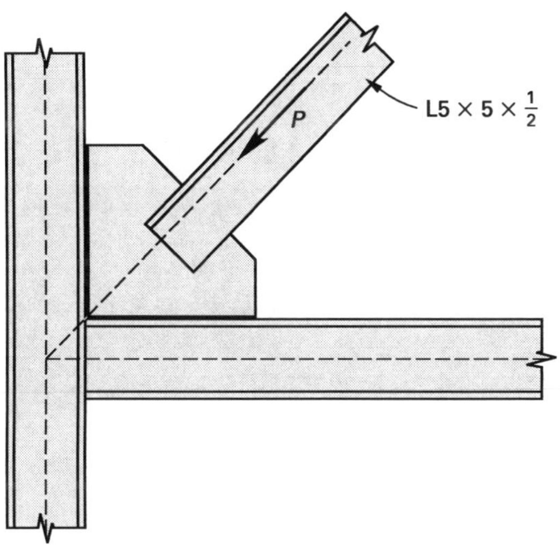

Section properties		Material properties
$A_g = 4.75\ \text{in}^2$	$r_x = r_y = 1.53$ in	ASTM A36
$I_x = I_y = 11.3\ \text{in}^4$	$Z_x = Z_y = 5.66\ \text{in}^3$	$F_y = 36$ ksi
$S_x = S_y = 3.15\ \text{in}^3$		$F_u = 58$ ksi

The gusset plate and bolts are not the governing limit states. Which of the following is most nearly the design strength (LRFD) or the allowable strength (ASD) of the angle? (LRFD options are in parentheses.)

(A) 49 kips (73 kips)

(B) 54 kips (81 kips)

(C) 62 kips (92 kips)

(D) 68 kips (102 kips)

Solution

Determine which equation to use for calculating the effective slenderness ratio.

$$\begin{aligned}\frac{L}{r_x} &= \frac{(8\ \text{ft})\left(12\ \frac{\text{in}}{\text{ft}}\right)}{1.53\ \text{in}}\\ &= 62.7\quad [< 80,\ \text{so use Eq. 7.25}]\end{aligned}$$

From Eq. 7.25, the effective slenderness ratio is

$$\begin{aligned}\frac{KL}{r_x} &= 72 + 0.75\left(\frac{L}{r_x}\right)\\ &= 72 + (0.75)(62.7)\\ &= 119\end{aligned}$$

Determine whether to use Eq. 7.6 or Eq. 7.7 for computing nominal strength.

$$\begin{aligned}4.71\sqrt{\frac{E}{F_y}} &= 4.71\sqrt{\frac{29{,}000\ \frac{\text{kips}}{\text{in}^2}}{36\ \frac{\text{kips}}{\text{in}^2}}}\\ &= 133\quad [> KL/r_x,\ \text{so use Eq. 7.6}]\end{aligned}$$

Use Eq. 7.8 to find the elastic critical buckling strength to use in Eq. 7.6.

$$\begin{aligned}F_e &= \frac{\pi^2 E}{\left(\frac{KL}{r_x}\right)^2}\\ &= \frac{\pi^2\left(29{,}000\ \frac{\text{kips}}{\text{in}^2}\right)}{(119)^2}\\ &= 20.21\ \text{ksi}\end{aligned}$$

From Eq. 7.6, the nominal strength is

$$\begin{aligned} F_{cr} &= 0.658^{F_y/F_e} F_y \\ &= (0.658)^{36 \frac{\text{kips}}{\text{in}^2} / 20.21 \frac{\text{kips}}{\text{in}^2}} \left(36 \ \frac{\text{kips}}{\text{in}^2}\right) \\ &= 17.1 \text{ ksi} \end{aligned}$$

Determine the design strength (LRFD) or allowable strength (ASD) of the angle.

LRFD	ASD
$P_u \le \phi_c P_n = \phi_c F_{cr} A_g$ $\le (0.90)\left(17.1 \ \frac{\text{kips}}{\text{in}^2}\right)(4.75 \text{ in}^2)$ $\le 73.1 \text{ kips} \quad (73 \text{ kips})$	$P_a \le \frac{P_n}{\Omega_c} = \frac{F_{cr} A_g}{\Omega}$ $\le \frac{\left(17.1 \ \frac{\text{kips}}{\text{in}}\right)(4.75 \text{ in}^2)}{1.67}$ $\le 48.6 \text{ kips} \quad (49 \text{ kips})$

The answer is (A).

PRACTICE PROBLEM 32

A HSS18 × 6 × $^5/_{16}$ beam spans 30 ft with a uniform dead load of 0.20 kips/ft and a live load of 0.60 kips/ft.

Sectional properties

$t = 0.291$ in

$A = 13.4 \text{ in}^2$

$b/t = 17.6$

$h/t = 58.9$

$I_x = 513 \text{ in}^4$

$S_x = 57 \text{ in}^3$

$r_x = 6.18$ in

$Z_x = 73.1 \text{ in}^3$

$I_y = 91.3 \text{ in}^4$

$S_y = 30.4 \text{ in}^3$

$r_y = 2.61$ in

$Z_y = 33.5 \text{ in}^3$

$J = 257 \text{ in}^4$

$C = 58.7 \text{ in}^3$

Material properties

ASTM A500, grade B

$F_y = 46$ ksi

$F_u = 65$ ksi

Which of the following is most nearly the torsional load that can be applied to the beam without exceeding its available strength? (LRFD options are in parentheses.)

(A) 38 ft-kips (57 ft-kips)

(B) 44 ft-kips (66 ft-kips)

(C) 49 ft-kips (74 ft-kips)

(D) 54 ft-kips (81 ft-kips)

Solution

Determine the required resistance for uniform loads.

LRFD	ASD
$w_u = 1.2D + 1.6L$ $= (1.2)\left(0.20\ \frac{\text{kips}}{\text{ft}}\right) + (1.6)\left(0.60\ \frac{\text{kips}}{\text{ft}}\right)$ $= 1.20\ \text{kips/ft}$	$w_a = D + L$ $= 0.20\ \frac{\text{kips}}{\text{ft}} + 0.60\ \frac{\text{kips}}{\text{ft}}$ $= 0.80\ \text{kips/ft}$
$V_u = \frac{w_u L}{2} = \frac{\left(1.2\ \frac{\text{kips}}{\text{ft}}\right)(30\ \text{ft})}{2}$ $= 18.0\ \text{kips}$	$V_a = \frac{w_a L}{2} = \frac{\left(0.80\ \frac{\text{kips}}{\text{ft}}\right)(30\ \text{ft})}{2}$ $= 12.0\ \text{kips}$
$M_u = \frac{w_u L^2}{8} = \frac{\left(1.2\ \frac{\text{kips}}{\text{ft}}\right)(30\ \text{ft})^2}{8}$ $= 135\ \text{ft-kips}$	$M_a = \frac{w_a L^2}{8} = \frac{\left(0.80\ \frac{\text{kips}}{\text{ft}}\right)(30\ \text{ft})^2}{8}$ $= 90\ \text{ft-kips}$

The effective height is

$$\begin{aligned} h_{\text{eff}} &= d - 3t \\ &= 18\ \text{in} - (3)(0.291\ \text{in}) \\ &= 17.1\ \text{in} \end{aligned}$$

Use Eq. 5.47 to determine available shear strength.

$$V_n = 0.6F_y A_w C_v$$

Determine the applicable formula to use for the web shear coefficient, C_v.

$$\begin{aligned} 1.10\sqrt{\frac{k_v E}{F_y}} &= 1.10\sqrt{\frac{(5)\left(29{,}000\ \frac{\text{kips}}{\text{in}^2}\right)}{50\ \frac{\text{kips}}{\text{in}^2}}} \\ &= 59.24\quad [> h/t_w = 58.9,\ \text{so use Eq. 5.48}] \end{aligned}$$

From Eq. 5.48,

$$C_v = 1.0$$

For two webs,

$$\begin{aligned} A_w &= 2ht = (2)(17.13 \text{ in})(0.291 \text{ in}) \\ &= 9.97 \text{ in}^2 \end{aligned}$$

From Eq. 5.47, the available shear strength is

$$\begin{aligned} V_n &= 0.6F_yA_wC_v \\ &= (0.60)\left(46\ \frac{\text{kips}}{\text{in}^2}\right)(9.97 \text{ in}^2)(1.0) \\ &= 275 \text{ kips} \end{aligned}$$

Use *AISC Manual* Table 3-12 to find the available flexural strengths.

LRFD	ASD
$\phi_b M_{nx} = 252$ ft-kips	$\dfrac{M_{nx}}{\Omega_b} = 168$ ft-kips

Use Eq. 8.8 to determine the nominal torsional strength.

$$T_n = F_{\text{cr}}C$$

Find the applicable formula for the critical stress.

$$\begin{aligned} 2.45\sqrt{\frac{E}{F_y}} &= 2.45\sqrt{\frac{29{,}000\ \frac{\text{kips}}{\text{in}^2}}{50\ \frac{\text{kips}}{\text{in}^2}}} \\ &= 59.00 \quad [> h/t = 58.9, \text{ so use Eq. 8.12}] \end{aligned}$$

From Eq. 8.12, the critical stress is

$$F_{\text{cr}} = 0.6F_y = (0.60)\left(46\ \frac{\text{kips}}{\text{in}^2}\right) = 27.60 \text{ ksi}$$

From Eq. 8.8, the nominal torsional strength is

$$\begin{aligned} T_n &= F_{\text{cr}}C \\ &= \frac{\left(27.60\frac{\text{kips}}{\text{in}^2}\right)(58.7 \text{ in}^3)}{12\ \frac{\text{in}}{\text{ft}}} \\ &= 135 \text{ ft-kips} \end{aligned}$$

Calculate the available torsional strength.

LRFD	ASD
$\phi_T T_n = (0.90)(135\text{ ft-kips}) = 121.5\text{ ft-kips}$	$\frac{T_n}{\Omega_T} = \frac{135\text{ ft-kips}}{1.67} = 80.8\text{ ft-kips}$

Use *AISC Specification* Eq. H3-6 to determine the available torsional strength for combined loading effects.

$$\left(\frac{P_r}{P_c}+\frac{M_r}{M_c}\right)+\left(\frac{V_r}{V_c}+\frac{T_r}{T_c}\right)^2 \le 1.0$$

$$T_r \le T_c\left(\sqrt{1.0-\frac{P_r}{P_c}-\frac{M_r}{M_c}-\frac{V_r}{V_c}}\right)$$

LRFD	ASD
$T_r \le (\phi_T T_n)\left(\sqrt{1.0-\frac{P_r}{P_c}-\frac{M_u}{\phi M_{px}}-\frac{V_u}{\phi V_n}}\right)$ $\le (121.5\text{ ft-kips}) \times \left(\sqrt{1.0-\frac{0\text{ kips}}{P_c}-\frac{135\text{ ft-kips}}{252\text{ ft-kips}}-\frac{18\text{ kips}}{(0.90)(275\text{ kips})}}\right)$ $= 74.03\text{ ft-kips}\quad(74\text{ ft-kips})$	$T_r \le \left(\frac{T_n}{\Omega_T}\right)\left(\sqrt{1.0-\frac{P_r}{P_c}-\frac{M_a}{\left(\frac{M_{px}}{\Omega}\right)}-\frac{V_a}{\frac{V_n}{\Omega_v}}}\right)$ $\le (80.8\text{ ft-kips}) \times \left(\sqrt{1.0-\frac{0\text{ kips}}{P_c}-\frac{90\text{ ft-kips}}{168\text{ ft-kips}}-\frac{12\text{ kips}}{\frac{275\text{ kips}}{1.67}}}\right)$ $= 49.17\text{ ft-kips}\quad(49\text{ ft-kips})$

This problem illustrates how effective a closed tubular section is at resisting torsional loads.

The answer is (C).

PRACTICE PROBLEM 33

Twelve $^3/_4$ in Group A bolts with the threads included in the shear plane are used in the beam-to-column moment connection shown. Assume that the plate-to-column welds are designed to develop the capacity of the bolts and that the limit states for column flange and web are satisfactory. The standard gage for a hole in the flange of a W18 × 35 is 3.5 in.

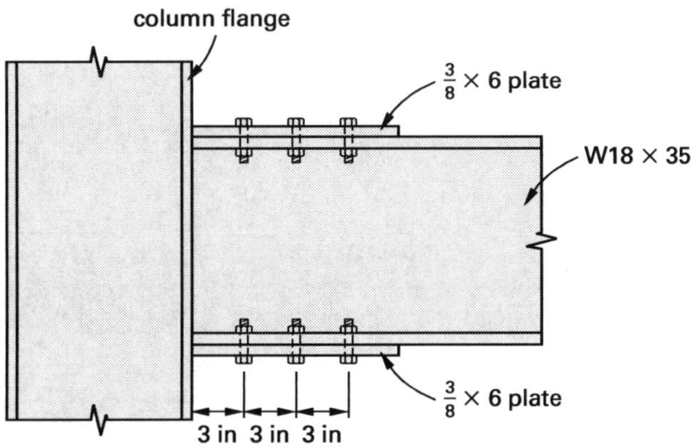

(not to scale)

Material properties

W18 × 35	plates
ASTM A992	ASTM A36
$F_y = 50$ ksi	$F_y = 36$ ksi
$F_u = 65$ ksi	$F_u = 58$ ksi

Which of the following is most nearly the moment capacity of the connection? (LRFD options are in parentheses.)

(A) 68 ft-kips (100 ft-kips)

(B) 83 ft-kips (120 ft-kips)

(C) 95 ft-kips (140 ft-kips)

(D) 110 ft-kips (160 ft-kips)

Solution

At each flange, six bolts in single shear bear on the flange of the beam and on the plate welded to the column flange. The flange is thicker than the plate, so the plate bearing is more critical than the flange bearing.

Determine the available strength of the bolts in single shear. From *AISC Manual* Table 7-1,

LRFD	ASD
$\phi_v r_n = 17.9 \text{ kips/bolt}$ $\phi_v R_n = n(\phi_v r_n)$ $= (6)\left(17.9 \ \dfrac{\text{kips}}{\text{bolt}}\right)$ $= 107.40 \text{ kips}$	$\dfrac{r_n}{\Omega_v} = 11.9 \text{ kips/bolt}$ $\dfrac{R_n}{\Omega_v} = \dfrac{nr_n}{\Omega_v} = n\left(\dfrac{r_n}{\Omega_v}\right)$ $= (6)\left(11.9 \ \dfrac{\text{kips}}{\text{bolt}}\right)$ $= 71.4 \text{ kips}$

Determine the available strength of the bolts bearing on moment plates. From *AISC Manual* Table 7-4,

LRFD	ASD
$\phi_v r_n = 78.3 \text{ kips/in of thickness}$ $\phi_v R_n = n(\phi_v r_n)t$ $= (6)\left(78.3 \ \dfrac{\text{kips}}{\text{in of thickness}}\right) \times (0.375 \text{ in})$ $= 176.18 \text{ kips}$	$\dfrac{r_n}{\Omega_v} = 52.2 \text{ kips/in of thickness}$ $\dfrac{R_n}{\Omega_v} = n\left(\dfrac{r_n}{\Omega_v}\right)t$ $= (6)\left(52.2 \ \dfrac{\text{kips}}{\text{in of thickness}}\right) \times (0.375 \text{ in})$ $= 117.45 \text{ kips}$

Use Eq. 4.2 to determine the available gross section yielding strength of the flange plates.

$$A_g = bt = (6 \text{ in})(0.375 \text{ in}) = 2.25 \text{ in}^2$$

$$P_n = F_y A_g = \left(36 \ \frac{\text{kips}}{\text{in}^2}\right)(2.25 \text{ in}^2) = 81.0 \text{ kips}$$

LRFD	ASD
$\phi_t P_n = (0.90)(81.0 \text{ kips})$ $= 72.90 \text{ kips}$	$\dfrac{P_n}{\Omega_t} = \dfrac{81.0 \text{ kips}}{1.67}$ $= 48.5 \text{ kips}$

Determine the available net section rupture strength of the flange plates. The area of the holes is

$$\begin{aligned} A_h &= n_{\text{holes}} t d_{\text{hole}} \\ &= n_{\text{holes}} t \left(d_{\text{bolt}} + 0.125 \text{ in}\right) \\ &= (2)(0.375 \text{ in})(0.75 \text{ in} + 0.125 \text{ in}) \\ &= 0.66 \text{ in}^2 \end{aligned}$$

The effective net area is

$$A_e = A_n = A_g - A_h = 2.25 \text{ in}^2 - 0.66 \text{ in}^2 = 1.59 \text{ in}^2$$

From Eq. 4.3,

LRFD	ASD
$\phi_t P_n = \phi_t F_u A_e$ $= (0.75)\left(58 \dfrac{\text{kips}}{\text{in}^2}\right)(1.59 \text{ in}^2)$ $= 69.2 \text{ kips}$	$\dfrac{P_n}{\Omega_t} = \dfrac{F_u A_e}{\Omega_t}$ $= \dfrac{\left(58 \dfrac{\text{kips}}{\text{in}^2}\right)(1.59 \text{ in}^2)}{2.00}$ $= 46.1 \text{ kips}$

Use Eq. 9.1 to determine the available block shear strength of the plates.

$$R_n = 0.60 F_u A_{nv} + U_{\text{bs}} F_u A_{nt} \le 0.60 F_y A_{gv} + U_{\text{bs}} F_u A_{nt}$$

Tension stress is uniform, so

$$U_{\text{bs}} = 1.0$$

Calculate the gross and net shear areas and the net tension area.

$$\begin{aligned} A_{gv} &= 2Lt = (2)(9 \text{ in})(0.375 \text{ in}) = 6.75 \text{ in}^2 \\ A_{nv} &= \left(2L - n d_{\text{hole}}\right) t \\ &= \left((2)(9 \text{ in}) - (2)(2.5)(0.875 \text{ in})\right)(0.375 \text{ in}) \\ &= 5.11 \text{ in}^2 \\ A_{nt} &= \left(b - n d_{\text{hole}}\right) t \\ &= \left(3.5 \text{ in} - (2)(0.5)(0.875 \text{ in})\right)(0.375 \text{ in}) \\ &= 0.98 \text{ in}^2 \end{aligned}$$

From Eq. 9.1,

$$R_n \le \begin{cases} 0.60F_uA_{nv} + U_{\text{bs}}F_uA_{nt} \\ \quad = (0.60)\left(58\ \dfrac{\text{kips}}{\text{in}^2}\right)(5.11\ \text{in}^2) + (1.0)\left(58\ \dfrac{\text{kips}}{\text{in}^2}\right)(0.98\ \text{in}^2) \\ \quad = 235\ \text{kips} \\ 0.60F_yA_{gv} + U_{\text{bs}}F_uA_{nt} \\ \quad = (0.60)\left(36\ \dfrac{\text{kips}}{\text{in}^2}\right)(6.75\ \text{in}^2) + (1.0)\left(58\ \dfrac{\text{kips}}{\text{in}^2}\right)(0.98\ \text{in}^2) \\ \quad = 203\ \text{kips} \quad [\text{controls}] \end{cases}$$

Calculate the available block shear strength.

LRFD	ASD
$\phi R_n = (0.90)(203\ \text{kips})$ $= 182.7\ \text{kips}$	$\dfrac{R_n}{\Omega_t} = \dfrac{203\ \text{kips}}{2.00}$ $= 101.5\ \text{kips}$

The controlling limit state is net section rupture with ϕ_tP_n = 69.2 kips and P_n/Ω_t = 46.1 kips. Determine the available moment capacity.

LRFD	ASD
$M_u = (\phi_tP_n)d$ $= \dfrac{(69.2\ \text{kips})(17.7\ \text{in})}{12\ \dfrac{\text{in}}{\text{ft}}}$ $= 102\ \text{ft-kips}\quad (100\ \text{ft-kips})$	$M_a = \left(\dfrac{P_n}{\Omega_t}\right)d$ $= \dfrac{(46.1\ \text{kips})(17.7\ \text{in})}{12\ \dfrac{\text{in}}{\text{ft}}}$ $= 68.0\ \text{ft-kips}\quad (68\ \text{ft-kips})$

The answer is (A).

PRACTICE PROBLEM 34

The top flange of the W-beam shown is laterally braced the entire length. The bottom flange is braced at the supports and at the end of the cantilever.

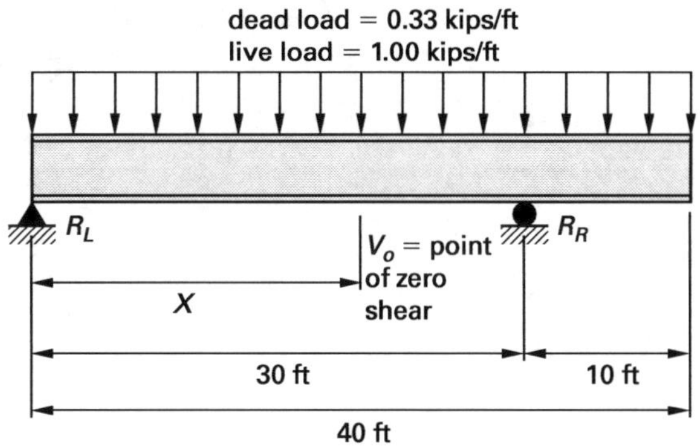

The lightest W16 section of ASTM A992 steel that meets the strength requirements is

(A) W16 × 26

(B) W16 × 31

(C) W16 × 36

(D) W16 × 40

Solution

The uniform load is

LRFD	ASD
$w_u = 1.2w_D + 1.6w_L$ $= (1.2)\left(0.33\ \frac{\text{kip}}{\text{ft}}\right) + (1.6)\left(1.00\ \frac{\text{kip}}{\text{ft}}\right)$ $= 2.0\ \text{kips/ft}$	$w_a = w_D + w_L$ $= 0.33\ \frac{\text{kip}}{\text{ft}} + 1.00\ \frac{\text{kip}}{\text{ft}}$ $= 1.33\ \text{kips/ft}$

The total load is

LRFD	ASD
$W_u = w_u L$ $= \left(2.0\ \frac{\text{kips}}{\text{ft}}\right)(40\ \text{ft})$ $= 80\ \text{kips}$	$W_a = w_a L$ $= \left(1.33\ \frac{\text{kips}}{\text{ft}}\right)(40\ \text{ft})$ $= 53.2\ \text{kips}$

The required flexural strength is

LRFD	ASD
$M_{\text{cantilever}} = \dfrac{w_u L^2_{\text{cantilever}}}{2}$ $= \dfrac{\left(2.0\ \dfrac{\text{kips}}{\text{ft}}\right)(10\ \text{ft})^2}{2}$ $= 100\ \text{ft-kips}$	$M_{\text{cantilever}} = \dfrac{w_a L^2_{\text{cantilever}}}{2}$ $= \dfrac{\left(1.33\ \dfrac{\text{kips}}{\text{ft}}\right)(10\ \text{ft})^2}{2}$ $= 66.5\ \text{ft-kips}$

The reactions at left and right are

LRFD	ASD
$R_{u,R} = \dfrac{\dfrac{wL^2}{2}}{\text{moment arm}}$ $= \dfrac{\dfrac{\left(2.0\ \dfrac{\text{kips}}{\text{ft}}\right)(40\ \text{ft})^2}{2}}{30\ \text{ft}}$ $= 53.3\ \text{kips}$ $R_{u,L} = W_u - R_{u,R}$ $= 80\ \text{kips} - 53.3\ \text{kips}$ $= 26.7\ \text{kips}$	$R_{a,R} = \dfrac{\dfrac{wL^2}{2}}{\text{moment arm}}$ $= \dfrac{\dfrac{\left(1.33\ \dfrac{\text{kips}}{\text{ft}}\right)(40\ \text{ft})^2}{2}}{30\ \text{ft}}$ $= 35.5\ \text{kips}$ $R_{a,L} = W_u - R_{a,R}$ $= 53.2\ \text{kips} - 35.5\ \text{kips}$ $= 17.7\ \text{kips}$

Determine the point of zero shear. (X is the distance from the left support.)

LRFD	ASD
$X = \dfrac{R_{u,L}}{w_u}$ $= \dfrac{26.7\ \text{kips}}{2.0\ \dfrac{\text{kips}}{\text{ft}}}$ $= 13.4\ \text{ft}$	$X = \dfrac{R_{u,L}}{w_a}$ $= \dfrac{17.7\ \text{kips}}{1.33\ \dfrac{\text{kips}}{\text{ft}}}$ $= 13.3\ \text{ft}$

Determine the maximum moment between supports.

LRFD	ASD
$M_u = R_{u,L}X - \dfrac{w_u X^2}{2} = (26.7 \text{ kips})(13.4 \text{ ft}) - \dfrac{\left(2.0 \ \dfrac{\text{kips}}{\text{ft}}\right)(13.4 \text{ ft})^2}{2} = 178 \text{ ft-kips}$	$M_a = R_{a,L}X - \dfrac{w_a X^2}{2} = (17.7 \text{ kips})(13.3 \text{ ft}) - \dfrac{\left(1.33 \ \dfrac{\text{kips}}{\text{ft}}\right)(13.3 \text{ ft})^2}{2} = 118 \text{ ft-kips}$

Check the design strength (LRFD) or allowable strength (ASD) and other qualities of each possible W16 section in *AISC Manual* Table 3-2 to see whether they are adequate. Start with the lightest of the options and check each in turn until an adequate one is found. For a W16 × 26,

$$L_p = 3.96 \text{ ft} \quad [< L_b = 10 \text{ ft, OK}]$$
$$L_r = 11.2 \text{ ft} \quad [> L_b = 10 \text{ ft, OK}]$$

LRFD	ASD
$\phi_b M_{px} = 166 \text{ ft-kips} \quad [< M_u = 178 \text{ ft-kips, not OK}]$	$\dfrac{M_{px}}{\Omega_b} = 110 \text{ ft-kips} \quad [< M_a = 118 \text{ ft-kips, not OK}]$
$\phi M_{rx} = 101 \text{ ft-kips} \quad [> M_{\text{cantilever}} = 100 \text{ ft-kips, OK}]$	$\dfrac{M_{rx}}{\Omega_b} = 67.1 \text{ ft-kips} \quad [> M_{\text{cantilever}} = 66.5 \text{ ft-kips, OK}]$
$\phi_v V_{nx} = 106 \text{ kips} \quad [> 33.33 \text{ kips, OK}]$	$\dfrac{V_{nx}}{\Omega_v} = 70.5 \text{ kips} \quad [> 22.17 \text{ kips, OK}]$

A W16 × 26 is not adequate. For a W16 × 31,

$$L_p = 4.13 \text{ ft} \quad [< L_b = 10 \text{ ft, OK}]$$
$$L_r = 11.9 \text{ ft} \quad [> L_b = 10 \text{ ft, OK}]$$

LRFD	ASD
$\phi_b M_{px} = 203 \text{ ft-kips} \quad [> M_u, \text{OK}]$ $\phi M_{rx} = 124 \text{ ft-kips} \quad [> M_{\text{cantilever}}, \text{OK}]$ $\phi_v V_{nx} = 131 \text{ kips} \quad [> 33.33 \text{ kips, OK}]$	$\frac{M_{px}}{\Omega_b} = 135 \text{ ft-kips} \quad [> M_a, \text{OK}]$ $\frac{M_{rx}}{\Omega_b} = 82.4 \text{ ft-kips} \quad [> M_{\text{cantilever}}, \text{OK}]$ $\frac{V_{nx}}{\Omega_v} = 87.3 \text{ kips} \quad [> 22.17 \text{ kips, OK}]$

The W16 × 31 is the lightest member with sufficient strength.

The answer is (B).

PRACTICE PROBLEM 35

A W10 × 60 A992 steel column is 16 ft tall with translation and rotation fixed at both ends of the column and for both axes. The column supports a concentric axial dead load of 13 kips and a concentric axial live load of 39 kips. There is no moment about the y-axis. Take the lateral-torsional buckling modification factor as $C_b = 1.0$. Which of the following is most nearly the maximum moment that can be placed on the x-axis? (LRFD options are in parentheses.)

(A) 133 ft-kips (220 ft-kips)

(B) 142 ft-kips (237 ft-kips)

(C) 150 ft-kips (257 ft-kips)

(D) 166 ft-kips (280 ft-kips)

Solution

Determine the required axial resistance.

LRFD	ASD
$P_u = 1.2D + 1.6L$ $= (1.2)(13 \text{ kips}) + (1.6)(39 \text{ kips})$ $= 78 \text{ kips}$	$P_a = D + L$ $= 13 \text{ kips} + 39 \text{ kips}$ $= 52 \text{ kips}$

From Table 7.1 , the effective length factor for a column with both ends restrained against rotation and translation is $K = 0.65$. The effective length of the column is

$$KL = (0.65)(16 \text{ ft}) = 10.4 \text{ ft}$$

Determine the effective slenderness ratio.

$$\frac{KL}{r_y} = \frac{(0.65)(16\ \text{ft})\left(12\ \frac{\text{in}}{\text{ft}}\right)}{2.57\ \text{in}} = 48.6$$

Determine which formula to use to compute critical stress.

$$4.71\sqrt{\frac{E}{F_y}} = 4.71\sqrt{\frac{29{,}000\ \frac{\text{kips}}{\text{in}^2}}{50\ \frac{\text{kips}}{\text{in}^2}}}$$
$$= 113 \quad \left[> KL/r_y,\ \text{so use Eq. 6.58}\right]$$

From Eq. 6.60, the elastic critical buckling stress is

$$F_e = \frac{\pi^2 E}{\left(\frac{KL}{r}\right)^2} = \frac{\pi^2\left(29{,}000\ \frac{\text{kips}}{\text{in}^2}\right)}{(48.6)^2} = 121\ \text{ksi}$$

From Eq. 6.58, the critical stress is

$$F_{cr} = 0.658^{F_y/F_e} F_y$$
$$= (0.658)^{50\ \frac{\text{kips}}{\text{in}^2}/121\ \frac{\text{kips}}{\text{in}^2}}\left(50\ \frac{\text{kips}}{\text{in}^2}\right)$$
$$= 42.1\ \text{ksi}$$

Use Eq. 6.57 to determine the nominal compressive strength. From *AISC Manual* Table 1-1, the area of a W10 × 60 is 17.7 in^2.

$$P_n = F_{cr}A_g$$
$$= \left(42.1\ \frac{\text{kips}}{\text{in}^2}\right)(17.7\ \text{in}^2)$$
$$= 745\ \text{kips}$$

The available compressive strength is

LRFD	ASD
$P_c = \phi_c P_n$ $= (0.90)(745\ \text{kips})$ $= 670\ \text{kips}$	$P_c = \frac{P_n}{\Omega_c} = \frac{745\ \text{kips}}{1.67}$ $= 446\ \text{kips}$

Calculate the ratio of required compressive strength to available strength to see whether Eq. 8.1 or 8.2 applies.

LRFD	ASD
$\frac{P_r}{P_c} = \frac{P_u}{\phi_c P_n} = \frac{78 \text{ kips}}{670 \text{ kips}}$ $= 0.12 \quad [< 0.2, \text{Eq. 8.2 applies}]$	$\frac{P_r}{P_c} = \frac{P_a}{\frac{P_n}{\Omega_c}} = \frac{52 \text{ kips}}{446 \text{ kips}}$ $= 0.12 \quad [< 0.2, \text{Eq. 8.2 applies}]$

From Eq. 8.2,

$$\frac{P_r}{2P_c} + \left(\frac{M_{rx}}{M_{cx}} + \frac{M_{ry}}{M_{cy}} \right) \le 1.0$$

Determine the available flexural strengths about the *x*- and *y*-axes. For the *x*-axis, from *AISC Manual* Table 3-2 and Eq 5.10 (LRFD) or Eq. 5.11 (ASD) and taking C_b as 1.0,

LRFD	ASD
$\phi_b M_n \le \begin{cases} C_b\left(\phi_b M_{px} - \text{BF}\left(L_b - L_p\right)\right) \\ \quad = (1.0) \\ \quad \times \begin{pmatrix} 280 \text{ ft-kips} \\ -(3.82 \text{ kips}) \\ \times(16 \text{ ft} - 9.08 \text{ ft}) \end{pmatrix} \\ \quad = 253.6 \text{ ft-kips} \quad [\text{controls}] \\ \phi_b M_{px} = 280 \text{ ft-kips} \end{cases}$	$\frac{M_n}{\Omega_b} \le \begin{cases} C_b\left(\frac{M_{px}}{\Omega_b} - \text{BF}\left(L_b - L_p\right)\right) \\ \quad = (1.0) \\ \quad \times \begin{pmatrix} 186 \text{ ft-kips} \\ -(2.54 \text{ kips}) \\ \times(16 \text{ ft} - 9.08 \text{ ft}) \end{pmatrix} \\ \quad = 168.4 \text{ ft-kips} \quad [\text{controls}] \\ \frac{M_{px}}{\Omega_b} = 186 \text{ ft-kips} \end{cases}$

For the *y*-axis, from *AISC Manual* Table 3-4,

LRFD	ASD
$\phi_b M_{py} = 131 \text{ ft-kips}$	$\frac{M_{py}}{\Omega_b} = 87.3 \text{ ft-kips}$

Determine the design strength (LRFD) or allowable strength (ASD) for the strong axis. From Eq. 8.2,

$$M_{rx} \le M_{cx}\left(1.0 - \frac{P_r}{2P_c} - \frac{M_{ry}}{M_{cy}}\right)$$

LRFD	ASD
$M_{rx} \le (253.7\text{ ft-kips}) \times \left(1.0 - \dfrac{78\text{ kips}}{(2)(667\text{ kips})} - \dfrac{0\text{ ft-kips}}{131\text{ ft-kips}}\right) = 239\text{ ft-kips} \quad (240\text{ ft-kips})$	$M_{rx} \le (168.5\text{ ft-kips}) \times \left(1.0 - \dfrac{52\text{ kips}}{(2)(444\text{ kips})} - \dfrac{0\text{ ft-kips}}{87.3\text{ ft-kips}}\right) = 159\text{ ft-kips} \quad (160\text{ ft-kips})$

The answer is (C).

PRACTICE PROBLEM 36

In the bolted bracket shown, bolt holes are spaced at 3 in centers.

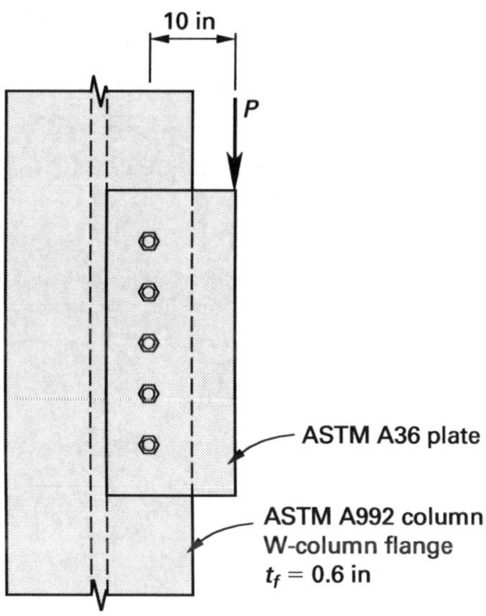

Material properties

column flange	bracket plate	bolts
ASTM A992	ASTM A36	Group A, threads excluded from shear plane
$F_y = 50$ ksi	$F_y = 36$ ksi	diameter = $^3/_4$ in
$F_u = 65$ ksi	$F_u = 58$ ksi	

Which of the following is most nearly the maximum load that the bracket can safely support and the minimum plate thickness? (LRFD options are in parentheses.)

(A) 22 kips and 0.250 in (33 kips and 0.250 in)

(B) 22 kips and 0.375 in (33 kips and 0.375 in)

(C) 33 kips and 0.375 in (49 kips and 0.375 in)

(D) 33 kips and 0.500 in (49 kips and 0.500 in)

Solution

From *AISC Manual* Table 7-6, the coefficient for a single row of five bolts with a bolt spacing of 3 in and an eccentricity of 10 in is $C = 1.66$. From the problem statement, the bolts are Group A with threads excluded from the shear plane. Determine the available strength based on single shear on the bolts. From *AISC Manual* Table 7-1,

LRFD	ASD
$\phi_v r_n = 22.5 \text{ kips}$ $C = \dfrac{P_u}{\phi_v r_n}$ $P_u = C(\phi_v r_n) = (1.66)(22.5 \text{ kips}) = 37.35 \text{ kips}$	$\dfrac{r_n}{\Omega_v} = 15.0 \text{ kips}$ $C = \dfrac{\Omega_v P_a}{r_n}$ $P_a = C\left(\dfrac{r_n}{\Omega_v}\right) = (1.66)(15.0 \text{ kips}) = 24.9 \text{ kips}$

Determine the minimum thickness for the bracket plate in order to develop strength based on bolt shear. From *AISC Manual* Table 7-5,

LRFD	ASD
$\phi_v r_n = 78.3 \dfrac{\text{kips}}{\text{in of thickness}}$ $t = \dfrac{P_u}{C(\phi_v r_n)} = \dfrac{37.35 \text{ kips}}{(1.66)\left(78.3 \dfrac{\text{kips}}{\text{in of thickness}}\right)} = 0.28 \text{ in} \quad (0.250 \text{ in})$	$\dfrac{r_n}{\Omega_v} = 52.2 \dfrac{\text{kips}}{\text{in of thickness}}$ $t = \dfrac{P_u}{C\left(\dfrac{r_n}{\Omega_v}\right)} = \dfrac{24.9 \text{ kips}}{(1.66)\left(52.2 \dfrac{\text{kips}}{\text{in of thickness}}\right)} = 0.28 \text{ in} \quad (0.250 \text{ in})$

The answer is (B).

PRACTICE PROBLEM 37

The slip-critical assembly shown is subject to shear and tension. The connection is designed for slip as a serviceability limit state. The bolts are Group A slip-critical class A bolts in standard holes. Assume that the beam and plates are adequate to transmit the loads.

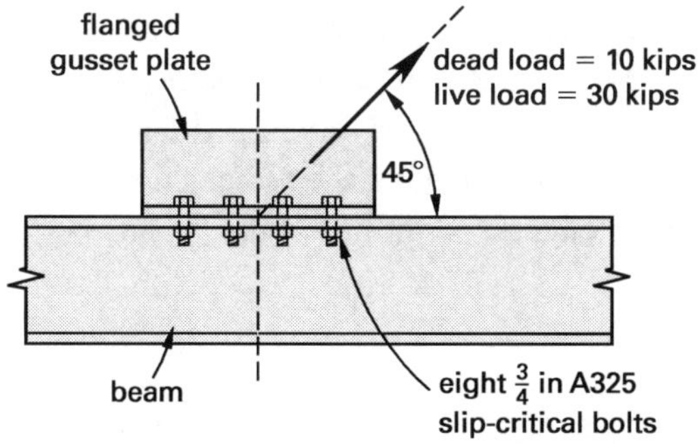

Which of the following is most nearly the design shear strength (LRFD) or allowable shear strength (ASD) per bolt? (LRFD options are in parentheses.)

(A) 4.7 kips/bolt (7.1 kips/bolt)

(B) 5.3 kips/bolt (7.9 kips/bolt)

(C) 5.8 kips/bolt (8.6 kips/bolt)

(D) 6.4 kips/bolt (9.4 kips/bolt)

Solution

Determine the tension and shear on the bolts. The load is

LRFD	ASD
$P_u = 1.2D + 1.6L$ $= (1.2)(10 \text{ kips}) + (1.6)(30 \text{ kips})$ $= 60 \text{ kips}$	$P_a = D + L$ $= 10 \text{ kips} + 30 \text{ kips}$ $= 40 \text{ kips}$

Due to the 45° angle, the tension and shear are equal.

LRFD	ASD
$T_u = V_u = \frac{P_u}{\sqrt{2}n_{\text{bolts}}}$ $= \frac{60 \text{ kips}}{\sqrt{2}(8 \text{ bolts})}$ $= 5.30 \text{ kips/bolt}$	$T_a = V_a = \frac{P_a}{\sqrt{2}n_{\text{bolts}}}$ $= \frac{40 \text{ kips}}{\sqrt{2}(8 \text{ bolts})}$ $= 3.54 \text{ kips/bolt}$

Check the tension on the bolts. From *AISC Manual* Table 7-2,

LRFD	ASD
$\phi r_n = 29.8 \text{ kips} \quad [> T_u = 5.30 \text{ kips, OK}]$	$\frac{r_n}{\Omega} = 19.9 \text{ kips} \quad [> T_a = 3.54 \text{ kips, OK}]$

Check the combined shear and tension in each bolt. The factor k_s is calculated from *AISC Specification* Eq. J3-5a (LRFD) or *AISC Specification* Eq. J3-5b (ASD). From *AISC Specification* Sec. J3.8, D_u = 1.13; from *AISC Specification* Table J3.1, T_b = 28 kips. N_b is the number of bolts carrying the indicated tension; because the value for the tension, T_u or T_a, was calculated per bolt, N_b equals one.

LRFD	ASD
$k_s = 1 - \frac{T_u}{D_u T_b N_b}$	$k_s = 1 - \frac{1.5T_a}{D_a T_b N_b}$
$= 1 - \frac{5.30 \text{ kips}}{(1.13)(28 \text{ kips})(1)}$	$= 1 - \frac{(1.5)(3.54 \text{ kips})}{(1.13)(28 \text{ kips})(1)}$
$= 0.83$	$= 0.83$

Determine the bolt shear capacity modified for slip-resistance. From *AISC Manual* Table 7-3,

LRFD	ASD
$\phi_v r_n = 9.49 \text{ kips/bolt}$	$\frac{r_n}{\Omega_v} = 6.33 \text{ kips/bolt}$
$[> 5.30 \text{ kips/bolt, OK}]$	$[> 3.54 \text{ kips/bolt, OK}]$
$k_s(\phi_v r_n) = (0.83)\left(9.49 \frac{\text{kips}}{\text{bolt}}\right)$	$k_s\left(\frac{r_n}{\Omega_v}\right) = (0.83)\left(6.33 \frac{\text{kips}}{\text{bolt}}\right)$
$= 7.87 \text{ kips/bolt}$	$= 5.25 \text{ kips/bolt}$
(7.9 kips/bolt)	(5.3 kips/bolt)

The answer is (B).

Index